Student Solutions Manual

Neil Wigley
University of Windsor

Dean Hickerson

to accompany

D0224675

CALCULUS

Single Variable

Ninth Edition

Howard Anton
Drexel University

Irl C. Bivens
Davidson College

Stephen L. Davis
Davidson College

WILEY

John Wiley & Sons, Inc.

ISBN-13   978-0-470-37962-2

Printed in the United States of America

10 9 8 7 6 5 4 3 2 1

Printed and bound by Bind-Rite Robbinsville

CONTENTS

CHAPTER 0
Before Calculus

EXERCISE SET 0.1

1. **(a)** $-2.9, -2.0, 2.35, 2.9$ **(b)** none   **(c)** $y = 0$
(d) $-1.75 \le x \le 2.15$ **(e)** $y_{\max} = 2.8$ at $x = -2.6$; $y_{\min} = -2.2$ at $x = 1.2$

3. **(a)** yes   **(b)** yes
(c) no (vertical line test fails)   **(d)** no (vertical line test fails)

5. **(a)** 1999, \$47,700   **(b)** 1993, \$41,600
(c) The slope between 2000 and 2001 is steeper than the slope between 2001 and 2002, so the median income was declining more rapidly during the first year of the 2-year period.

7. **(a)** $f(0) = 3(0)^2 - 2 = -2$; $f(2) = 3(2)^2 - 2 = 10$; $f(-2) = 3(-2)^2 - 2 = 10$; $f(3) = 3(3)^2 - 2 = 25$; $f(\sqrt{2}) = 3(\sqrt{2})^2 - 2 = 4$; $f(3t) = 3(3t)^2 - 2 = 27t^2 - 2$
(b) $f(0) = 2(0) = 0$; $f(2) = 2(2) = 4$; $f(-2) = 2(-2) = -4$; $f(3) = 2(3) = 6$; $f(\sqrt{2}) = 2\sqrt{2}$; $f(3t) = 1/3t$ for $t > 1$ and $f(3t) = 6t$ for $t \le 1$.

9. **(a)** Natural domain: $x \ne 3$. Range: $y \ne 0$.
(b) Natural domain: $x \ne 0$. Range: $\{1, -1\}$.
(c) Natural domain: $x \le -\sqrt{3}$ or $x \ge \sqrt{3}$. Range: $y \ge 0$.
(d) $x^2 - 2x + 5 = (x-1)^2 + 4 \ge 4$. So $G(x)$ is defined for all x, and is $\ge \sqrt{4} = 2$. Natural domain: all x. Range: $y \ge 2$.
(e) Natural domain: $\sin x \ne 1$, so $x \ne (2n + \frac{1}{2})\pi$, $n = 0, \pm 1, \pm 2, \ldots$. For such x, $-1 \le \sin x < 1$, so $0 < 1 - \sin x \le 2$, and $\frac{1}{1 - \sin x} \ge \frac{1}{2}$. Range: $y \ge \frac{1}{2}$.
(f) Division by 0 occurs for $x = 2$. For all other x, $\frac{x^2 - 4}{x - 2} = x + 2$, which is nonnegative for $x \ge -2$. Natural domain: $[-2, 2) \cup (2, +\infty)$. The range of $\sqrt{x + 2}$ is $[0, +\infty)$. But we must exclude $x = 2$, for which $\sqrt{x + 2} = 2$. Range: $[0, 2) \cup (2, +\infty)$.

11. **(a)** The curve is broken whenever someone is born or someone dies.
(b) C decreases for eight hours, increases rapidly (but continuously), and then repeats.

13.

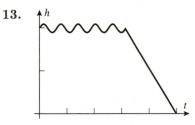

15. Yes. $y = \sqrt{25 - x^2}$ **17.** Yes. $y = \begin{cases} \sqrt{25 - x^2}, & -5 \le x \le 0 \\ -\sqrt{25 - x^2}, & 0 < x \le 5 \end{cases}$

19. False. E.g. the graph of $x^2 - 1$ crosses the x-axis at $x = 1$ and $x = -1$.

21. False. The range also includes 0.

23. **(a)** $x = 2, 4$ **(b)** none   **(c)** $x \le 2$; $4 \le x$ **(d)** $y_{\min} = -1$; no maximum value

25. The cosine of θ is $(L - h)/L$ (side adjacent over hypotenuse), so $h = L(1 - \cos\theta)$.

27. (a) If $x < 0$, then $|x| = -x$ so $f(x) = -x + 3x + 1 = 2x + 1$. If $x \geq 0$, then $|x| = x$ so
$f(x) = x + 3x + 1 = 4x + 1$;

$$f(x) = \begin{cases} 2x + 1, & x < 0 \\ 4x + 1, & x \geq 0 \end{cases}$$

(b) If $x < 0$, then $|x| = -x$ and $|x - 1| = 1 - x$ so $g(x) = -x + (1 - x) = 1 - 2x$. If $0 \leq x < 1$,
then $|x| = x$ and $|x - 1| = 1 - x$ so $g(x) = x + (1 - x) = 1$. If $x \geq 1$, then $|x| = x$ and
$|x - 1| = x - 1$ so $g(x) = x + (x - 1) = 2x - 1$;

$$g(x) = \begin{cases} 1 - 2x, & x < 0 \\ 1, & 0 \leq x < 1 \\ 2x - 1, & x \geq 1 \end{cases}$$

29. (a) $V = (8 - 2x)(15 - 2x)x$

(b) $0 < x < 4$

(c) $0 < V \leq 91$, approximately

(d) As x increases, V increases and then decreases; the
maximum value occurs when x is about 1.7.

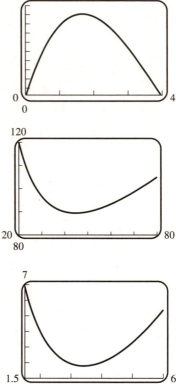

31. (a) The side adjacent to the building
has length x, so $L = x + 2y$.

(b) $A = xy = 1000$, so $L = x + 2000/x$.

(c) $0 < x \leq 100$

(d) $x \approx 44.72$ ft, $y \approx 22.36$ ft

33. (a) $V = 500 = \pi r^2 h$ so $h = \dfrac{500}{\pi r^2}$. Then

$$C = (0.02)(2)\pi r^2 + (0.01)2\pi rh = 0.04\pi r^2 + 0.02\pi r\frac{500}{\pi r^2}$$

$$= 0.04\pi r^2 + \frac{10}{r};$$

$C_{\min} \approx 4.39$ cents at $r \approx 3.4$ cm, $h \approx 13.7$ cm

(b) $C = (0.02)(2)(2r)^2 + (0.01)2\pi rh = 0.16r^2 + \dfrac{10}{r}$. Since
$0.04\pi < 0.16$, the top and bottom now get more weight.
Since they cost more, we diminish their sizes in the solution, and the cans become taller.

(c) $r \approx 3.1$ cm, $h \approx 16.0$ cm, $C \approx 4.76$ cents

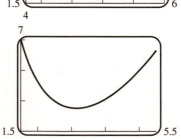

35. (i) $x = 1, -2$ causes division by zero   **(ii)** $g(x) = x + 1$, all x

37. (a) 25°F   **(b)** 13°F   **(c)** 5°F

39. If $v = 48$ then $-60 = $ WCT $\approx 1.4157T - 30.6763$; thus $T \approx 15$°F when WCT $= -10$.

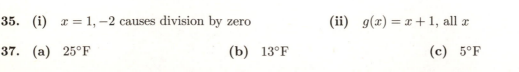

EXERCISE SET 0.2

1. (a)

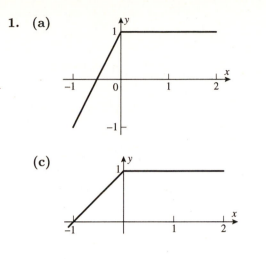

(b)

(c)

(d)

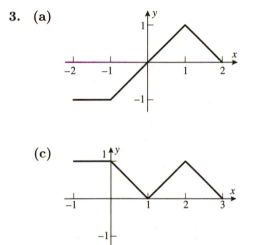

3. (a)

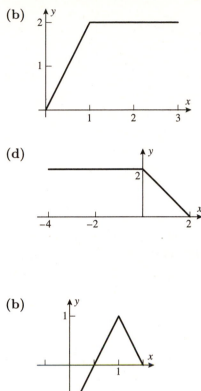

(b)

(c)

(d)

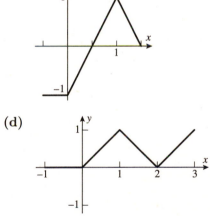

5. Translate left 1 unit, stretch vertically by a factor of 2, reflect over x-axis, translate down 3 units.

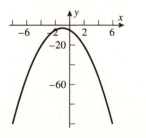

7. $y = -(x-1)^2 + 2$; translate right 1 unit, reflect over x-axis, translate up 2 units.

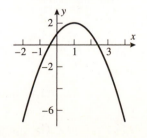

9. Translate left 1 unit, reflect over x-axis, translate up 3 units.

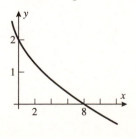

11. Compress vertically by a factor of $\frac{1}{2}$, translate up 1 unit.

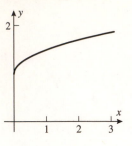

13. Translate right 3 units.

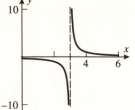

15. Translate left 1 unit, reflect over x-axis, translate up 2 units.

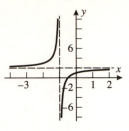

17. Translate left 2 units and down 2 units.

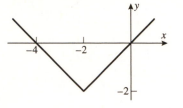

19. Stretch vertically by a factor of 2, translate right 1/2 unit and up 1 unit.

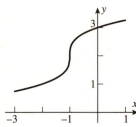

21. Stretch vertically by a factor of 2, reflect over x-axis, translate up 1 unit.

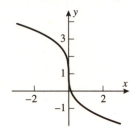

23. Translate left 1 unit and up 2 units.

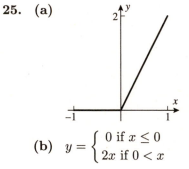

25. (a)

(b) $y = \begin{cases} 0 \text{ if } x \leq 0 \\ 2x \text{ if } 0 < x \end{cases}$

27. $(f + g)(x) = 3\sqrt{x - 1}$, $x \geq 1$; $(f - g)(x) = \sqrt{x - 1}$, $x \geq 1$; $(fg)(x) = 2x - 2$, $x \geq 1$; $(f/g)(x) = 2$, $x > 1$

29. (a) 3
(d) 2
(b) 9
(e) $\sqrt{2 + h}$
(c) 2
(f) $(3 + h)^3 + 1$

31. $(f \circ g)(x) = 1 - x$, $x \leq 1$; $(g \circ f)(x) = \sqrt{1 - x^2}$, $|x| \leq 1$

33. $(f \circ g)(x) = \dfrac{1}{1 - 2x}$, $x \neq \dfrac{1}{2}, 1$; $(g \circ f)(x) = -\dfrac{1}{2x} - \dfrac{1}{2}$, $x \neq 0, 1$

35. (a) $g(x) = \sqrt{x}$, $h(x) = x + 2$
(b) $g(x) = |x|$, $h(x) = x^2 - 3x + 5$

37. (a) $g(x) = x^2$, $h(x) = \sin x$
(b) $g(x) = 3/x$, $h(x) = 5 + \cos x$

39. **(a)** $g(x) = (1 + x)^3$, $h(x) = \sin(x^2)$　　　**(b)** $g(x) = \sqrt{1 - x}$, $h(x) = \sqrt[3]{x}$

41. True, by Definition 0.2.1.

43. True, by Theorem 0.2.3(a).

45.

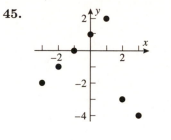

47. Note that
$$f(g(-x)) = f(-g(x)) = f(g(x)),$$
so $f(g(x))$ is even.

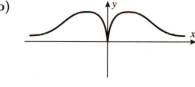

49. $f(g(x)) = 0$ when $g(x) = \pm 2$, so $x \approx \pm 1.5$; $g(f(x)) = 0$ when $f(x) = 0$, so $x = \pm 2$.

51. $\dfrac{3(x + h)^2 - 5 - (3x^2 - 5)}{h} = \dfrac{6xh + 3h^2}{h} = 6x + 3h;$

$\dfrac{3w^2 - 5 - (3x^2 - 5)}{w - x} = \dfrac{3(w - x)(w + x)}{w - x} = 3w + 3x$

53. $\dfrac{1/(x + h) - 1/x}{h} = \dfrac{x - (x + h)}{xh(x + h)} = \dfrac{-1}{x(x + h)};\ \dfrac{1/w - 1/x}{w - x} = \dfrac{x - w}{wx(w - x)} = -\dfrac{1}{xw}$

55. neither; odd; even

57. **(a)**　　　　　　　　　　　　　　　**(b)**

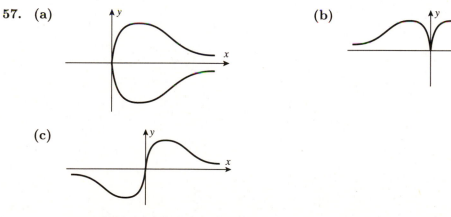

(c)

59. **(a)** $f(-x) = (-x)^2 = x^2 = f(x)$, even　　　**(b)** $f(-x) = (-x)^3 = -x^3 = -f(x)$, odd
　　(c) $f(-x) = |-x| = |x| = f(x)$, even　　　**(d)** $f(-x) = -x + 1$, neither

　　(e) $f(-x) = \dfrac{(-x)^5 - (-x)}{1 + (-x)^2} = -\dfrac{x^5 - x}{1 + x^2} = -f(x)$, odd

　　(f) $f(-x) = 2 = f(x)$, even

61. In Exercise 60 it was shown that g is an even function, and h is odd. Moreover by inspection $f(x) = g(x) + h(x)$ for all x, so f is the sum of an even function and an odd function.

63. (a) y-axis, because $(-x)^4 = 2y^3 + y$ gives $x^4 = 2y^3 + y$

(b) origin, because $(-y) = \dfrac{(-x)}{3 + (-x)^2}$ gives $y = \dfrac{x}{3 + x^2}$

(c) x-axis, y-axis, and origin because $(-y)^2 = |x| - 5$, $y^2 = |-x| - 5$, and $(-y)^2 = |-x| - 5$ all give $y^2 = |x| - 5$

65.

67.

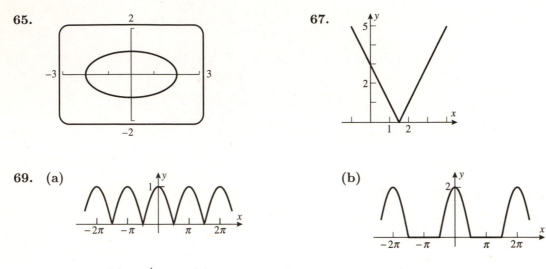

69. (a)

(b)

71. Yes, e.g. $f(x) = x^k$ and $g(x) = x^n$ where k and n are integers.

EXERCISE SET 0.3

1. (a) $y = 3x + b$ **(b)** $y = 3x + 6$ **(c)**

3. (a) $y = mx + 2$

(b) $m = \tan\phi = \tan 135° = -1$, so $y = -x + 2$ **(c)**

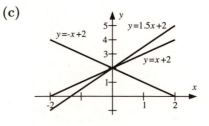

5. Let the line be tangent to the circle at the point (x_0, y_0) where $x_0^2 + y_0^2 = 9$. The slope of the tangent line is the negative reciprocal of y_0/x_0 (why?), so $m = -x_0/y_0$ and $y = -(x_0/y_0)x + b$.

Substituting the point (x_0, y_0) as well as $y_0 = \pm\sqrt{9 - x_0^2}$ we get $y = \pm\dfrac{9 - x_0 x}{\sqrt{9 - x_0^2}}$.

7. The x-intercept is $x = 10$ so that with depreciation at 10% per year the final value is always zero, and hence $y = m(x - 10)$. The y-intercept is the original value.

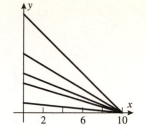

9. **(a)** The slope is -1.

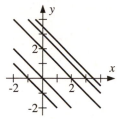

(b) The y-intercept is $y = -1$.

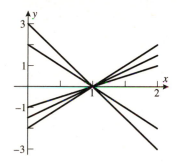

(c) They pass through the point $(-4, 2)$.

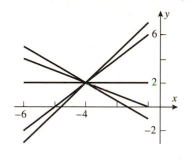

(d) The x-intercept is $x = 1$.

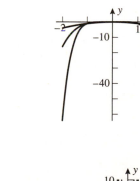
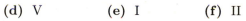

11. **(a)** VI   **(b)** IV   **(c)** III   **(d)** V   **(e)** I   **(f)** II

13. **(a)**

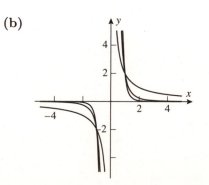

(b)

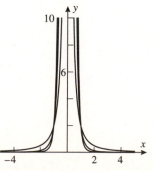

(c)

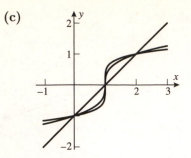 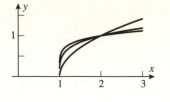

15. (a)

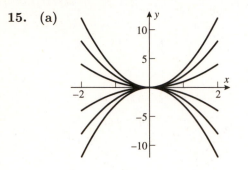

(b)

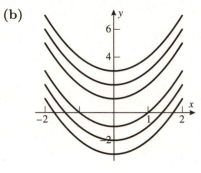

(c)

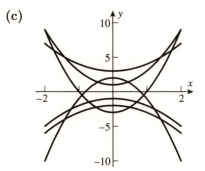

17. (a)

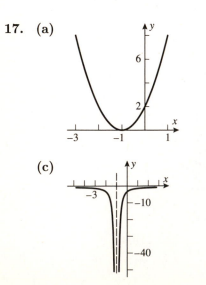

(b)

(c)

(d)

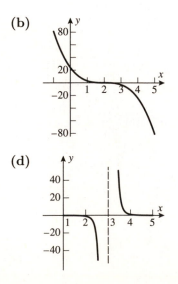

19. The part of the graph of $y = \sqrt{|x|}$ with $x \geq 0$ is the same as the graph of $y = \sqrt{x}$. The part with $x \leq 0$ is the reflection of the graph of $y = \sqrt{x}$ across the y-axis.

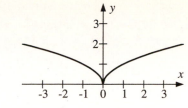

21. **(a)** N·m **(b)** k = 20 N·m

(c)

$V(\text{L})$	0.25	0.5	1.0	1.5	2.0
$P\ (\text{N/m}^2)$	80×10^3	40×10^3	20×10^3	13.3×10^3	10×10^3

(d)

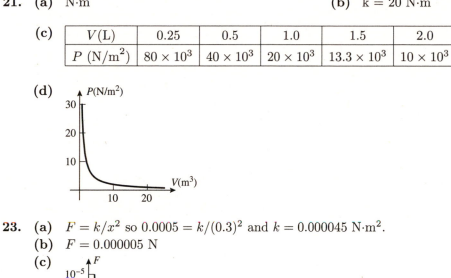

23. **(a)** $F = k/x^2$ so $0.0005 = k/(0.3)^2$ and $k = 0.000045$ N·m^2.
(b) $F = 0.000005$ N
(c)

(d) When they approach one another, the force increases without bound; when they get far apart it tends to zero.

25. True. The graph of $y = 2x + b$ is obtained by translating the graph of $y = 2x$ up b units (or down $-b$ units if $b < 0$).

27. False. The curve's equation is $y = 12/x$, so the constant of proportionality is 12.

29. **(a)** II; $y = 1$, $x = -1, 2$ **(b)** I; $y = 0$, $x = -2, 3$
    (c) IV; $y = 2$ **(d)** III; $y = 0$, $x = -2$

31. **(a)** $y = 3\sin(x/2)$ **33.** **(a)** $y = \sin(x + \pi/2)$
    (b) $y = 4\cos 2x$ **(b)** $y = 3 + 3\sin(2x/9)$
    (c) $y = -5\sin 4x$ **(c)** $y = 1 + 2\sin(2x - \pi/2)$

35. **(a)** $3, \pi/2$ **(b)** $2, 2$ **(c)** $1, 4\pi$

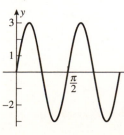 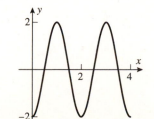 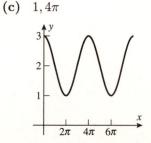

37. Let $\omega = 2\pi$. Then $A\sin(\omega t + \theta) = A(\cos\theta\sin 2\pi t + \sin\theta\cos 2\pi t) = (A\cos\theta)\sin 2\pi t + (A\sin\theta)\cos 2\pi t$, so for the two equations for x to be equivalent, we need $A\cos\theta = \sqrt{2}$ and $A\sin\theta = \sqrt{6}$. These imply that $A^2 = (A\cos\theta)^2 + (A\sin\theta)^2 = 2 + 6 = 8$ and

$$\tan\theta = \frac{A\sin\theta}{A\cos\theta} = \frac{\sqrt{6}}{\sqrt{2}} = \sqrt{3}.\ \text{So take }\theta = \pi/3\text{ and}$$

$$A = \sqrt{8} = 2\sqrt{2}: \qquad x = 2\sqrt{2}\sin(2\pi t + \pi/3).$$

EXERCISE SET 0.4

1. **(a)** $f(g(x)) = 4(x/4) = x$, $g(f(x)) = (4x)/4 = x$, f and g are inverse functions

(b) $f(g(x)) = 3(3x - 1) + 1 = 9x - 2 \neq x$ so f and g are not inverse functions

(c) $f(g(x)) = \sqrt[3]{(x^3 + 2) - 2} = x$, $g(f(x)) = (x - 2) + 2 = x$, f and g are inverse functions

(d) $f(g(x)) = (x^{1/4})^4 = x$, $g(f(x)) = (x^4)^{1/4} = |x| \neq x$, f and g are not inverse functions

3. **(a)** yes   **(b)** yes   **(c)** no   **(d)** yes   **(e)** no   **(f)** no

5. **(a)** yes; all outputs (the elements of row two) are distinct

(b) no; $f(1) = f(6)$

7. **(a)** f has an inverse because the graph passes the horizontal line test. To compute $f^{-1}(2)$ start at 2 on the y-axis and go to the curve and then down, so $f^{-1}(2) = 8$; similarly, $f^{-1}(-1) = -1$ and $f^{-1}(0) = 0$.

(b) domain of f^{-1} is $[-2, 2]$, range is $[-8, 8]$ **(c)**

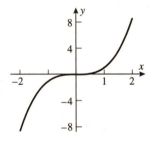

9. $y = f^{-1}(x)$, $x = f(y) = 7y - 6$, $y = \dfrac{1}{7}(x + 6) = f^{-1}(x)$

11. $y = f^{-1}(x)$, $x = f(y) = 3y^3 - 5$, $y = \sqrt[3]{(x + 5)/3} = f^{-1}(x)$

13. $y = f^{-1}(x)$, $x = f(y) = 3/y^2$, $y = -\sqrt{3/x} = f^{-1}(x)$

15. $y = f^{-1}(x), x = f(y) = \begin{cases} 5/2 - y, & y < 2 \\ 1/y, & y \geq 2 \end{cases}$, $y = f^{-1}(x) = \begin{cases} 5/2 - x, & x > 1/2 \\ 1/x, & 0 < x \leq 1/2 \end{cases}$

17. $y = f^{-1}(x)$, $x = f(y) = (y + 2)^4$ for $y \geq 0$, $y = f^{-1}(x) = x^{1/4} - 2$ for $x \geq 16$

19. $y = f^{-1}(x)$, $x = f(y) = -\sqrt{3 - 2y}$ for $y \leq 3/2$, $y = f^{-1}(x) = (3 - x^2)/2$ for $x \leq 0$

21. False. $f^{-1}(2) = f^{-1}(f(2)) = 2$

23. True. Both terms have the same definition; see the paragraph before Theorem 0.4.3.

25. $y = f^{-1}(x)$, $x = f(y) = ay^2 + by + c$, $ay^2 + by + c - x = 0$, use the quadratic formula to get

$$y = \frac{-b \pm \sqrt{b^2 - 4a(c - x)}}{2a};$$

(a) $f^{-1}(x) = \dfrac{-b + \sqrt{b^2 - 4a(c - x)}}{2a}$ (b) $f^{-1}(x) = \dfrac{-b - \sqrt{b^2 - 4a(c - x)}}{2a}$

27. (a) $y = f(x) = (6.214 \times 10^{-4})x$ (b) $x = f^{-1}(y) = \dfrac{10^4}{6.214}y$

(c) how many meters in y miles

29. (a) $f(f(x)) = \dfrac{3 - \dfrac{3 - x}{1 - x}}{1 - \dfrac{3 - x}{1 - x}} = \dfrac{3 - 3x - 3 + x}{1 - x - 3 + x} = x$ so $f = f^{-1}$

(b) symmetric about the line $y = x$

31. if $f^{-1}(x) = 1$, then $x = f(1) = 2(1)^3 + 5(1) + 3 = 10$

33. $f(f(x)) = x$ thus $f = f^{-1}$ so the graph is symmetric about $y = x$.

35. Suppose that g and h are both inverses of f. Then $f(g(x)) = x$, $h[f(g(x))] = h(x)$; but $h[f(g(x))] = g(x)$ because h is an inverse of f so $g(x) = h(x)$.

REVIEW EXERCISES, CHAPTER 0

1.

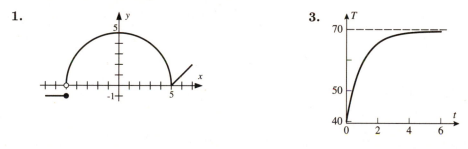

3.

5. (a) If the side has length x and height h, then $V = 8 = x^2 h$, so $h = 8/x^2$. Then the cost $C = 5x^2 + 2(4)(xh) = 5x^2 + 64/x$.

(b) The domain of C is $(0, +\infty)$ because x can be very large (just take h very small).

7. (a) The base has sides $(10 - 2x)/2$ and $6 - 2x$, and the height is x, so $V = (6 - 2x)(5 - x)x$ ft^3.

(b) From the picture we see that $x < 5$ and $2x < 6$, so $0 < x < 3$.

(c) 3.57 ft $\times 3.79$ ft $\times 1.21$ ft

9.

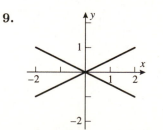

11.

x	-4	-3	-2	-1	0	1	2	3	4
$f(x)$	0	-1	2	1	3	-2	-3	4	-4
$g(x)$	3	2	1	-3	-1	-4	4	-2	0
$(f \circ g)(x)$	4	-3	-2	-1	1	0	-4	2	3
$(g \circ f)(x)$	-1	-3	4	-4	-2	1	2	0	3

13. $f(g(x)) = (3x+2)^2 + 1, g(f(x)) = 3(x^2+1)+2$, so $9x^2 + 12x + 5 = 3x^2 + 5, 6x^2 + 12x = 0, x = 0, -2$

15. For $g(h(x))$ to be defined, we require $h(x) \neq 0$, i.e. $x \neq \pm 1$. For $f(g(h(x)))$ to be defined, we also require $g(h(x)) \neq 1$, i.e. $x \neq \pm\sqrt{2}$. So the domain of $f \circ g \circ h$ consists of all x except ± 1 and $\pm\sqrt{2}$. For all x in the domain, $(f \circ g \circ h)(x) = 1/(2 - x^2)$.

17. (a) even × odd = odd
    (c) even + odd is neither
    (b) odd × odd = even
    (d) odd × odd = even

19. (a) The circle of radius 1 centered at (a, a^2); therefore, the family of all circles of radius 1 with centers on the parabola $y = x^2$.

    (b) All translates of the parabola $y = x^2$ with vertex on the line $y = x/2$.

21. (a)

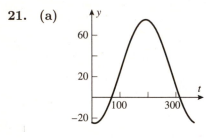

    (b) when $\dfrac{2\pi}{365}(t - 101) = \dfrac{3\pi}{2}$, or $t = 374.75$, which is the same date as $t = 9.75$, so during the night of January 10th-11th

    (c) from $t = 0$ to $t = 70.58$ and from $t = 313.92$ to $t = 365$ (the same date as $t = 0$), for a total of about 122 days

23. When $x = 0$ the value of the green curve is higher than that of the blue curve, therefore the blue curve is given by $y = 1 + 2\sin x$.
The points A, B, C, D are the points of intersection of the two curves, i.e. where
$1 + 2\sin x = 2\sin(x/2) + 2\cos(x/2)$. Let $\sin(x/2) = p, \cos(x/2) = q$. Then $2\sin x = 4\sin(x/2)\cos(x/2)$ (basic trigonometric identity), so the equation which yields the points of intersection becomes $1 + 4pq = 2p + 2q, 4pq - 2p - 2q + 1 = 0, (2p-1)(2q-1) = 0$; thus whenever either $\sin(x/2) = 1/2$ or $\cos(x/2) = 1/2$, i.e. when $x/2 = \pi/6, 5\pi/6, \pm\pi/3$. Thus A has coordinates $(-2\pi/3, 1 - \sqrt{3})$, B has coordinates $(\pi/3, 1 + \sqrt{3})$, C has coordinates $(2\pi/3, 1 + \sqrt{3})$, and D has coordinates $(5\pi/3, 1 - \sqrt{3})$.

25. (a) $f(g(x)) = x$ for all x in the domain of g, and $g(f(x)) = x$ for all x in the domain of f.

    (b) They are reflections of each other through the line $y = x$.

    (c) The domain of one is the range of the other and vice versa.

    (d) The equation $y = f(x)$ can always be solved for x as a function of y. Functions with no inverses include $y = x^2$, $y = \sin x$.

27. (a) $x = f(y) = 8y^3 - 1$; $f^{-1}(x) = y = \left(\dfrac{x+1}{8}\right)^{1/3} = \dfrac{1}{2}(x+1)^{1/3}$

(b) $f(x) = (x-1)^2$; f does not have an inverse because f is not one-to-one, for example $f(0) = f(2) = 1$.

(c) $x = f(y) = \dfrac{3}{y+1}$; $f^{-1}(x) = y = \dfrac{3}{x} - 1$

(d) $x = f(y) = \dfrac{y+2}{y-1}$; $f^{-1}(x) = y = \dfrac{x+2}{x-1}$

EXERCISE SET 1.1

1. (a) 3   (b) 3   (c) 3   (d) 3

3. (a) −1   (b) 3   (c) does not exist   (d) 1

5. (a) 0   (b) 0   (c) 0   (d) 3

7. (a) $-\infty$ (b) $-\infty$ (c) $-\infty$ (d) 1

9. (a) 1   (b) $-\infty$ (c) does not exist   (d) −2

11. (i)

−0.1	−0.01	−0.001	0.001	0.01	0.1
1.9866933	1.9998667	1.9999987	1.9999987	1.9998667	1.9866933

(ii)

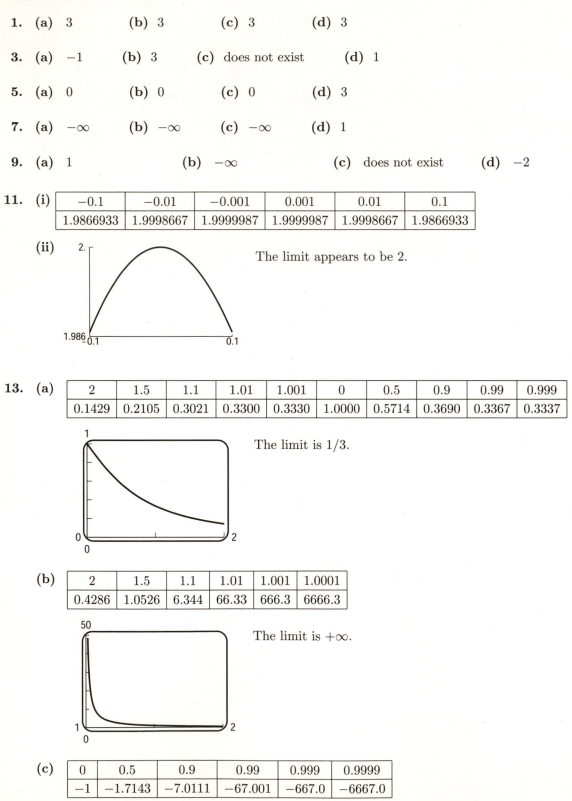

The limit appears to be 2.

13. (a)

2	1.5	1.1	1.01	1.001	0	0.5	0.9	0.99	0.999
0.1429	0.2105	0.3021	0.3300	0.3330	1.0000	0.5714	0.3690	0.3367	0.3337

The limit is 1/3.

(b)

2	1.5	1.1	1.01	1.001	1.0001
0.4286	1.0526	6.344	66.33	666.3	6666.3

The limit is $+\infty$.

(c)

0	0.5	0.9	0.99	0.999	0.9999
−1	−1.7143	−7.0111	−67.001	−667.0	−6667.0

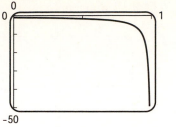

The limit is $-\infty$.

15. **(a)**

-0.25	-0.1	-0.001	-0.0001	0.0001	0.001	0.1	0.25
2.7266	2.9552	3.0000	3.0000	3.0000	3.0000	2.9552	2.7266

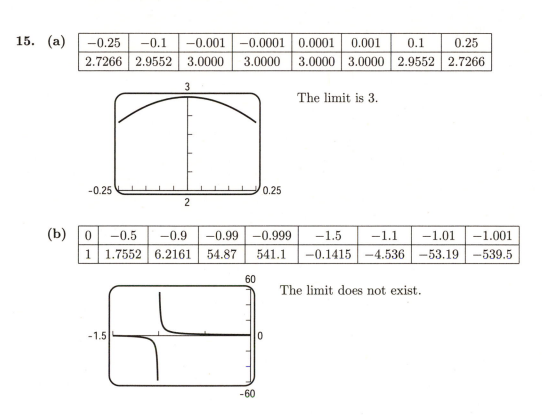

The limit is 3.

(b)

0	-0.5	-0.9	-0.99	-0.999	-1.5	-1.1	-1.01	-1.001
1	1.7552	6.2161	54.87	541.1	-0.1415	-4.536	-53.19	-539.5

The limit does not exist.

17. false; define $f(x) = x$ for $x \neq a$ and $f(a) = a + 1$. Then $\lim_{x \to a} f(x) = a \neq f(a)$.

19. false; define $f(x) = 0$ for $x < 0$ and $f(x) = x + 1$ for $x \geq 0$. Then the left and right limits exist but are unequal.

27. $m_{\text{sec}} = \dfrac{x^2 - 1}{x + 1} = x - 1$ which gets close to -2 as x gets close to -1, thus $y - 1 = -2(x + 1)$ or $y = -2x - 1$

29. $m_{\text{sec}} = \dfrac{x^4 - 1}{x - 1} = x^3 + x^2 + x + 1$ which gets close to 4 as x gets close to 1, thus $y - 1 = 4(x - 1)$ or $y = 4x - 3$

31. **(a)** The length of the rod while at rest.

(b) The limit is zero. The length of the rod approaches zero as its speed approaches c.

33. **(a)** The limit appears to be 3.   **(b)** The limit appears to be 3.

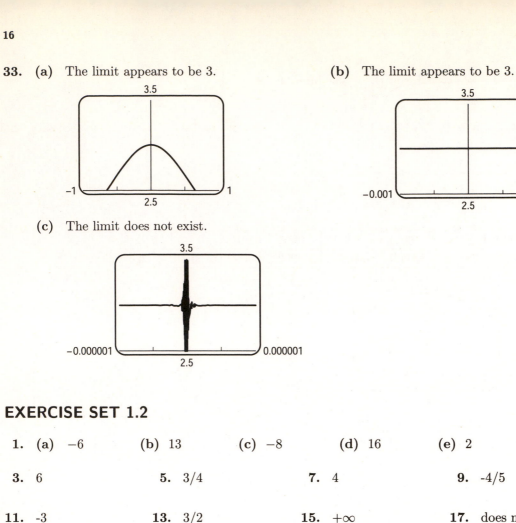

(c) The limit does not exist.

EXERCISE SET 1.2

1. **(a)** -6 **(b)** 13   **(c)** -8 **(d)** 16   **(e)** 2   **(f)** $-1/2$

3. 6   **5.** $3/4$ **7.** 4   **9.** -4/5

11. -3   **13.** $3/2$ **15.** $+\infty$ **17.** does not exist

19. $-\infty$ **21.** $+\infty$ **23.** does not exist   **25.** $+\infty$

27. $+\infty$ **29.** 6

31. **(a)** 2   **(b)** 2   **(c)** 2

33. true, Thm. 1.2.2

35. false; e.g. $f(x) = 2x, g(x) = x$, so $\lim_{x\to 0} f(x) = \lim_{x\to 0} g(x)$

37. $\displaystyle\lim_{x\to 0} \frac{x}{x\left(\sqrt{x+4}+2\right)} = \frac{1}{4}$

39. **(a)** 3   **(b)**

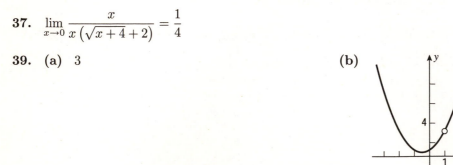

41. **(a)** Theorem 1.2.2(a) doesn't apply; moreover one cannot add/subtract infinities.

(b) $\displaystyle\lim_{x\to 0^+}\left(\frac{1}{x}-\frac{1}{x^2}\right)=\lim_{x\to 0^+}\left(\frac{x-1}{x^2}\right)=-\infty$

43. For $x\neq 1$, $\dfrac{1}{x-1}-\dfrac{a}{x^2-1}=\dfrac{x+1-a}{x^2-1}$ and for this to have a limit it is necessary that $\lim_{x\to 1}(x+1-a)=0$, i.e. $a=2$.

45. The left and/or right limits could be plus or minus infinity; or the limit could exist, or equal any preassigned real number. For example, let $q(x)=x-x_0$ and let $p(x)=a(x-x_0)^n$ where n takes on the values $0,1,2$.

47. $g(x)=[f(x)+g(x)]-f(x),\lim g(x)=\lim[f(x)+g(x)-f(x)]=\lim[f(x)+g(x)]-\lim f(x)$ by Theorem 1.2.2.

EXERCISE SET 1.3

1. **(a)** $-\infty$ **(b)** $+\infty$

3. **(a)** 0 **(b)** -1

5. **(a)** -12 **(b)** 21 **(c)** -15 **(d)** 25
(e) 2 **(f)** $-3/5$ **(g)** 0
(h) The limit doesn't exist because the denominator tends to zero but the numerator doesn't.

7.

x	0.1	0.01	0.001	0.0001	0.00001	0.000001
$f(x)$	0.3015	0.09950	0.03161	0.01000	0.00316	0.00100

The limit appears to be 0.

9. $-\infty$ **11.** $+\infty$ **13.** $3/2$ **15.** 0

17. 0 **19.** $-\infty$ **21.** $-1/7$ **23.** $-5^{1/3}/2$

25. $-\sqrt{5}$ **27.** $1/\sqrt{6}$ **29.** $\sqrt{3}$

31. $\displaystyle\lim_{x\to+\infty}(\sqrt{x^2+3}-x)\frac{\sqrt{x^2+3}+x}{\sqrt{x^2+3}+x}=\lim_{x\to+\infty}\frac{3}{\sqrt{x^2+3}+x}=0$

33. false; if $x/2>1000$ then $1000x<x^2/2,\,x^2-1000x>x^2/2$, so the limit is $+\infty$.

35. true: for example $f(x)=\sin x/x$ crosses the x-axis infinitely many times at $x=n\pi,n=1,2,\ldots$

37. It appears that $\displaystyle\lim_{t\to+\infty}n(t)=+\infty$, and $\displaystyle\lim_{t\to+\infty}e(t)=c$.

39. **(a)** $+\infty$ **(b)** -5

41. $\displaystyle\lim_{x\to+\infty}p(x)=(-1)^n\infty$ and $\displaystyle\lim_{x\to-\infty}p(x)=+\infty$

43. **(a)** no
(b) yes, $\tan x$ and $\sec x$ at $x=n\pi+\pi/2$ and $\cot x$ and $\csc x$ at $x=n\pi,n=0,\pm 1,\pm 2,\ldots$

45. (a) If $f(t) \to +\infty$ (resp $f(t) \to -\infty$) then $f(t)$ can be made arbitrarily large (resp small) by taking t large enough. But by considering the values $g(x)$ where $g(x) > t$, we see that $f(g(x))$ has the limit $+\infty$ too (resp limit $-\infty$).

If $f(t)$ has the limit L as $t \to +\infty$ the values $f(t)$ can be made arbitrarily close L by taking t large enough. But if x is large enough then $g(x) > t$ and hence $f(g(x))$ is also arbitrarily close to L.

(b) For $\lim\limits_{x \to -\infty}$ the same argument holds with the substitutiion "x decreases without bound" instead of "x increases without bound". For $\lim\limits_{x \to c^-}$ substitute "x close enough to $c, x < c$", etc.

47. $t = 1/x, \ \lim\limits_{t \to +\infty} f(t) = +\infty$ **49.** $t = \csc x, \ \lim\limits_{t \to +\infty} f(t) = +\infty$

51. $f(x) = x + 2 + \dfrac{2}{x - 2}$,

so $\lim\limits_{x \to \pm\infty} (f(x) - (x + 2)) = 0$

and $f(x)$ is asymptotic to $y = x + 2$.

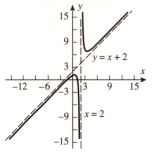

53. $f(x) = -x^2 + 1 + 2/(x - 3)$

so $\lim\limits_{x \to \pm\infty} [f(x) - (-x^2 + 1)] = 0$

and $f(x)$ is asymptotic to $y = -x^2 + 1$.

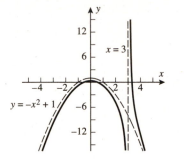

55. $\lim\limits_{x \to +\infty} (f(x) - \sin x) = 0$ and $f(x)$ is asymptotic to $y = \sin x$.

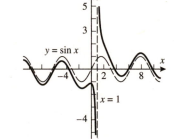

EXERCISE SET 1.4

1. (a) $|f(x) - f(0)| = |x + 2 - 2| = |x| < 0.1$ if and only if $|x| < 0.1$

(b) $|f(x) - f(3)| = |(4x - 5) - 7| = 4|x - 3| < 0.1$ if and only if $|x - 3| < (0.1)/4 = 0.0025$

(c) $|f(x) - f(4)| = |x^2 - 16| < \epsilon$ if $|x - 4| < \delta$. We get $f(x) = 16 + \epsilon = 16.001$ at $x = 4.000124998$, which corresponds to $\delta = 0.000124998$; and $f(x) = 16 - \epsilon = 15.999$ at $x = 3.999874998$, for which $\delta = 0.000125002$. Use the smaller δ: thus $|f(x) - 16| < \epsilon$ provided $|x - 4| < 0.000125$ (to six decimals).

3. (a) $x_0 = (1.95)^2 = 3.8025, x_2 = (2.05)^2 = 4.2025$

(b) $\delta = \min(|4 - 3.8025|, |4 - 4.2025|) = 0.1975$

5. $|(x^3 - 4x + 5) - 2| < 0.05$, $-0.05 < (x^3 - 4x + 5) - 2 < 0.05$, $1.95 < x^3 - 4x + 5 < 2.05$; $x^3 - 4x + 5 = 1.95$ at $x = 1.0616$, $x^3 - 4x + 5 = 2.05$ at $x = 0.9558$; $\delta = \min(1.0616 - 1, 1 - 0.9558) = 0.0442$

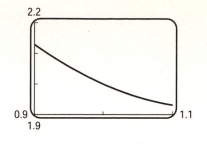

7. With the TRACE feature of a calculator we discover that (to five decimal places) $(0.87000, 1.80274)$ and $(1.13000, 2.19301)$ belong to the graph. Set $x_0 = 0.87$ and $x_1 = 1.13$. Since the graph of $f(x)$ rises from left to right, we see that if $x_0 < x < x_1$ then $1.80274 < f(x) < 2.19301$, and therefore $1.8 < f(x) < 2.2$. So we can take $\delta = 0.13$.

9. We have only shown that, for one specific epsilon, there is a delta; for other, smaller epsilons there may not be a delta.

11. $3x - 6, x - 2, \delta$

13. Let $\epsilon > 0$ be given. Then $|f(x) - 3| = |3 - 3| = 0 < \epsilon$ regardless of x, and hence any $\delta > 0$ will work.

15. $|3x - 15| = 3|x - 5| < \epsilon$ if $|x - 5| < \frac{1}{3}\epsilon$, $\delta = \frac{1}{3}\epsilon$

17. $\left|\dfrac{2x^2 + x}{x} - 1\right| = |2x| < \epsilon$ if $|x| < \frac{1}{2}\epsilon$, $\delta = \frac{1}{2}\epsilon$

19. $|f(x) - 3| = |x + 2 - 3| = |x - 1| < \epsilon$ if $0 < |x - 1| < \epsilon$, $\delta = \epsilon$

21. If $\epsilon > 0$ is given, then take $\delta = \epsilon$; if $|x - 0| = |x| < \delta$, then $|x - 0| = |x| < \epsilon$.

23. True. Clearly $L = ma + b$; then $f(x) - f(a) = mx + b - (ma + b) = m(x - a)$ so $|f(x) - f(a)| < m\delta = \epsilon$.

25. True only for constant functions. Given $\epsilon > 0$, in order to make $|f(x) - f(a)| < \epsilon$ it is sufficient to have $|x - a|$ smaller than a prescribed but constant δ. Since ϵ is arbitrary, let it take on the values $1/n, n = 1, 2, \ldots$. Then $|f(x) - f(a)| = 0$ and f is a constant function within the region $|x - a| < \delta$.

27. For the first part, let $\epsilon > 0$. Then there exists $\delta > 0$ such that if $a < x < a + \delta$ then $|f(x) - L| < \epsilon$. For the left limit replace $a < x < a + \delta$ with $a - \delta < x < a$.

29. (a) $|(3x^2 + 2x - 20 - 300| = |3x^2 + 2x - 320| = |(3x + 32)(x - 10)| = |3x + 32| \cdot |x - 10|$
  (b) If $|x - 10| < 1$ then $|3x + 32| < 65$, since clearly $x < 11$
  (c) $\delta = \min(1, \epsilon/65)$; $|3x + 32| \cdot |x - 10| < 65 \cdot |x - 10| < 65 \cdot \epsilon/65 = \epsilon$

31. if $\delta < 1$ then $|2x^2 - 2| = 2|x - 1||x + 1| < 6|x - 1| < \epsilon$ if $|x - 1| < \frac{1}{6}\epsilon$, $\delta = \min(1, \frac{1}{6}\epsilon)$

33. If $\delta < \frac{1}{2}$ and $|x - (-2)| < \delta$ then $-\frac{5}{2} < x < -\frac{3}{2}$, $x + 1 < -\frac{1}{2}$, $|x + 1| > \frac{1}{2}$; then
$$\left|\frac{1}{x + 1} - (-1)\right| = \frac{|x + 2|}{|x + 1|} < 2|x + 2| < \epsilon \text{ if } |x + 2| < \frac{1}{2}\epsilon, \delta = \min\left(\frac{1}{2}, \frac{1}{2}\epsilon\right)$$

35. $|\sqrt{x} - 2| = \left|(\sqrt{x} - 2)\dfrac{\sqrt{x} + 2}{\sqrt{x} + 2}\right| = \left|\dfrac{x - 4}{\sqrt{x} + 2}\right| < \frac{1}{2}|x - 4| < \epsilon$ if $|x - 4| < 2\epsilon$, $\delta = 2\epsilon$

37. Let $\epsilon > 0$ be given and take $\delta = \epsilon$. If $|x| < \delta$, then $|f(x) - 0| = 0$ if x is rational, and equals $|x| < \delta = \epsilon$.

39. (a) $\dfrac{x_1^2}{1 + x_1^2} = 1 - \epsilon, \ x_1 = -\sqrt{\dfrac{1 - \epsilon}{\epsilon}}; \quad \dfrac{x_2^2}{1 + x_2^2} = 1 - \epsilon, \ x_2 = \sqrt{\dfrac{1 - \epsilon}{\epsilon}}$

   (b) $N = \sqrt{\dfrac{1 - \epsilon}{\epsilon}}$ $\qquad\qquad\qquad\qquad$ (c) $N = -\sqrt{\dfrac{1 - \epsilon}{\epsilon}}$

41. $\dfrac{1}{x^2} < 0.01$ if $|x| > 10$, $N = 10$

43. $\left| \dfrac{x}{x + 1} - 1 \right| = \left| \dfrac{1}{x + 1} \right| < 0.001$ if $|x + 1| > 1000$, $x > 999$, $N = 999$

45. $\left| \dfrac{1}{x + 2} - 0 \right| < 0.005$ if $|x + 2| > 200$, $-x - 2 > 200$, $x < -202$, $N = -202$

47. $\left| \dfrac{4x - 1}{2x + 5} - 2 \right| = \left| \dfrac{11}{2x + 5} \right| < 0.1$ if $|2x + 5| > 110$, $-2x - 5 > 110$, $2x < -115$, $x < -57.5$, $N = -57.5$

49. $\left| \dfrac{1}{x^2} \right| < \epsilon$ if $|x| > \dfrac{1}{\sqrt{\epsilon}}$, $N = \dfrac{1}{\sqrt{\epsilon}}$

51. $\left| \dfrac{4x - 1}{2x + 5} - 2 \right| = \left| \dfrac{11}{2x + 5} \right| < \epsilon$ if $|2x + 5| > \dfrac{11}{\epsilon}$, $-2x - 5 > \dfrac{11}{\epsilon}$, $2x < -\dfrac{11}{\epsilon} - 5$, $x < -\dfrac{11}{2\epsilon} - \dfrac{5}{2}$,

$N = -\dfrac{5}{2} - \dfrac{11}{2\epsilon}$

53. $\left| \dfrac{2\sqrt{x}}{\sqrt{x} - 1} - 2 \right| = \left| \dfrac{2}{\sqrt{x} - 1} \right| < \epsilon$ if $\sqrt{x} - 1 > \dfrac{2}{\epsilon}$, $\sqrt{x} > 1 + \dfrac{2}{\epsilon}$, $x > \left(1 + \dfrac{2}{\epsilon} \right)^2$, $N > \left(1 + \dfrac{2}{\epsilon} \right)^2$

55. (a) $\dfrac{1}{x^2} > 100$ if $|x| < \dfrac{1}{10}$ $\qquad\qquad$ (b) $\dfrac{1}{|x - 1|} > 1000$ if $|x - 1| < \dfrac{1}{1000}$

   (c) $\dfrac{-1}{(x - 3)^2} < -1000$ if $|x - 3| < \dfrac{1}{10\sqrt{10}}$ $\qquad$ (d) $-\dfrac{1}{x^4} < -10000$ if $x^4 < \dfrac{1}{10000}$, $|x| < \dfrac{1}{10}$

57. if $M > 0$ then $\dfrac{1}{(x - 3)^2} > M$, $0 < (x - 3)^2 < \dfrac{1}{M}$, $0 < |x - 3| < \dfrac{1}{\sqrt{M}}$, $\delta = \dfrac{1}{\sqrt{M}}$

59. if $M > 0$ then $\dfrac{1}{|x|} > M$, $0 < |x| < \dfrac{1}{M}$, $\delta = \dfrac{1}{M}$

61. if $M < 0$ then $-\dfrac{1}{x^4} < M$, $0 < x^4 < -\dfrac{1}{M}$, $|x| < \dfrac{1}{(-M)^{1/4}}$, $\delta = \dfrac{1}{(-M)^{1/4}}$

63. if $x > 2$ then $|x + 1 - 3| = |x - 2| = x - 2 < \epsilon$ if $2 < x < 2 + \epsilon$, $\delta = \epsilon$

65. if $x > 4$ then $\sqrt{x - 4} < \epsilon$ if $x - 4 < \epsilon^2$, $4 < x < 4 + \epsilon^2$, $\delta = \epsilon^2$

67. if $x > 2$ then $|f(x) - 2| = |x - 2| = x - 2 < \epsilon$ if $2 < x < 2 + \epsilon$, $\delta = \epsilon$

69. (a) Definition: For every $M < 0$ there corresponds $\delta > 0$ such that if $1 < x < 1 + \delta$ then $f(x) < M$. In our case we want $\dfrac{1}{1 - x} < M$, $1 - x > \dfrac{1}{M}$, $x < 1 - \dfrac{1}{M}$, take $\delta = -\dfrac{1}{M}$.

(b) Definition: For every $M > 0$ there corresponds $\delta > 0$ such that if $1 - \delta < x < 1$ then $f(x) > M$. In our case we want $\dfrac{1}{1 - x} > M, 1 - x < \dfrac{1}{M}, x > 1 - \dfrac{1}{M}$, take $\delta = \dfrac{1}{M}$.

71. (a) Given any $M > 0$ there corresponds $N > 0$ such that if $x > N$ then $f(x) > M$, $x + 1 > M$, $x > M - 1$, $N = M - 1$.

(b) Given any $M < 0$ there corresponds $N < 0$ such that if $x < N$ then $f(x) < M$, $x + 1 < M$, $x < M - 1$, $N = M - 1$.

73. (a) 0.4 amperes   **(b)** $[0.3947, 0.4054]$ **(c)** $\left[\dfrac{3}{7.5 + \delta}, \dfrac{3}{7.5 - \delta}\right]$

(d) 0.0187   **(e)** It becomes infinite.

EXERCISE SET 1.5

1. (a) $\lim\limits_{x \to 2} f(x)$ does not exist   **(b)** $\lim\limits_{x \to 2} f(x)$ does not exist
  (c) $\lim\limits_{x \to 2} f(x) \neq f(2)$ **(d)** yes
  (e) yes   **(f)** yes

3. (a) $f(1)$ is not defined   **(b)** yes
  (c) $f(1)$ is not defined   **(d)** yes
  (e) $\lim\limits_{x \to 2} f(x) \neq f(2)$ **(f)** yes

5. (a) no, $x = 4$ **(b)** no, $x = 4$ **(c)** no, $x = 4$ **(d)** yes
  (e) yes   **(f)** yes   **(g)** yes

7. (a)
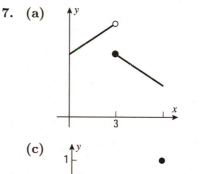
      (b)
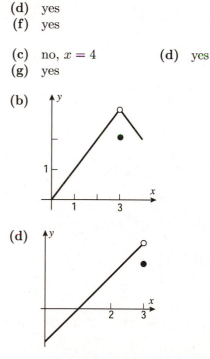

  (c)
      (d)

9. (a)
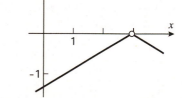
      (b) One second could cost you one dollar.

11. none   **13.** none   **15.** $x = 0, -1/2$

17. $x = -1, 0, 1$

19. none

21. none; $f(x) = 2x + 3$ is continuous on $x < 4$ and $f(x) = 7 + \dfrac{16}{x}$ is continuous on $4 < x$;
$\lim\limits_{x \to 4^-} f(x) = \lim\limits_{x \to 4^+} f(x) = f(4) = 11$ so f is continuous at $x = 4$

23. true; Theorem 1.5.5

25. false; e.g. $f(x) = g(x) = 2$ if $x \neq 3$, $f(3) = -7, g(3) = 9$

27. true; use Theorem 1.5.3 with $f(x) = g(x) = \sqrt{x}$

29. (a) f is continuous for $x < 1$, and for $x > 1$; $\lim\limits_{x \to 1^-} f(x) = 5$, $\lim\limits_{x \to 1^+} f(x) = k$, so if $k = 5$ then f is continuous for all x

    (b) f is continuous for $x < 2$, and for $x > 2$; $\lim\limits_{x \to 2^-} f(x) = 4k$, $\lim\limits_{x \to 2^+} f(x) = 4 + k$, so if $4k = 4 + k$, $k = 4/3$ then f is continuous for all x

31. f is continuous for $x < -1$, $-1 < x < 2$ and $x > 2$; $\lim\limits_{x \to -1^-} f(x) = 4$, $\lim\limits_{x \to -1^+} f(x) = k$, so $k = 4$ is required. Next, $\lim\limits_{x \to 2^-} f(x) = 3m + k = 3m + 4$, $\lim\limits_{x \to 2^+} f(x) = 9$, so $3m + 4 = 9, m = 5/3$ and f is continuous everywhere if $k = 4, m = 5/3$

33. (a)   (b)

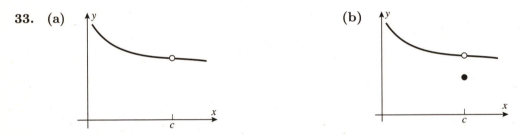

35. (a) $x = 0$, $\lim\limits_{x \to 0^-} f(x) = -1 \neq +1 = \lim\limits_{x \to 0^+} f(x)$ so the discontinuity is not removable

    (b) $x = -3$; define $f(-3) = -3 = \lim\limits_{x \to -3} f(x)$, then the discontinuity is removable

    (c) f is undefined at $x = \pm 2$; at $x = 2$, $\lim\limits_{x \to 2} f(x) = 1$, so define $f(2) = 1$ and f becomes continuous there; at $x = -2$, $\lim\limits_{x \to -2}$ does not exist, so the discontinuity is not removable

37. (a) discontinuity at $x = 1/2$, not removable;   (b) $2x^2 + 5x - 3 = (2x - 1)(x + 3)$
           at $x = -3$, removable

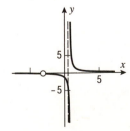

39. Write $f(x) = x^{3/5} = (x^3)^{1/5}$ as the composition (Theorem 1.5.6) of the two continuous functions $g(x) = x^3$ and $h(x) = x^{1/5}$; it is thus continuous.

41. Since f and g are continuous at $x = c$ we know that $\lim_{x \to c} f(x) = f(c)$ and $\lim_{x \to c} g(x) = g(c)$. In the following we use Theorem 1.2.2.

   (a) $f(c) + g(c) = \lim_{x \to c} f(x) + \lim_{x \to c} g(x) = \lim_{x \to c} (f(x) + g(x))$ so $f + g$ is continuous at $x = c$.

   (b) same as (a) except the $+$ sign becomes a $-$ sign

   (c) $\dfrac{f(c)}{g(c)} = \dfrac{\lim_{x \to c} f(x)}{\lim_{x \to c} g(x)} = \lim_{x \to c} \dfrac{f(x)}{g(x)}$ so $\dfrac{f}{g}$ is continuous at $x = c$

43. **(a)** Let $h = x - c, x = h + c$. Then by Theorem 1.5.5,
   $$\lim_{h \to 0} f(h + c) = f(\lim_{h \to 0} (h + c)) = f(c)$$

   (b) With $g(h) = f(c + h)$, $\lim_{h \to 0} g(h) = \lim_{h \to 0} f(c + h) = f(c) = g(0)$, so $g(h)$ is continuous at $h = 0$. That is, $f(c + h)$ is continuous at $h = 0$, so f is continuous at $x = c$.

45. Of course such a function must be discontinuous. Let $f(x) = 1$ on $0 \leq x < 1$, and $f(x) = -1$ on $1 \leq x \leq 2$.

47. If $f(x) = x^3 + x^2 - 2x - 1$ then $f(-1) = 1$, $f(1) = -1$. –Use the Intermediate Value Theorem.

49. For the negative root, use intervals on the x-axis as follows: $[-2, -1]$; since $f(-1.3) < 0$ and $f(-1.2) > 0$, the midpoint $x = -1.25$ of $[-1.3, -1.2]$ is the required approximation of the root. For the positive root use the interval $[0, 1]$; since $f(0.7) < 0$ and $f(0.8) > 0$, the midpoint $x = 0.75$ of $[0.7, 0.8]$ is the required approximation.

51. For the positive root, use intervals on the x-axis as follows: $[2, 3]$; since $f(2.2) < 0$ and $f(2.3) > 0$, use the interval $[2.2, 2.3]$. Since $f(2.23) < 0$ and $f(2.24) > 0$ the midpoint $x = 2.235$ of $[2.23, 2.24]$ is the required approximation of the root. For the negative root use, of course, $x = -2.235$.

53. Note that T has period $2\pi, T(\theta + 2\pi) = T(\theta)$, so that
   $f(\theta + \pi) = T(\theta + 2\pi) - T(\theta + \pi) = -(T(\theta + \pi) - T(\theta)) = -f(\theta)$. Thus f changes sign from $\theta = 0$ to $\theta = \pi$, and thus there is t_0 between 0 and π such that $f(t_0) = 0$, so that $T(t_0 + \pi) = T(t_0)$.

55. Since R and L are arbitrary, we can introduce coordinates so that L is the x-axis. Let $f(z)$ be as in Exercise 54. Then for large z, $f(z) = $ area of ellipse, and for small z, $f(z) = 0$. By the Intermediate Value Theorem there is z_1 such that $f(z_1) = $ half of the area of the ellipse.

57. For $x \geq 0, f$ is increasing and so is $1 - 1$. It is continuous everywhere and thus by Theorem 1.5.7 it has an inverse defined on its range $[5, +\infty)$ which is continuous there.

EXERCISE SET 1.6

1. none

3. $n\pi, n = 0, \pm 1, \pm 2, \ldots$

5. $x = n\pi, n = 0, \pm 1, \pm 2, \ldots$

7. $2n\pi + \pi/6, 2n\pi + 5\pi/6, n = 0, \pm 1, \pm 2, \ldots$

9. **(a)** $\sin x, x^3 + 7x + 1$ **(b)** $|x|, \sin x$ **(c)** $x^3, \cos x, x + 1$

11. $\cos \left(\lim_{x \to +\infty} \dfrac{1}{x} \right) = \cos 0 = 1$

13. $3 \lim_{\theta \to 0} \dfrac{\sin 3\theta}{3\theta} = 3$

15. $\lim\limits_{x\to0} x - 3\lim\limits_{x\to0}\dfrac{\sin x}{x} = -3$

17. $\left(\lim\limits_{\theta\to0^+}\dfrac{1}{\theta}\right)\lim\limits_{\theta\to0^+}\dfrac{\sin\theta}{\theta} = +\infty$

19. $\dfrac{\tan 7x}{\sin 3x} = \dfrac{7}{3\cos 7x}\dfrac{\sin 7x}{7x}\dfrac{3x}{\sin 3x}$ so $\lim\limits_{x\to0}\dfrac{\tan 7x}{\sin 3x} = \dfrac{7}{3(1)}(1)(1) = \dfrac{7}{3}$

21. $\dfrac{1}{5}\lim\limits_{x\to0^+}\sqrt{x}\lim\limits_{x\to0^+}\dfrac{\sin x}{x} = 0$

23. $\left(\lim\limits_{x\to0} x\right)\left(\lim\limits_{x\to0}\dfrac{\sin x^2}{x^2}\right) = 0$

25. $\dfrac{t^2}{1-\cos^2 t} = \left(\dfrac{t}{\sin t}\right)^2$, so $\lim\limits_{t\to0}\dfrac{t^2}{1-\cos^2 t} = 1$

27. $\dfrac{\theta^2}{1-\cos\theta}\dfrac{1+\cos\theta}{1+\cos\theta} = \dfrac{\theta^2(1+\cos\theta)}{1-\cos^2\theta} = \left(\dfrac{\theta}{\sin\theta}\right)^2(1+\cos\theta)$ so $\lim\limits_{\theta\to0}\dfrac{\theta^2}{1-\cos\theta} = (1)^2 2 = 2$

29. $\lim\limits_{x\to0^+}\sin\left(\dfrac{1}{x}\right) = \lim\limits_{t\to+\infty}\sin t$; limit does not exist

31. $\lim\limits_{x\to0}\dfrac{\tan ax}{\sin bx} = \lim\limits_{\to0}\dfrac{a}{b}\dfrac{\sin ax}{ax}\dfrac{1}{\cos ax}\dfrac{bx}{\sin bx} = a/b$

33. **(a)**

4	4.5	4.9	5.1	5.5	5.01
0.093497	0.100932	0.100842	0.098845	0.091319	0.0998984

The limit appears to be 0.1.

(b) Let $t = x - 5$. Then $t \to 0$ as $x \to 5$ and

$$\lim\limits_{x\to5}\dfrac{\sin(x-5)}{x^2-25} = \lim\limits_{x\to5}\dfrac{1}{x+5}\lim\limits_{t\to0}\dfrac{\sin t}{t} = \dfrac{1}{10}\cdot 1 = \dfrac{1}{10}$$

35. Let $\epsilon > 0$ and $\delta = \epsilon$. Then if $|x+1| < \delta$ then $|f(x)+5| < \epsilon$.

37. true: the functions $f(x) = x, g(x) = \sin x$, and $h(x) = 1/x$ are continuous everywhere except possibly at $x = 0$, so by Theorem 1.5.6 the given function is continuous everywhere except possibly at $x = 0$. We prove that $\lim\limits_{x\to0} x\sin(1/x) = 0$. Let $\epsilon > 0$. Then with $\delta = \epsilon$, if $|x| < \delta$ then $|x\sin(1/x)| \le |x| < \delta = \epsilon$, and hence f is continuous everywhere.

39. **(a)** The student calculated x in degrees rather than radians.

(b) $\sin x = \sin t$ where x is measured in degrees, t is measured in radians and $t = \dfrac{\pi x}{180}$. Thus

$$\lim\limits_{x\to0}\dfrac{\sin x}{x} = \lim\limits_{t\to0}\dfrac{\sin t}{(180t/\pi)} = \dfrac{\pi}{180}.$$

41. Denote θ by x in accordance with Figure 1.6.4. Let P have coordinates $(\cos x, \sin x)$ and Q coordinates $(1,0)$ so that $c^2(x) = (1-\cos x)^2 + \sin^2 x = 2(1-\cos x)$. Since $s = r\theta = 1\cdot x = x$ we have

$$\lim\limits_{x\to0^+}\dfrac{c^2(x)}{s^2(x)} = \lim\limits_{x\to0^+} 2\dfrac{1-\cos x}{x^2} = \lim\limits_{x\to0^+} 2\dfrac{1-\cos x}{x^2}\dfrac{1+\cos x}{1+\cos x} = \lim\limits_{x\to0^+}\left(\dfrac{\sin x}{x}\right)^2\dfrac{2}{1+\cos x} = 1$$

43. $\lim\limits_{x\to0^-} f(x) = k\lim\limits_{x\to0}\dfrac{\sin kx}{kx\cos kx} = k$, $\lim\limits_{x\to0^+} f(x) = 2k^2$, so $k = 2k^2$, $k = \dfrac{1}{2}$

45. (a) $\lim\limits_{t \to 0^+} \dfrac{\sin t}{t} = 1$

(b) $\lim\limits_{t \to 0^-} \dfrac{1 - \cos t}{t} = 0$ (Theorem 1.6.3)

(c) $\sin(\pi - t) = \sin t$, so $\lim\limits_{x \to \pi} \dfrac{\pi - x}{\sin x} = \lim\limits_{t \to 0} \dfrac{t}{\sin t} = 1$

47. Let $t = x - 1$. Then $\sin(\pi(t+1)) = -\sin \pi t$, so $\lim\limits_{x \to 1} \dfrac{\sin(\pi x)}{x - 1} = \lim\limits_{t \to 0} \dfrac{-\pi \sin \pi t}{\pi t} = -\pi$

49. Since $\lim\limits_{x \to 0} \sin(1/x)$ does not exist, no conclusions can be drawn.

51. $\lim\limits_{x \to +\infty} f(x) = 0$ by the Squeezing Theorem

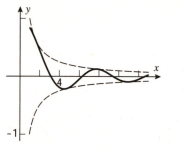

53. (a) Let $f(x) = x - \cos x$; $f(0) = -1$, $f(\pi/2) = \pi/2$. By the IVT there must be a solution of $f(x) = 0$.

(b)

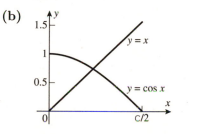

(c) 0.739

55. (a) Gravity is strongest at the poles and weakest at the equator.

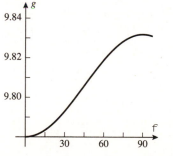

(b) Let $g(\phi)$ be the given function. Then $g(38) < 9.8$ and $g(39) > 9.8$, so by the Intermediate Value Theorem there is a value c between 38 and 39 for which $g(c) = 9.8$ exactly.

REVIEW EXERCISES, CHAPTER 1

1. (a) 1   **(b)** no limit   **(c)** no limit

    (d) 1   **(e)** 3   **(f)** 0

    (g) 0   **(h)** 2   **(i)** 1/2

3. 1

5. If $x \neq -3$ then $\dfrac{3x+9}{x^2+4x+3} = \dfrac{3}{x+1}$ with limit $-\dfrac{3}{2}$

7. $\dfrac{2^5}{3} = \dfrac{32}{3}$

9. (a) $y = 0$ (b) none   (c) $y = 2$

11. 1   **13.** $3 - k$

15. Let $t = x + 1$. Then $\displaystyle\lim_{t \to 0} \dfrac{\sin t}{t(t-2)} = \dfrac{1}{2}$

17. (a) $f(x) = 2x/(x-1)$ (b)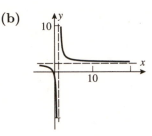

19. (a) $\displaystyle\lim_{x \to 2} f(x) = 5$
  (b) 0.0045

21. (a) $|4x - 7 - 1| < 0.01, 4|x - 2| < 0.01, |x - 2| < 0.0025$, let $\delta = 0.0025$
  (b) $\left|\dfrac{4x^2 - 9}{2x - 3} - 6\right| < 0.05, |2x + 3 - 6| < 0.05, |x - 1.5| < 0.025$, take $\delta = 0.025$
  (c) $|x^2 - 16| < 0.001$; if $\delta < 1$ then $|x + 4| < 9$ if $|x - 4| < 1$;   then $|x^2 - 16| = |x - 4||x + 4| \leq$
      $9|x - 4| < 0.001$ provided $|x - 4| < 0.001/9 = 1/9000$, take $\delta = 1/9000$, then $|x^2 - 16| <$
      $9|x - 4| < 9(1/9000) = 1/1000 = 0.001$

23. Let $\epsilon = f(x_0)/2 > 0$; then there corresponds $\delta > 0$ such that if $|x - x_0| < \delta$ then $|f(x) - f(x_0)| < \epsilon$,
    $-\epsilon < f(x) - f(x_0) < \epsilon, f(x) > f(x_0) - \epsilon = f(x_0)/2 > 0$ for $x_0 - \delta < x < x_0 + \delta$.

25. (a) f is not defined at $x = \pm 1$, continuous elsewhere
  (b) none
  (c) f is not defined at $x = 0, -3$

27. For $x < 2$ f is a polynomial and is continuous; for $x > 2$ f is a polynomial and is continuous. At
    $x = 2$, $f(2) = -13 \neq 13 = \displaystyle\lim_{x \to 2^+} f(x)$ so f is not continuous there.

29. $f(x) = -1$ for $a \leq x < \dfrac{a+b}{2}$ and $f(x) = 1$ for $\dfrac{a+b}{2} \leq x \leq b$

31. $f(-6) = 185, f(0) = -1, f(2) = 65$; apply Theorem 1.5.8 twice, once on $[-6, 0]$ and once on $[0, 2]$

MAKING CONNECTIONS, CHAPTER 1

1. Let $P(x, x^2)$ be an arbitrary point on the curve, let $Q(-x, x^2)$ be its reflection through the y-axis, let $O(0,0)$ be the origin. The perpendicular bisector of the line which connects P with O meets the y-axis at a point $C(0, \lambda(x))$, whose ordinate is as yet unknown. A segment of the bisector is also the altitude of the triangle ΔOPC which is isosceles, so that $CP = CO$.
   Using the symmetrically opposing point Q in the second quadrant, we see that $\overline{OP} = \overline{OQ}$ too, and thus C is equidistant from the three points O, P, Q and is thus the center of the unique circle that passes through the three points.

3. Replace the parabola with the general curve $y = f(x)$ which passes through $P(x, f(x))$ and $S(0, f(0))$. Let the perpendicular bisector of the line through S and P meet the y-axis at $C(0, \lambda)$, and let $R(x/2, (f(x) - \lambda)/2)$ be the midpoint of P and S. By the Pythagorean Theorem,
   $$\overline{CS}^2 = \overline{RS}^2 + \overline{CR}^2, \text{ or } (\lambda - f(0))^2 = x^2/4 + \left[\frac{f(x) + f(0)}{2} - f(0)\right]^2 + x^2/4 + \left[\frac{f(x) + f(0)}{2} - \lambda\right]^2,$$
   which yields $\lambda = \frac{1}{2}\left[f(0) + f(x) + \frac{x^2}{f(x) - f(0)}\right].$

CHAPTER 2
The Derivative

EXERCISE SET 2.1

1. **(a)** $m_{\text{tan}} = (50 - 10)/(15 - 5)$
$= 40/10$
$= 4 \text{ m/s}$

(b)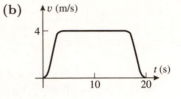

3. **(a)** $(10 - 10)/(3 - 0) = 0 \text{ cm/s}$

(b) $t = 0$, $t = 2$, $t = 4.2$, and $t = 8$ (horizontal tangent line)

(c) maximum: $t = 1$ (slope > 0) minimum: $t = 3$ (slope < 0)

(d) $(3 - 18)/(4 - 2) = -7.5 \text{ cm/s}$ (slope of estimated tangent line to curve at $t = 3$)

5. It is a straight line with slope equal to the velocity.

7.

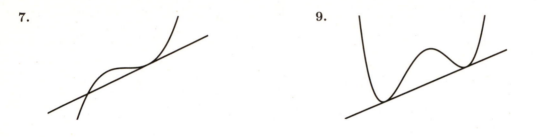

9.

11. **(a)** $m_{\text{sec}} = \dfrac{f(1) - f(0)}{1 - 0} = \dfrac{2}{1} = 2$

(b) $m_{\text{tan}} = \lim\limits_{x_1 \to 0} \dfrac{f(x_1) - f(0)}{x_1 - 0} = \lim\limits_{x_1 \to 0} \dfrac{2x_1^2 - 0}{x_1 - 0} = \lim\limits_{x_1 \to 0} 2x_1 = 0$

(c) $m_{\text{tan}} = \lim\limits_{x_1 \to x_0} \dfrac{f(x_1) - f(x_0)}{x_1 - x_0}$

$= \lim\limits_{x_1 \to x_0} \dfrac{2x_1^2 - 2x_0^2}{x_1 - x_0}$

$= \lim\limits_{x_1 \to x_0} (2x_1 + 2x_0)$

$= 4x_0$

(d) The tangent line is the x-axis.

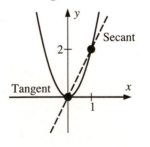

13. **(a)** $m_{\text{sec}} = \dfrac{f(3) - f(2)}{3 - 2} = \dfrac{1/3 - 1/2}{1} = -\dfrac{1}{6}$

(b) $m_{\text{tan}} = \lim\limits_{x_1 \to 2} \dfrac{f(x_1) - f(2)}{x_1 - 2} = \lim\limits_{x_1 \to 2} \dfrac{1/x_1 - 1/2}{x_1 - 2}$

$= \lim\limits_{x_1 \to 2} \dfrac{2 - x_1}{2x_1(x_1 - 2)} = \lim\limits_{x_1 \to 2} \dfrac{-1}{2x_1} = -\dfrac{1}{4}$

(c) $\quad m_{\text{tan}} = \displaystyle\lim_{x_1 \to x_0} \frac{f(x_1) - f(x_0)}{x_1 - x_0}$

$\qquad\qquad = \displaystyle\lim_{x_1 \to x_0} \frac{1/x_1 - 1/x_0}{x_1 - x_0}$

$\qquad\qquad = \displaystyle\lim_{x_1 \to x_0} \frac{x_0 - x_1}{x_0 x_1 (x_1 - x_0)}$

$\qquad\qquad = \displaystyle\lim_{x_1 \to x_0} \frac{-1}{x_0 x_1} = -\frac{1}{x_0^2}$

(d)

15. (a) $\quad m_{\text{tan}} = \displaystyle\lim_{x_1 \to x_0} \frac{f(x_1) - f(x_0)}{x_1 - x_0} = \lim_{x_1 \to x_0} \frac{(x_1^2 - 1) - (x_0^2 - 1)}{x_1 - x_0}$

$\qquad\qquad = \displaystyle\lim_{x_1 \to x_0} \frac{(x_1^2 - x_0^2)}{x_1 - x_0} = \lim_{x_1 \to x_0} (x_1 + x_0) = 2x_0$

(b) $\quad m_{\text{tan}} = 2(-1) = -2$

17. (a) $\quad m_{\text{tan}} = \displaystyle\lim_{x_1 \to x_0} \frac{f(x_1) - f(x_0)}{x_1 - x_0} = \lim_{x_1 \to x_0} \frac{(x_1 + \sqrt{x_1}) - (x_0 + \sqrt{x_0})}{x_1 - x_0}$

$\qquad\qquad = \displaystyle\lim_{x_1 \to x_0} \left(1 + \frac{1}{\sqrt{x_1} + \sqrt{x_0}}\right) = 1 + \frac{1}{2\sqrt{x_0}}$

(b) $\quad m_{\text{tan}} = 1 + \dfrac{1}{2\sqrt{1}} = \dfrac{3}{2}$

19. True. Let $x = 1 + h$.

21. False. Velocity represents the <u>rate</u> at which position changes.

23. (a) 72°F at about 4:30 P.M. $\qquad\qquad$ **(b)** about $(67 - 43)/6 = 4°\text{F/h}$

$\quad$ **(c)** decreasing most rapidly at about 9 P.M.; rate of change of temperature is about $-7°\text{F/h}$ (slope of estimated tangent line to curve at 9 P.M.)

25. (a) during the first year after birth

$\quad$ **(b)** about 6 cm/year (slope of estimated tangent line at age 5)

$\quad$ **(c)** the growth rate is greatest at about age 14; about 10 cm/year

$\quad$ **(d)**

27. (a) $0.3 \cdot 40^3 = 19{,}200 \text{ ft}$ $\qquad\qquad$ **(b)** $v_{\text{ave}} = 19{,}200/40 = 480 \text{ ft/s}$

$\quad$ **(c)** Solve $s = 0.3t^3 = 1000; t \approx 14.938$ so $v_{\text{ave}} \approx 1000/14.938 \approx 66.943 \text{ ft/s}$.

$\quad$ **(d)** $v_{\text{inst}} = \displaystyle\lim_{h \to 0} \frac{0.3(40 + h)^3 - 0.3 \cdot 40^3}{h} = \lim_{h \to 0} \frac{0.3(4800h + 120h^2 + h^3)}{h}$

$\qquad\qquad = \displaystyle\lim_{h \to 0} 0.3(4800 + 120h + h^2) = 1440 \text{ ft/s}$

EXERCISE SET 2.2

1. $f'(1) = 2$, $f'(3) = 0$, $f'(5) = -2$, $f'(6) = -1$

3. **(a)** $f'(a)$ is the slope of the tangent line.

    (b) $f'(2) = m = 3$ **(c)** the same, $f'(2) = 3$

5.

7. $y - (-1) = 5(x - 3)$, $y = 5x - 16$

9. $f'(x) = \lim\limits_{h \to 0} \dfrac{f(x+h) - f(x)}{h} = \lim\limits_{h \to 0} \dfrac{2(x+h)^2 - 2x^2}{h} = \lim\limits_{h \to 0} \dfrac{4xh + 2h^2}{h} = 4x$;

$f'(1) = 4$ so $y - 2 = 4(x - 1)$, $y = 4x - 2$

11. $f'(x) = \lim\limits_{h \to 0} \dfrac{f(x+h) - f(x)}{h} = \lim\limits_{h \to 0} \dfrac{(x+h)^3 - x^3}{h} = \lim\limits_{h \to 0} (3x^2 + 3xh + h^2) = 3x^2$;

$f'(0) = 0$ so $y - 0 = 0(x - 0)$, $y = 0$

13. $f'(x) = \lim\limits_{h \to 0} \dfrac{f(x+h) - f(x)}{h} = \lim\limits_{h \to 0} \dfrac{\sqrt{x+1+h} - \sqrt{x+1}}{h}$

$\qquad = \lim\limits_{h \to 0} \dfrac{\sqrt{x+1+h} - \sqrt{x+1}}{h} \cdot \dfrac{\sqrt{x+1+h} + \sqrt{x+1}}{\sqrt{x+1+h} + \sqrt{x+1}}$

$\qquad = \lim\limits_{h \to 0} \dfrac{h}{h(\sqrt{x+1+h} + \sqrt{x+1})} = \dfrac{1}{2\sqrt{x+1}}$;

$\qquad f(8) = \sqrt{8+1} = 3$ and $f'(8) = \dfrac{1}{6}$ so $y - 3 = \dfrac{1}{6}(x - 8)$, $y = \dfrac{1}{6}x + \dfrac{5}{3}$

15. $f'(x) = \lim\limits_{\Delta x \to 0} \dfrac{\dfrac{1}{x + \Delta x} - \dfrac{1}{x}}{\Delta x} = \lim\limits_{\Delta x \to 0} \dfrac{\dfrac{x - (x + \Delta x)}{x(x + \Delta x)}}{\Delta x}$

$\qquad = \lim\limits_{\Delta x \to 0} \dfrac{-\Delta x}{x \Delta x(x + \Delta x)} = \lim\limits_{\Delta x \to 0} -\dfrac{1}{x(x + \Delta x)} = -\dfrac{1}{x^2}$

17. $f'(x) = \lim\limits_{\Delta x \to 0} \dfrac{(x + \Delta x)^2 - (x + \Delta x) - (x^2 - x)}{\Delta x} = \lim\limits_{\Delta x \to 0} \dfrac{2x\Delta x + (\Delta x)^2 - \Delta x}{\Delta x}$

$\qquad = \lim\limits_{\Delta x \to 0} (2x - 1 + \Delta x) = 2x - 1$

19. $f'(x) = \lim\limits_{\Delta x \to 0} \dfrac{\dfrac{1}{\sqrt{x + \Delta x}} - \dfrac{1}{\sqrt{x}}}{\Delta x} = \lim\limits_{\Delta x \to 0} \dfrac{\sqrt{x} - \sqrt{x + \Delta x}}{\Delta x \sqrt{x}\sqrt{x + \Delta x}}$

$\qquad = \lim\limits_{\Delta x \to 0} \dfrac{x - (x + \Delta x)}{\Delta x \sqrt{x}\sqrt{x + \Delta x}(\sqrt{x} + \sqrt{x + \Delta x})} = \lim\limits_{\Delta x \to 0} \dfrac{-1}{\sqrt{x}\sqrt{x + \Delta x}(\sqrt{x} + \sqrt{x + \Delta x})} = -\dfrac{1}{2x^{3/2}}$

21. $f'(t) = \lim\limits_{h \to 0} \dfrac{f(t+h) - f(t)}{h} = \lim\limits_{h \to 0} \dfrac{[4(t+h)^2 + (t+h)] - [4t^2 + t]}{h}$

$= \lim\limits_{h \to 0} \dfrac{4t^2 + 8th + 4h^2 + t + h - 4t^2 - t}{h}$

$= \lim\limits_{h \to 0} \dfrac{8th + 4h^2 + h}{h} = \lim\limits_{h \to 0} (8t + 4h + 1) = 8t + 1$

23. **(a)** D $\qquad$ **(b)** F $\qquad$ **(c)** B $\qquad$ **(d)** C $\qquad$ **(e)** A $\qquad$ **(f)** E

25. **(a)** $\qquad\qquad\qquad$ **(b)** $\qquad\qquad\qquad$ **(c)**

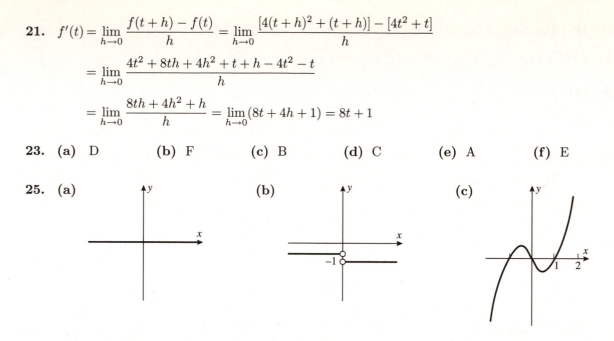

27. False. If the tangent line is horizontal then $f'(a) = 0$.

29. False. E.g. $|x|$ is continuous but not differentiable at $x = 0$.

31. **(a)** $f(x) = \sqrt{x}$ and $a = 1$ $\qquad\qquad\qquad$ **(b)** $f(x) = x^2$ and $a = 3$

33. $\dfrac{dy}{dx} = \lim\limits_{h \to 0} \dfrac{(1 - (x+h)^2) - (1 - x^2)}{h} = \lim\limits_{h \to 0} \dfrac{-2xh - h^2}{h} = \lim\limits_{h \to 0} (-2x - h) = -2x,$

and $\dfrac{dy}{dx}\bigg|_{x=1} = -2$

35. $y = -2x + 1$

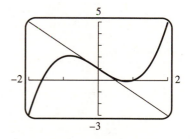

37. **(b)**

w	1.5	1.1	1.01	1.001	1.0001	1.00001
$\dfrac{f(w) - f(1)}{w - 1}$	1.6569	1.4355	1.3911	1.3868	1.3863	1.3863

39. **(a)** $\dfrac{f(3) - f(1)}{3 - 1} = \dfrac{2.2 - 2.12}{2} = 0.04;\quad \dfrac{f(2) - f(1)}{2 - 1} = \dfrac{2.34 - 2.12}{1} = 0.22;$

$\dfrac{f(2) - f(0)}{2 - 0} = \dfrac{2.34 - 0.58}{2} = 0.88$

(b) The tangent line at $x = 1$ appears to have slope about 0.8, so $\dfrac{f(2) - f(0)}{2 - 0}$ gives the best approximation and $\dfrac{f(3) - f(1)}{3 - 1}$ gives the worst.

41. (a) dollars/ft

(b) $f'(x)$ is roughly the price per additional foot.

(c) If each additional foot costs extra money (this is to be expected) then $f'(x)$ remains positive.

(d) From the approximation $1000 = f'(300) \approx \dfrac{f(301) - f(300)}{301 - 300}$

we see that $f(301) \approx f(300) + 1000$, so the extra foot will cost around $1000.

43. (a) $F \approx 200$ lb, $dF/d\theta \approx 50$ lb/rad $\qquad$ **(b)** $\mu = (dF/d\theta)/F \approx 50/200 = 0.25$

45. (a) $T \approx 115°\mathrm{F}$, $dT/dt \approx -3.35°\mathrm{F/min}$

(b) $k = (dT/dt)/(T - T_0) \approx (-3.35)/(115 - 75) = -0.084$

47. $\displaystyle\lim_{x \to 1^-} f(x) = \lim_{x \to 1^+} f(x) = f(1)$, so f is continuous at $x = 1$.

$$\lim_{h \to 0^-} \frac{f(1+h) - f(1)}{h} = \lim_{h \to 0^-} \frac{[(1+h)^2 + 1] - 2}{h} = \lim_{h \to 0^-} (2 + h) = 2;$$

$$\lim_{h \to 0^+} \frac{f(1+h) - f(1)}{h} = \lim_{h \to 0^+} \frac{2(1+h) - 2}{h} = \lim_{h \to 0^+} 2 = 2, \text{ so}$$

$$f'(1) = 2.$$

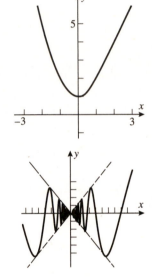

49. Since $-|x| \le x\sin(1/x) \le |x|$ it follows by the Squeezing Theorem (Theorem 1.6.4) that $\displaystyle\lim_{x \to 0} x\sin(1/x) = 0$. The derivative cannot exist: consider $\dfrac{f(x) - f(0)}{x} = \sin(1/x)$. This function oscillates between -1 and $+1$ and does not tend to zero as x tends to zero.

51. Let $\epsilon = |f'(x_0)/2|$. Then there exists $\delta > 0$ such that if $0 < |x - x_0| < \delta$, then

$$\left| \frac{f(x) - f(x_0)}{x - x_0} - f'(x_0) \right| < \epsilon. \text{ Since } f'(x_0) > 0 \text{ and } \epsilon = f'(x_0)/2 \text{ it follows that}$$

$$\frac{f(x) - f(x_0)}{x - x_0} > \epsilon > 0. \text{ If } x = x_1 < x_0 \text{ then } f(x_1) < f(x_0) \text{ and if } x = x_2 > x_0 \text{ then } f(x_2) > f(x_0).$$

53. (a) Let $\epsilon = |m|/2$. Since $m \ne 0$, $\epsilon > 0$. Since $f(0) = f'(0) = 0$ we know there exists $\delta > 0$ such that $\left| \dfrac{f(0+h) - f(0)}{h} \right| < \epsilon$ whenever $0 < |h| < \delta$. It follows that $|f(h)| < \frac{1}{2}|hm|$ for $0 < |h| < \delta$. Replace h with x to get the result.

(b) For $0 < |x| < \delta, |f(x)| < \frac{1}{2}|mx|$. Moreover $|mx| = |mx - f(x) + f(x)| \le |f(x) - mx| + |f(x)|$, which yields $|f(x) - mx| \ge |mx| - |f(x)| > \frac{1}{2}|mx| > |f(x)|$, i.e. $|f(x) - mx| > |f(x)|$.

(c) If any straight line $y = mx + b$ is to approximate the curve $y = f(x)$ for small values of x, then $b = 0$ since $f(0) = 0$. The inequality $|f(x) - mx| > |f(x)|$ can also be interpreted as $|f(x) - mx| > |f(x) - 0|$, i.e. the line $y = 0$ is a better approximation than is $y = mx$.

EXERCISE SET 2.3

1. $28x^6$ **3.** $24x^7 + 2$ **5.** 0 **7.** $-\dfrac{1}{3}(7x^6 + 2)$

9. $-3x^{-4} - 7x^{-8}$ **11.** $24x^{-9} + 1/\sqrt{x}$

13. $f(x) = x^\pi + x^{-\sqrt{10}}$, so $f'(x) = \pi x^{\pi - 1} - \sqrt{10}\, x^{-1-\sqrt{10}}$.

15. $3ax^2 + 2bx + c$ **17.** $y' = 10x - 3,\ y'(1) = 7$ **19.** $2t - 1$

21. $dy/dx = 1 + 2x + 3x^2 + 4x^3 + 5x^4,\ dy/dx|_{x=1} = 15$

23. $y = (1 - x^2)(1 + x^2)(1 + x^4) = (1 - x^4)(1 + x^4) = 1 - x^8$

$\dfrac{dy}{dx} = -8x^7,\ dy/dx|_{x=1} = -8$

25. $f'(1) \approx \dfrac{f(1.01) - f(1)}{0.01} = \dfrac{-0.999699 - (-1)}{0.01} = 0.0301$, and by differentiation, $f'(1) = 3(1)^2 - 3 = 0$

27. The estimate will depend on your graphing utility and on how far you zoom in. Since $f'(x) = 1 - \dfrac{1}{x^2}$, the exact value is $f'(1) = 0$.

29. $32t$ **31.** $3\pi r^2$

33. True. In general, $\dfrac{d}{dx}[f(x) - 8g(x)] = f'(x) - 8g'(x)$; substitute $x = 2$ to get the result.

35. False. $\dfrac{d}{dx}[4f(x) + x^3]\Big|_{x=2} = (4f'(x) + 3x^2)\big|_{x=2} = 4f'(2) + 3 \cdot 2^2 = 32$

37. **(a)** $\dfrac{dV}{dr} = 4\pi r^2$ **(b)** $\dfrac{dV}{dr}\Big|_{r=5} = 4\pi(5)^2 = 100\pi$

39. $y - 2 = 5(x + 3),\ y = 5x + 17$

41. **(a)** $dy/dx = 21x^2 - 10x + 1,\ d^2y/dx^2 = 42x - 10$
(b) $dy/dx = 24x - 2,\ d^2y/dx^2 = 24$
(c) $dy/dx = -1/x^2,\ d^2y/dx^2 = 2/x^3$
(d) $y = 35x^5 - 16x^3 - 3x,\ dy/dx = 175x^4 - 48x^2 - 3,\ d^2y/dx^2 = 700x^3 - 96x$

43. **(a)** $y' = -5x^{-6} + 5x^4,\ y'' = 30x^{-7} + 20x^3,\ y''' = -210x^{-8} + 60x^2$
(b) $y = x^{-1},\ y' = -x^{-2},\ y'' = 2x^{-3},\ y''' = -6x^{-4}$
(c) $y' = 3ax^2 + b,\ y'' = 6ax,\ y''' = 6a$

45. **(a)** $f'(x) = 6x,\ f''(x) = 6,\ f'''(x) = 0,\ f'''(2) = 0$

(b) $\dfrac{dy}{dx} = 30x^4 - 8x,\ \dfrac{d^2y}{dx^2} = 120x^3 - 8,\ \dfrac{d^2y}{dx^2}\Big|_{x=1} = 112$

(c) $\dfrac{d}{dx}[x^{-3}] = -3x^{-4},\ \dfrac{d^2}{dx^2}[x^{-3}] = 12x^{-5},\ \dfrac{d^3}{dx^3}[x^{-3}] = -60x^{-6},\ \dfrac{d^4}{dx^4}[x^{-3}] = 360x^{-7},$

$\dfrac{d^4}{dx^4}[x^{-3}]\Big|_{x=1} = 360$

47. $y' = 3x^2 + 3$, $y'' = 6x$, and $y''' = 6$ so
$y''' + xy'' - 2y' = 6 + x(6x) - 2(3x^2 + 3) = 6 + 6x^2 - 6x^2 - 6 = 0$

49. The graph has a horizontal tangent at points where $\dfrac{dy}{dx} = 0$,

but $\dfrac{dy}{dx} = x^2 - 3x + 2 = (x-1)(x-2) = 0$ if $x = 1, 2$. The
corresponding values of y are 5/6 and 2/3 so the tangent
line is horizontal at $(1, 5/6)$ and $(2, 2/3)$.

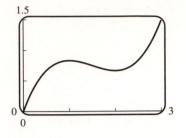

51. The y-intercept is -2 so the point $(0, -2)$ is on the graph; $-2 = a(0)^2 + b(0) + c$, $c = -2$. The
x-intercept is 1 so the point $(1, 0)$ is on the graph; $0 = a + b - 2$. The slope is $dy/dx = 2ax + b$; at
$x = 0$ the slope is b so $b = -1$, thus $a = 3$. The function is $y = 3x^2 - x - 2$.

53. The points $(-1, 1)$ and $(2, 4)$ are on the secant line so its slope is $(4 - 1)/(2 + 1) = 1$. The slope
of the tangent line to $y = x^2$ is $y' = 2x$ so $2x = 1$, $x = 1/2$.

55. $y' = -2x$, so at any point (x_0, y_0) on $y = 1 - x^2$ the tangent line is $y - y_0 = -2x_0(x - x_0)$, or
$y = -2x_0 x + x_0^2 + 1$. The point $(2, 0)$ is to be on the line, so $0 = -4x_0 + x_0^2 + 1$, $x_0^2 - 4x_0 + 1 = 0$.
Use the quadratic formula to get $x_0 = \dfrac{4 \pm \sqrt{16 - 4}}{2} = 2 \pm \sqrt{3}$. The points are $(2 + \sqrt{3}, -6 - 4\sqrt{3})$
and $(2 - \sqrt{3}, -6 + 4\sqrt{3})$.

57. $y' = 3ax^2 + b$; the tangent line at $x = x_0$ is $y - y_0 = (3ax_0^2 + b)(x - x_0)$ where $y_0 = ax_0^3 + bx_0$.
Solve with $y = ax^3 + bx$ to get

$$\begin{aligned}
(ax^3 + bx) - (ax_0^3 + bx_0) &= (3ax_0^2 + b)(x - x_0) \\
ax^3 + bx - ax_0^3 - bx_0 &= 3ax_0^2 x - 3ax_0^3 + bx - bx_0 \\
x^3 - 3x_0^2 x + 2x_0^3 &= 0 \\
(x - x_0)(x^2 + xx_0 - 2x_0^2) &= 0 \\
(x - x_0)^2 (x + 2x_0) &= 0, \text{ so } x = -2x_0.
\end{aligned}$$

59. $y' = -\dfrac{1}{x^2}$; the tangent line at $x = x_0$ is $y - y_0 = -\dfrac{1}{x_0^2}(x - x_0)$, or $y = -\dfrac{x}{x_0^2} + \dfrac{2}{x_0}$. The tangent line

crosses the x-axis at $2x_0$, the y-axis at $2/x_0$, so that the area of the triangle is $\dfrac{1}{2}(2/x_0)(2x_0) = 2$.

61. $F = GmMr^{-2}$, $\dfrac{dF}{dr} = -2GmMr^{-3} = -\dfrac{2GmM}{r^3}$

63. $f'(x) = 1 + 1/x^2 > 0$ for all $x \neq 0$

65. f is continuous at 1 because $\lim\limits_{x\to1^-} f(x) = \lim\limits_{x\to1^+} f(x) = f(1)$;

also $\lim\limits_{x\to1^-} f'(x) = \lim\limits_{x\to1^-} (2x+1) = 3$

and $\lim\limits_{x\to1^+} f'(x) = \lim\limits_{x\to1^+} 3 = 3$ so f is differentiable at 1,

and the derivative equals 3.

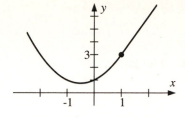

67. f is continuous at 1 because $\lim\limits_{x\to1^-} f(x) = \lim\limits_{x\to1^+} f(x) = f(1)$. Also, $\lim\limits_{x\to1^-} \dfrac{f(x)-f(1)}{x-1}$ equals the

derivative of x^2 at $x=1$, namely $2x\big|_{x=1} = 2$, while $\lim\limits_{x\to1^+} \dfrac{f(x)-f(1)}{x-1}$ equals the derivative of

$\sqrt{x}$ at $x=1$, namely $\dfrac{1}{2\sqrt{x}}\Big|_{x=1} = \dfrac{1}{2}$. Since these are not equal, f is not differentiable at $x=1$.

69. (a) $f(x) = 3x-2$ if $x \geq 2/3$, $f(x) = -3x+2$ if $x < 2/3$ so f is differentiable everywhere
except perhaps at 2/3. f is continuous at 2/3, also $\lim\limits_{x\to2/3^-} f'(x) = \lim\limits_{x\to2/3^-} (-3) = -3$ and
$\lim\limits_{x\to2/3^+} f'(x) = \lim\limits_{x\to2/3^+} (3) = 3$ so f is not differentiable at $x = 2/3$.

(b) $f(x) = x^2 - 4$ if $|x| \geq 2$, $f(x) = -x^2 + 4$ if $|x| < 2$ so f is differentiable everywhere
except perhaps at ±2. f is continuous at -2 and 2, also $\lim\limits_{x\to2^-} f'(x) = \lim\limits_{x\to2^-} (-2x) = -4$
and $\lim\limits_{x\to2^+} f'(x) = \lim\limits_{x\to2^+} (2x) = 4$ so f is not differentiable at $x = 2$. Similarly, f is not
differentiable at $x = -2$.

71. (a) $\dfrac{d^2}{dx^2}[cf(x)] = \dfrac{d}{dx}\left[\dfrac{d}{dx}[cf(x)]\right] = \dfrac{d}{dx}\left[c\dfrac{d}{dx}[f(x)]\right] = c\dfrac{d}{dx}\left[\dfrac{d}{dx}[f(x)]\right] = c\dfrac{d^2}{dx^2}[f(x)]$

$\dfrac{d^2}{dx^2}[f(x)+g(x)] = \dfrac{d}{dx}\left[\dfrac{d}{dx}[f(x)+g(x)]\right] = \dfrac{d}{dx}\left[\dfrac{d}{dx}[f(x)] + \dfrac{d}{dx}[g(x)]\right]$

$\qquad = \dfrac{d^2}{dx^2}[f(x)] + \dfrac{d^2}{dx^2}[g(x)]$

(b) yes, by repeated application of the procedure illustrated in part (a)

73. (a) $f'(x) = nx^{n-1}$, $f''(x) = n(n-1)x^{n-2}$, $f'''(x) = n(n-1)(n-2)x^{n-3}, \ldots,$
$f^{(n)}(x) = n(n-1)(n-2)\cdots 1$

(b) from part (a), $f^{(k)}(x) = k(k-1)(k-2)\cdots 1$ so $f^{(k+1)}(x) = 0$ thus $f^{(n)}(x) = 0$ if $n > k$

(c) from parts (a) and (b), $f^{(n)}(x) = a_n n(n-1)(n-2)\cdots 1$

75. Let $g(x) = x^n$, $f(x) = (mx+b)^n$. Use Exercise 52 in Section 2.2, but with f and g permuted. If
$x_0 = mx_1 + b$ then Exercise 52 says that f is differentiable at x_1 and $f'(x_1) = mg'(x_0)$. Since
$g'(x_0) = nx_0^{n-1}$, the result follows.

77. $f(x) = 27x^3 - 27x^2 + 9x - 1$ so $f'(x) = 81x^2 - 54x + 9 = 3\cdot 3(3x-1)^2$, as predicted by
Exercise 75.

79. $f(x) = 3(2x+1)^{-2}$ so $f'(x) = 3(-2)2(2x+1)^{-3} = -12/(2x+1)^3$

81. $f(x) = \dfrac{2x^2 + 4x + 2 + 1}{(x+1)^2} = 2 + (x+1)^{-2}$, so $f'(x) = -2(x+1)^{-3} = -2/(x+1)^3$

EXERCISE SET 2.4

1. **(a)** $f(x) = 2x^2 + x - 1$, $f'(x) = 4x + 1$

    (b) $f'(x) = (x+1) \cdot (2) + (2x-1) \cdot (1) = 4x + 1$

3. **(a)** $f(x) = x^4 - 1$, $f'(x) = 4x^3$

    (b) $f'(x) = (x^2 + 1) \cdot (2x) + (x^2 - 1) \cdot (2x) = 4x^3$

5. $f'(x) = (3x^2 + 6)\dfrac{d}{dx}\left(2x - \dfrac{1}{4}\right) + \left(2x - \dfrac{1}{4}\right)\dfrac{d}{dx}(3x^2 + 6) = (3x^2 + 6)(2) + \left(2x - \dfrac{1}{4}\right)(6x)$

$$= 18x^2 - \frac{3}{2}x + 12$$

7. $f'(x) = (x^3 + 7x^2 - 8)\dfrac{d}{dx}(2x^{-3} + x^{-4}) + (2x^{-3} + x^{-4})\dfrac{d}{dx}(x^3 + 7x^2 - 8)$

$$= (x^3 + 7x^2 - 8)(-6x^{-4} - 4x^{-5}) + (2x^{-3} + x^{-4})(3x^2 + 14x)$$

$$= -15x^{-2} - 14x^{-3} + 48x^{-4} + 32x^{-5}$$

9. $f(x) = x^3 - 8$ so $f'(x) = 3x^2$

11. $f'(x) = \dfrac{(x^2 + 1)\frac{d}{dx}(3x + 4) - (3x + 4)\frac{d}{dx}(x^2 + 1)}{(x^2 + 1)^2} = \dfrac{(x^2 + 1) \cdot 3 - (3x + 4) \cdot 2x}{(x^2 + 1)^2} = \dfrac{-3x^2 - 8x + 3}{(x^2 + 1)^2}$

13. $f'(x) = \dfrac{(3x - 4)\frac{d}{dx}(x^2) - x^2\frac{d}{dx}(3x - 4)}{(3x - 4)^2} = \dfrac{(3x - 4) \cdot 2x - x^2 \cdot 3}{(3x - 4)^2} = \dfrac{3x^2 - 8x}{(3x - 4)^2}$

15. $f(x) = \dfrac{2x^{3/2} + x - 2x^{1/2} - 1}{x + 3}$, so

$$f'(x) = \frac{(x + 3)\frac{d}{dx}(2x^{3/2} + x - 2x^{1/2} - 1) - (2x^{3/2} + x - 2x^{1/2} - 1)\frac{d}{dx}(x + 3)}{(x + 3)^2}$$

$$= \frac{(x + 3) \cdot (3x^{1/2} + 1 - x^{-1/2}) - (2x^{3/2} + x - 2x^{1/2} - 1) \cdot 1}{(x + 3)^2} = \frac{x^{3/2} + 10x^{1/2} + 4 - 3x^{-1/2}}{(x + 3)^2}$$

17. This could be computed by two applications of the product rule, but it's simpler to expand $f(x)$:

$f(x) = 14x + 21 + 7x^{-1} + 2x^{-2} + 3x^{-3} + x^{-4}$, so $f'(x) = 14 - 7x^{-2} - 4x^{-3} - 9x^{-4} - 4x^{-5}$.

19. In general, $\frac{d}{dx}\left[g(x)^2\right] = 2g(x)g'(x)$ and

$$\frac{d}{dx}\left[g(x)^3\right] = \frac{d}{dx}\left[g(x)^2 g(x)\right] = g(x)^2 g'(x) + g(x)\frac{d}{dx}\left[g(x)^2\right] = g(x)^2 g'(x) + g(x) \cdot 2g(x)g'(x)$$

$$= 3g(x)^2 g'(x).$$

Letting $g(x) = x^7 + 2x - 3$, we have $f'(x) = 3(x^7 + 2x - 3)^2(7x^6 + 2)$.

21. $\dfrac{dy}{dx} = \left(\dfrac{3x + 2}{x}\right)\dfrac{d}{dx}(x^{-5} + 1) + (x^{-5} + 1)\dfrac{d}{dx}\left(\dfrac{3x + 2}{x}\right)$

$$= \left(\frac{3x + 2}{x}\right)(-5x^{-6}) + (x^{-5} + 1)\left[\frac{x(3) - (3x + 2)(1)}{x^2}\right]$$

$$= \left(\frac{3x + 2}{x}\right)(-5x^{-6}) + (x^{-5} + 1)\left(-\frac{2}{x^2}\right);$$

$$\left.\frac{dy}{dx}\right|_{x=1} = 5(-5) + 2(-2) = -29$$

23. $f'(1) = 0$

25. **(a)** $g'(x) = \sqrt{x}f'(x) + \dfrac{1}{2\sqrt{x}}f(x)$, $g'(4) = (2)(-5) + \dfrac{1}{4}(3) = -37/4$

(b) $g'(x) = \dfrac{xf'(x) - f(x)}{x^2}$, $g'(4) = \dfrac{(4)(-5) - 3}{16} = -23/16$

27. **(a)** $F'(x) = 5f'(x) + 2g'(x)$, $F'(2) = 5(4) + 2(-5) = 10$
(b) $F'(x) = f'(x) - 3g'(x)$, $F'(2) = 4 - 3(-5) = 19$
(c) $F'(x) = f(x)g'(x) + g(x)f'(x)$, $F'(2) = (-1)(-5) + (1)(4) = 9$
(d) $F'(x) = [g(x)f'(x) - f(x)g'(x)]/g^2(x)$, $F'(2) = [(1)(4) - (-1)(-5)]/(1)^2 = -1$

29. $\dfrac{dy}{dx} = \dfrac{2x(x+2) - (x^2 - 1)}{(x+2)^2}$,

$\dfrac{dy}{dx} = 0$ if $x^2 + 4x + 1 = 0$. By the quadratic formula,

$x = \dfrac{-4 \pm \sqrt{16 - 4}}{2} = -2 \pm \sqrt{3}$. The tangent line is horizontal at $x = -2 \pm \sqrt{3}$.

31. The tangent line is parallel to the line $y = x$ when it has slope 1.
$\dfrac{dy}{dx} = \dfrac{2x(x+1) - (x^2 + 1)}{(x+1)^2} = \dfrac{x^2 + 2x - 1}{(x+1)^2} = 1$ if $x^2 + 2x - 1 = (x+1)^2$, which reduces to $-1 = +1$,
impossible. Thus the tangent line is never parallel to the line $y = x$.

33. Fix x_0. The slope of the tangent line to the curve $y = \dfrac{1}{x+4}$ at the point $(x_0, 1/(x_0 + 4))$ is
given by $\dfrac{dy}{dx} = \dfrac{-1}{(x+4)^2}\bigg|_{x=x_0} = \dfrac{-1}{(x_0+4)^2}$. The tangent line to the curve at (x_0, y_0) thus has the
equation $y - y_0 = \dfrac{-1}{(x_0+4)^2}(x - x_0)$, and this line passes through the origin if its constant term
$y_0 - x_0\dfrac{-1}{(x_0+4)^2}$ is zero. Then $\dfrac{1}{x_0+4} = \dfrac{-x_0}{(x_0+4)^2}$, $x_0 + 4 = -x_0$, $x_0 = -2$.

35. **(a)** Their tangent lines at the intersection point must be perpendicular.
(b) They intersect when $\dfrac{1}{x} = \dfrac{1}{2-x}$, $x = 2 - x$, $x = 1$, $y = 1$. The first curve has derivative
$y = -\dfrac{1}{x^2}$, so the slope when $x = 1$ is -1. Second curve has derivative $y = \dfrac{1}{(2-x)^2}$ so the
slope when $x = 1$ is 1. Since the two slopes are negative reciprocals of each other, the tangent
lines are perpendicular at the point $(1, 1)$.

37. $F'(x) = xf'(x) + f(x)$, $F''(x) = xf''(x) + f'(x) + f'(x) = xf''(x) + 2f'(x)$

39. $R'(p) = p \cdot f'(p) + f(p) \cdot 1 = f(p) + pf'(p)$, so $R'(120) = 9000 + 120 \cdot (-60) = 1800$.
Increasing the price by a small amount Δp dollars would increase the revenue by about $1800\Delta p$
dollars.

41. $f(x) = \dfrac{1}{x^n}$ so $f'(x) = \dfrac{x^n \cdot (0) - 1 \cdot (nx^{n-1})}{x^{2n}} = -\dfrac{n}{x^{n+1}} = -nx^{-n-1}$

EXERCISE SET 2.5

1. $f'(x) = -4\sin x + 2\cos x$

3. $f'(x) = 4x^2 \sin x - 8x \cos x$

5. $f'(x) = \dfrac{\sin x(5 + \sin x) - \cos x(5 - \cos x)}{(5 + \sin x)^2} = \dfrac{1 + 5(\sin x - \cos x)}{(5 + \sin x)^2}$

7. $f'(x) = \sec x \tan x - \sqrt{2} \sec^2 x$

9. $f'(x) = -4\csc x \cot x + \csc^2 x$

11. $f'(x) = \sec x(\sec^2 x) + (\tan x)(\sec x \tan x) = \sec^3 x + \sec x \tan^2 x$

13. $f'(x) = \dfrac{(1 + \csc x)(-\csc^2 x) - \cot x(0 - \csc x \cot x)}{(1 + \csc x)^2} = \dfrac{\csc x(-\csc x - \csc^2 x + \cot^2 x)}{(1 + \csc x)^2}$ but

$1 + \cot^2 x = \csc^2 x$ (identity) thus $\cot^2 x - \csc^2 x = -1$ so

$f'(x) = \dfrac{\csc x(-\csc x - 1)}{(1 + \csc x)^2} = -\dfrac{\csc x}{1 + \csc x}$

15. $f(x) = \sin^2 x + \cos^2 x = 1$ (identity) so $f'(x) = 0$

17. $f(x) = \dfrac{\tan x}{1 + x \tan x}$ (because $\sin x \sec x = (\sin x)(1/\cos x) = \tan x$),

$f'(x) = \dfrac{(1 + x \tan x)(\sec^2 x) - \tan x[x(\sec^2 x) + (\tan x)(1)]}{(1 + x \tan x)^2}$

$= \dfrac{\sec^2 x - \tan^2 x}{(1 + x \tan x)^2} = \dfrac{1}{(1 + x \tan x)^2}$ (because $\sec^2 x - \tan^2 x = 1$)

19. $dy/dx = -x \sin x + \cos x$, $d^2y/dx^2 = -x \cos x - \sin x - \sin x = -x \cos x - 2\sin x$

21. $dy/dx = x(\cos x) + (\sin x)(1) - 3(-\sin x) = x \cos x + 4\sin x$,

$d^2y/dx^2 = x(-\sin x) + (\cos x)(1) + 4\cos x = -x \sin x + 5\cos x$

23. $dy/dx = (\sin x)(-\sin x) + (\cos x)(\cos x) = \cos^2 x - \sin^2 x$,

$d^2y/dx^2 = (\cos x)(-\sin x) + (\cos x)(-\sin x) - [(\sin x)(\cos x) + (\sin x)(\cos x)] = -4\sin x \cos x$

25. Let $f(x) = \tan x$, then $f'(x) = \sec^2 x$.

   (a) $f(0) = 0$ and $f'(0) = 1$ so $y - 0 = (1)(x - 0)$, $y = x$.

   (b) $f\left(\dfrac{\pi}{4}\right) = 1$ and $f'\left(\dfrac{\pi}{4}\right) = 2$ so $y - 1 = 2\left(x - \dfrac{\pi}{4}\right)$, $y = 2x - \dfrac{\pi}{2} + 1$.

   (c) $f\left(-\dfrac{\pi}{4}\right) = -1$ and $f'\left(-\dfrac{\pi}{4}\right) = 2$ so $y + 1 = 2\left(x + \dfrac{\pi}{4}\right)$, $y = 2x + \dfrac{\pi}{2} - 1$.

27. **(a)** If $y = x \sin x$ then $y' = \sin x + x \cos x$ and $y'' = 2\cos x - x \sin x$ so $y'' + y = 2\cos x$.

   (b) Differentiate the result of part (a) twice more to get $y^{(4)} + y'' = -2\cos x$.

29. **(a)** $f'(x) = \cos x = 0$ at $x = \pm\pi/2, \pm 3\pi/2$.

   (b) $f'(x) = 1 - \sin x = 0$ at $x = -3\pi/2, \pi/2$.

   (c) $f'(x) = \sec^2 x \geq 1$ always, so no horizontal tangent line.

   (d) $f'(x) = \sec x \tan x = 0$ when $\sin x = 0$, $x = \pm 2\pi, \pm\pi, 0$

31. $x = 10\sin\theta$, $dx/d\theta = 10\cos\theta$; if $\theta = 60°$, then

$dx/d\theta = 10(1/2) = 5$ ft/rad $= \pi/36$ ft/deg ≈ 0.087 ft/deg

33. $D = 50 \tan \theta$, $dD/d\theta = 50 \sec^2 \theta$; if $\theta = 45°$, then
$dD/d\theta = 50(\sqrt{2})^2 = 100$ m/rad $= 5\pi/9$ m/deg ≈ 1.75 m/deg

35. False. $g'(x) = f(x)\cos x + f'(x)\sin x$

37. True. $f(x) = \dfrac{\sin x}{\cos x} = \tan x$, so $f'(x) = \sec^2 x$.

39. $\dfrac{d^4}{dx^4}\sin x = \sin x$, so $\dfrac{d^{4k}}{dx^{4k}}\sin x = \sin x$; $\dfrac{d^{87}}{dx^{87}}\sin x = \dfrac{d^3}{dx^3}\dfrac{d^{4\cdot21}}{dx^{4\cdot21}}\sin x = \dfrac{d^3}{dx^3}\sin x = -\cos x$

41. $f'(x) = -\sin x$, $f''(x) = -\cos x$, $f'''(x) = \sin x$, and $f^{(4)}(x) = \cos x$ with higher order derivatives repeating this pattern, so $f^{(n)}(x) = \sin x$ for $n = 3, 7, 11, \ldots$

43. **(a)** all x **(b)** all x
      (c) $x \neq \pi/2 + n\pi$, $n = 0, \pm1, \pm2, \ldots$ **(d)** $x \neq n\pi$, $n = 0, \pm1, \pm2, \ldots$
      (e) $x \neq \pi/2 + n\pi$, $n = 0, \pm1, \pm2, \ldots$ **(f)** $x \neq n\pi$, $n = 0, \pm1, \pm2, \ldots$
      (g) $x \neq (2n+1)\pi$, $n = 0, \pm1, \pm2, \ldots$ **(h)** $x \neq n\pi/2$, $n = 0, \pm1, \pm2, \ldots$
      (i) all x

45. $\dfrac{d}{dx}\sin x = \lim\limits_{w\to x}\dfrac{\sin w - \sin x}{w - x} = \lim\limits_{w\to x}\dfrac{2\sin\frac{w-x}{2}\cos\frac{w+x}{2}}{w - x}$

$\qquad = \lim\limits_{w\to x}\dfrac{\sin\frac{w-x}{2}}{\frac{w-x}{2}}\cos\dfrac{w+x}{2} = 1 \cdot \cos x = \cos x$

47. **(a)** $\lim\limits_{h\to 0}\dfrac{\tan h}{h} = \lim\limits_{h\to 0}\dfrac{\left(\frac{\sin h}{\cos h}\right)}{h} = \lim\limits_{h\to 0}\dfrac{\left(\frac{\sin h}{h}\right)}{\cos h} = \dfrac{1}{1} = 1$

      (b) $\dfrac{d}{dx}[\tan x] = \lim\limits_{h\to 0}\dfrac{\tan(x+h) - \tan x}{h} = \lim\limits_{h\to 0}\dfrac{\frac{\tan x + \tan h}{1 - \tan x \tan h} - \tan x}{h}$

$\qquad\qquad = \lim\limits_{h\to 0}\dfrac{\tan x + \tan h - \tan x + \tan^2 x \tan h}{h(1 - \tan x \tan h)} = \lim\limits_{h\to 0}\dfrac{\tan h(1 + \tan^2 x)}{h(1 - \tan x \tan h)}$

$\qquad\qquad = \lim\limits_{h\to 0}\dfrac{\tan h \sec^2 x}{h(1 - \tan x \tan h)} = \sec^2 x \lim\limits_{h\to 0}\dfrac{\frac{\tan h}{h}}{1 - \tan x \tan h}$

$\qquad\qquad = \sec^2 x \dfrac{\lim\limits_{h\to 0}\frac{\tan h}{h}}{\lim\limits_{h\to 0}(1 - \tan x \tan h)} = \sec^2 x$

49. By Exercises 49 and 50 of Section 1.6, we have $\lim\limits_{h\to 0}\dfrac{\sin h}{h} = \dfrac{\pi}{180}$ and $\lim\limits_{h\to 0}\dfrac{\cos h - 1}{h} = 0$. Therefore:

      (a) $\dfrac{d}{dx}[\sin x] = \lim\limits_{h\to 0}\dfrac{\sin(x+h) - \sin x}{h} = \sin x \lim\limits_{h\to 0}\dfrac{\cos h - 1}{h} + \cos x \lim\limits_{h\to 0}\dfrac{\sin h}{h}$

$\qquad\qquad = (\sin x)(0) + (\cos x)(\pi/180) = \dfrac{\pi}{180}\cos x$

      (b) $\dfrac{d}{dx}[\cos x] = \lim\limits_{h\to 0}\dfrac{\cos(x+h) - \cos x}{h} = \lim\limits_{h\to 0}\dfrac{\cos x \cos h - \sin x \sin h - \cos x}{h}$

$\qquad\qquad = \cos x \lim\limits_{h\to 0}\dfrac{\cos h - 1}{h} - \sin x \lim\limits_{h\to 0}\dfrac{\sin h}{h} = 0 \cdot \cos x - \dfrac{\pi}{180}\cdot\sin x = -\dfrac{\pi}{180}\sin x$

EXERCISE SET 2.6

1. $(f \circ g)'(x) = f'(g(x))g'(x)$ so $(f \circ g)'(0) = f'(g(0))g'(0) = f'(0)(3) = (2)(3) = 6$

3. **(a)** $(f \circ g)(x) = f(g(x)) = (2x-3)^5$ and $(f \circ g)'(x) = f'(g(x))g'(x) = 5(2x-3)^4(2) = 10(2x-3)^4$

   (b) $(g \circ f)(x) = g(f(x)) = 2x^5 - 3$ and $(g \circ f)'(x) = g'(f(x))f'(x) = 2(5x^4) = 10x^4$

5. **(a)** $F'(x) = f'(g(x))g'(x)$, $F'(3) = f'(g(3))g'(3) = -1(7) = -7$

   (b) $G'(x) = g'(f(x))f'(x)$, $G'(3) = g'(f(3))f'(3) = 4(-2) = -8$

7. $f'(x) = 37(x^3 + 2x)^{36} \dfrac{d}{dx}(x^3 + 2x) = 37(x^3 + 2x)^{36}(3x^2 + 2)$

9. $f'(x) = -2\left(x^3 - \dfrac{7}{x}\right)^{-3} \dfrac{d}{dx}\left(x^3 - \dfrac{7}{x}\right) = -2\left(x^3 - \dfrac{7}{x}\right)^{-3}\left(3x^2 + \dfrac{7}{x^2}\right)$

11. $f(x) = 4(3x^2 - 2x + 1)^{-3}$,

$f'(x) = -12(3x^2 - 2x + 1)^{-4}\dfrac{d}{dx}(3x^2 - 2x + 1) = -12(3x^2 - 2x + 1)^{-4}(6x - 2) = \dfrac{24(1 - 3x)}{(3x^2 - 2x + 1)^4}$

13. $f'(x) = \dfrac{1}{2\sqrt{4 + \sqrt{3x}}} \dfrac{d}{dx}(4 + \sqrt{3x}) = \dfrac{\sqrt{3}}{4\sqrt{x}\sqrt{4 + \sqrt{3x}}}$

15. $f'(x) = \cos(1/x^2)\dfrac{d}{dx}(1/x^2) = -\dfrac{2}{x^3}\cos(1/x^2)$

17. $f'(x) = 20\cos^4 x\dfrac{d}{dx}(\cos x) = 20\cos^4 x(-\sin x) = -20\cos^4 x \sin x$

19. $f'(x) = 2\cos(3\sqrt{x})\dfrac{d}{dx}[\cos(3\sqrt{x})] = -2\cos(3\sqrt{x})\sin(3\sqrt{x})\dfrac{d}{dx}(3\sqrt{x}) = -\dfrac{3\cos(3\sqrt{x})\sin(3\sqrt{x})}{\sqrt{x}}$

21. $f'(x) = 4\sec(x^7)\dfrac{d}{dx}[\sec(x^7)] = 4\sec(x^7)\sec(x^7)\tan(x^7)\dfrac{d}{dx}(x^7) = 28x^6\sec^2(x^7)\tan(x^7)$

23. $f'(x) = \dfrac{1}{2\sqrt{\cos(5x)}}\dfrac{d}{dx}[\cos(5x)] = -\dfrac{5\sin(5x)}{2\sqrt{\cos(5x)}}$

25. $f'(x) = -3\left[x + \csc(x^3 + 3)\right]^{-4}\dfrac{d}{dx}\left[x + \csc(x^3 + 3)\right]$

$= -3\left[x + \csc(x^3 + 3)\right]^{-4}\left[1 - \csc(x^3 + 3)\cot(x^3 + 3)\dfrac{d}{dx}(x^3 + 3)\right]$

$= -3\left[x + \csc(x^3 + 3)\right]^{-4}\left[1 - 3x^2\csc(x^3 + 3)\cot(x^3 + 3)\right]$

27. $\dfrac{dy}{dx} = x^3(2\sin 5x)\dfrac{d}{dx}(\sin 5x) + 3x^2\sin^2 5x = 10x^3\sin 5x\cos 5x + 3x^2\sin^2 5x$

29. $\dfrac{dy}{dx} = x^5\sec\left(\dfrac{1}{x}\right)\tan\left(\dfrac{1}{x}\right)\dfrac{d}{dx}\left(\dfrac{1}{x}\right) + \sec\left(\dfrac{1}{x}\right)(5x^4)$

$= x^5\sec\left(\dfrac{1}{x}\right)\tan\left(\dfrac{1}{x}\right)\left(-\dfrac{1}{x^2}\right) + 5x^4\sec\left(\dfrac{1}{x}\right)$

$= -x^3\sec\left(\dfrac{1}{x}\right)\tan\left(\dfrac{1}{x}\right) + 5x^4\sec\left(\dfrac{1}{x}\right)$

31. $\dfrac{dy}{dx} = -\sin(\cos x)\dfrac{d}{dx}(\cos x) = -\sin(\cos x)(-\sin x) = \sin(\cos x)\sin x$

33. $\dfrac{dy}{dx} = 3\cos^2(\sin 2x)\dfrac{d}{dx}[\cos(\sin 2x)] = 3\cos^2(\sin 2x)[-\sin(\sin 2x)]\dfrac{d}{dx}(\sin 2x)$

$\qquad = -6\cos^2(\sin 2x)\sin(\sin 2x)\cos 2x$

35. $\dfrac{dy}{dx} = (5x+8)^7\dfrac{d}{dx}(1-\sqrt{x})^6 + (1-\sqrt{x})^6\dfrac{d}{dx}(5x+8)^7$

$\qquad = 6(5x+8)^7(1-\sqrt{x})^5\dfrac{-1}{2\sqrt{x}} + 7\cdot 5(1-\sqrt{x})^6(5x+8)^6$

$\qquad = \dfrac{-3}{\sqrt{x}}(5x+8)^7(1-\sqrt{x})^5 + 35(1-\sqrt{x})^6(5x+8)^6$

37. $\dfrac{dy}{dx} = 3\left[\dfrac{x-5}{2x+1}\right]^2\dfrac{d}{dx}\left[\dfrac{x-5}{2x+1}\right] = 3\left[\dfrac{x-5}{2x+1}\right]^2\cdot\dfrac{11}{(2x+1)^2} = \dfrac{33(x-5)^2}{(2x+1)^4}$

39. $\dfrac{dy}{dx} = \dfrac{(4x^2-1)^8(3)(2x+3)^2(2) - (2x+3)^3(8)(4x^2-1)^7(8x)}{(4x^2-1)^{16}}$

$\qquad = \dfrac{2(2x+3)^2(4x^2-1)^7[3(4x^2-1) - 32x(2x+3)]}{(4x^2-1)^{16}} = -\dfrac{2(2x+3)^2(52x^2+96x+3)}{(4x^2-1)^9}$

41. $\dfrac{dy}{dx} = 5\left[x\sin 2x + \tan^4(x^7)\right]^4\dfrac{d}{dx}\left[x\sin 2x\tan^4(x^7)\right]$

$\qquad = 5\left[x\sin 2x + \tan^4(x^7)\right]^4\left[x\cos 2x\dfrac{d}{dx}(2x) + \sin 2x + 4\tan^3(x^7)\dfrac{d}{dx}\tan(x^7)\right]$

$\qquad = 5\left[x\sin 2x + \tan^4(x^7)\right]^4\left[2x\cos 2x + \sin 2x + 28x^6\tan^3(x^7)\sec^2(x^7)\right]$

43. $\dfrac{dy}{dx} = \cos 3x - 3x\sin 3x$; if $x = \pi$ then $\dfrac{dy}{dx} = -1$ and $y = -\pi$, so the equation of the tangent line is $y + \pi = -(x-\pi)$, $y = -x$

45. $\dfrac{dy}{dx} = -3\sec^3(\pi/2 - x)\tan(\pi/2 - x)$; if $x = -\pi/2$ then $\dfrac{dy}{dx} = 0$, $y = -1$ so the equation of the tangent line is $y + 1 = 0$, $y = -1$

47. $\dfrac{dy}{dx} = \sec^2(4x^2)\dfrac{d}{dx}(4x^2) = 8x\sec^2(4x^2)$, $\left.\dfrac{dy}{dx}\right|_{x=\sqrt{\pi}} = 8\sqrt{\pi}\sec^2(4\pi) = 8\sqrt{\pi}$. When $x = \sqrt{\pi}$, $y = \tan(4\pi) = 0$, so the equation of the tangent line is $y = 8\sqrt{\pi}(x - \sqrt{\pi}) = 8\sqrt{\pi}x - 8\pi$.

49. $\dfrac{dy}{dx} = 2x\sqrt{5-x^2} + \dfrac{x^2}{2\sqrt{5-x^2}}(-2x)$, $\left.\dfrac{dy}{dx}\right|_{x=1} = 4 - 1/2 = 7/2$. When $x = 1$, $y = 2$, so the equation of the tangent line is $y - 2 = (7/2)(x-1)$, or $y = \dfrac{7}{2}x - \dfrac{3}{2}$.

51. $\dfrac{dy}{dx} = x(-\sin(5x))\dfrac{d}{dx}(5x) + \cos(5x) - 2\sin x\dfrac{d}{dx}(\sin x)$

$\qquad = -5x\sin(5x) + \cos(5x) - 2\sin x\cos x = -5x\sin(5x) + \cos(5x) - \sin(2x),$

$$\frac{d^2y}{dx^2} = -5x\cos(5x)\frac{d}{dx}(5x) - 5\sin(5x) - \sin(5x)\frac{d}{dx}(5x) - \cos(2x)\frac{d}{dx}(2x)$$

$$= -25x\cos(5x) - 10\sin(5x) - 2\cos(2x)$$

53. $\dfrac{dy}{dx} = \dfrac{(1-x)+(1+x)}{(1-x)^2} = \dfrac{2}{(1-x)^2} = 2(1-x)^{-2}$ and $\dfrac{d^2y}{dx^2} = -2(2)(-1)(1-x)^{-3} = 4(1-x)^{-3}$

55. $y = \cot^3(\pi - \theta) = -\cot^3\theta$ so $dy/dx = 3\cot^2\theta\csc^2\theta$

57. $\dfrac{d}{d\omega}[a\cos^2\pi\omega + b\sin^2\pi\omega] = -2\pi a\cos\pi\omega\sin\pi\omega + 2\pi b\sin\pi\omega\cos\pi\omega$

$$= \pi(b-a)(2\sin\pi\omega\cos\pi\omega) = \pi(b-a)\sin 2\pi\omega$$

59. **(a)**

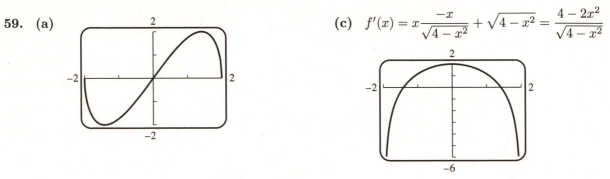

(c) $f'(x) = x\dfrac{-x}{\sqrt{4-x^2}} + \sqrt{4-x^2} = \dfrac{4-2x^2}{\sqrt{4-x^2}}$

(d) $f(1) = \sqrt{3}$ and $f'(1) = \dfrac{2}{\sqrt{3}}$ so the tangent line has the equation $y - \sqrt{3} = \dfrac{2}{\sqrt{3}}(x-1)$.

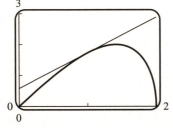

61. False. $\dfrac{d}{dx}[\sqrt{y}] = \dfrac{1}{2\sqrt{y}}\dfrac{dy}{dx} = \dfrac{f'(x)}{2\sqrt{f(x)}}$

63. False. $dy/dx = -\sin[g(x)]\,g'(x)$

65. **(a)** $dy/dt = -A\omega\sin\omega t, d^2y/dt^2 = -A\omega^2\cos\omega t = -\omega^2 y$

(b) one complete oscillation occurs when ωt increases over an interval of length 2π, or if t increases over an interval of length $2\pi/\omega$

(c) $f = 1/T$

(d) amplitude $= 0.6$ cm, $T = 2\pi/15$ s/oscillation, $f = 15/(2\pi)$ oscillations/s

67. By the chain rule, $\dfrac{d}{dx}\left[\sqrt{x+f(x)}\right] = \dfrac{1+f'(x)}{2\sqrt{x+f(x)}}$. From the graph, $f(x) = \dfrac{4}{3}x + 5$ for $x < 0$, so

$$f(-1) = \dfrac{11}{3},\ f'(-1) = \dfrac{4}{3},\ \text{and}\ \dfrac{d}{dx}\left[\sqrt{x+f(x)}\right]\Big|_{x=-1} = \dfrac{7/3}{2\sqrt{8/3}} = \dfrac{7\sqrt{6}}{24}.$$

69. **(a)** $p \approx 10$ lb/in^2, $dp/dh \approx -2$ lb/in^2/mi

(b) $\dfrac{dp}{dt} = \dfrac{dp}{dh}\dfrac{dh}{dt} \approx (-2)(0.3) = -0.6$ lb/in^2/s

71. With $u = \sin x$, $\dfrac{d}{dx}(|\sin x|) = \dfrac{d}{dx}(|u|) = \dfrac{d}{du}(|u|)\dfrac{du}{dx} = \dfrac{d}{du}(|u|)\cos x = \begin{cases} \cos x, & u > 0 \\ -\cos x, & u < 0 \end{cases}$

$$= \begin{cases} \cos x, & \sin x > 0 \\ -\cos x, & \sin x < 0 \end{cases} = \begin{cases} \cos x, & 0 < x < \pi \\ -\cos x, & -\pi < x < 0 \end{cases}$$

73. (a) For $x \neq 0$, $|f(x)| \leq |x|$, and $\lim\limits_{x\to 0} |x| = 0$, so by the Squeezing Theorem, $\lim\limits_{x\to 0} f(x) = 0$.

(b) If $f'(0)$ were to exist, then the limit $\dfrac{f(x) - f(0)}{x - 0} = \sin(1/x)$ would have to exist, but it doesn't.

(c) For $x \neq 0$, $f'(x) = x\left(\cos\dfrac{1}{x}\right)\left(-\dfrac{1}{x^2}\right) + \sin\dfrac{1}{x} = -\dfrac{1}{x}\cos\dfrac{1}{x} + \sin\dfrac{1}{x}$

(d) If $x = \dfrac{1}{2\pi n}$ for an integer $n \neq 0$, then $f'(x) = -2\pi n\cos(2\pi n) + \sin(2\pi n) = -2\pi n$. This approaches $+\infty$ as $n \to -\infty$, so there are points x arbitrarily close to 0 where $f'(x)$ becomes arbitrarily large. Hence $\lim\limits_{x\to 0} f'(x)$ does not exist.

75. (a) $g'(x) = 3[f(x)]^2 f'(x)$, $g'(2) = 3[f(2)]^2 f'(2) = 3(1)^2(7) = 21$

(b) $h'(x) = f'(x^3)(3x^2)$, $h'(2) = f'(8)(12) = (-3)(12) = -36$

77. $F'(x) = f'(g(x))g'(x) = f'(\sqrt{3x-1})\dfrac{3}{2\sqrt{3x-1}} = \dfrac{\sqrt{3x-1}}{(3x-1)+1}\dfrac{3}{2\sqrt{3x-1}} = \dfrac{1}{2x}$

79. $\dfrac{d}{dx}[f(3x)] = f'(3x)\dfrac{d}{dx}(3x) = 3f'(3x) = 6x$, so $f'(3x) = 2x$. Let $u = 3x$ to get $f'(u) = \dfrac{2}{3}u$;

$\dfrac{d}{dx}[f(x)] = f'(x) = \dfrac{2}{3}x$.

81. For an even function, the graph is symmetric about the y-axis; the slope of the tangent line at $(a, f(a))$ is the negative of the slope of the tangent line at $(-a, f(-a))$. For an odd function, the graph is symmetric about the origin; the slope of the tangent line at $(a, f(a))$ is the same as the slope of the tangent line at $(-a, f(-a))$.

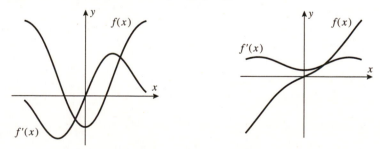

83. $\dfrac{d}{dx}[f(g(h(x)))] = \dfrac{d}{dx}[f(g(u))]$, $\quad u = h(x)$

$$= \dfrac{d}{du}[f(g(u))]\dfrac{du}{dx} = f'(g(u))g'(u)\dfrac{du}{dx} = f'(g(h(x)))g'(h(x))h'(x)$$

EXERCISE SET 2.7

1. **(a)** $1 + y + x\dfrac{dy}{dx} - 6x^2 = 0$, $\dfrac{dy}{dx} = \dfrac{6x^2 - y - 1}{x}$

   (b) $y = \dfrac{2 + 2x^3 - x}{x} = \dfrac{2}{x} + 2x^2 - 1$, $\dfrac{dy}{dx} = -\dfrac{2}{x^2} + 4x$

   (c) From Part (a), $\dfrac{dy}{dx} = 6x - \dfrac{1}{x} - \dfrac{1}{x}y = 6x - \dfrac{1}{x} - \dfrac{1}{x}\left(\dfrac{2}{x} + 2x^2 - 1\right) = 4x - \dfrac{2}{x^2}$

3. $2x + 2y\dfrac{dy}{dx} = 0$ so $\dfrac{dy}{dx} = -\dfrac{x}{y}$

5. $x^2\dfrac{dy}{dx} + 2xy + 3x(3y^2)\dfrac{dy}{dx} + 3y^3 - 1 = 0$

   $(x^2 + 9xy^2)\dfrac{dy}{dx} = 1 - 2xy - 3y^3$ so $\dfrac{dy}{dx} = \dfrac{1 - 2xy - 3y^3}{x^2 + 9xy^2}$

7. $-\dfrac{1}{2x^{3/2}} - \dfrac{\frac{dy}{dx}}{2y^{3/2}} = 0$, $\dfrac{dy}{dx} = -\dfrac{y^{3/2}}{x^{3/2}}$

9. $\cos(x^2 y^2)\left[x^2(2y)\dfrac{dy}{dx} + 2xy^2\right] = 1$, $\dfrac{dy}{dx} = \dfrac{1 - 2xy^2\cos(x^2 y^2)}{2x^2 y\cos(x^2 y^2)}$

11. $3\tan^2(xy^2 + y)\sec^2(xy^2 + y)\left(2xy\dfrac{dy}{dx} + y^2 + \dfrac{dy}{dx}\right) = 1$

    so $\dfrac{dy}{dx} = \dfrac{1 - 3y^2\tan^2(xy^2 + y)\sec^2(xy^2 + y)}{3(2xy + 1)\tan^2(xy^2 + y)\sec^2(xy^2 + y)}$

13. $4x - 6y\dfrac{dy}{dx} = 0$, $\dfrac{dy}{dx} = \dfrac{2x}{3y}$, $4 - 6\left(\dfrac{dy}{dx}\right)^2 - 6y\dfrac{d^2 y}{dx^2} = 0$,

    $\dfrac{d^2 y}{dx^2} = -\dfrac{3\left(\frac{dy}{dx}\right)^2 - 2}{3y} = \dfrac{2(3y^2 - 2x^2)}{9y^3} = -\dfrac{8}{9y^3}$

15. $\dfrac{dy}{dx} = -\dfrac{y}{x}$, $\dfrac{d^2 y}{dx^2} = -\dfrac{x(dy/dx) - y(1)}{x^2} = -\dfrac{x(-y/x) - y}{x^2} = \dfrac{2y}{x^2}$

17. $\dfrac{dy}{dx} = (1 + \cos y)^{-1}$, $\dfrac{d^2 y}{dx^2} = -(1 + \cos y)^{-2}(-\sin y)\dfrac{dy}{dx} = \dfrac{\sin y}{(1 + \cos y)^3}$

19. By implicit differentiation, $2x + 2y(dy/dx) = 0$, $\dfrac{dy}{dx} = -\dfrac{x}{y}$; at $(1/2, \sqrt{3}/2)$, $\dfrac{dy}{dx} = -\sqrt{3}/3$; at

    $(1/2, -\sqrt{3}/2)$, $\dfrac{dy}{dx} = +\sqrt{3}/3$. Directly, at the upper point $y = \sqrt{1 - x^2}$, $\dfrac{dy}{dx} = \dfrac{-x}{\sqrt{1 - x^2}} = $

    $-\dfrac{1/2}{\sqrt{3/4}} = -1/\sqrt{3}$ and at the lower point $y = -\sqrt{1 - x^2}$, $\dfrac{dy}{dx} = \dfrac{x}{\sqrt{1 - x^2}} = +1/\sqrt{3}$.

21. false; $x = y^2$ defines two functions $y = \pm\sqrt{x}$. See Definition 2.7.1.

23. false; the equation is equivalent to $x^2 = y^2$ which is satisfied by $y = |x|$.

25. $x^m x^{-m} = 1$, $x^{-m}\dfrac{d}{dx}(x^m) - mx^{-m-1}x^m = 0$, $\dfrac{d}{dx}(x^m) = x^m(mx^{-m-1})x^m = mx^{m-1}$

27. $4x^3 + 4y^3 \dfrac{dy}{dx} = 0$, so $\dfrac{dy}{dx} = -\dfrac{x^3}{y^3} = -\dfrac{1}{15^{3/4}} \approx -0.1312$.

29. $4(x^2 + y^2)\left(2x + 2y\dfrac{dy}{dx}\right) = 25\left(2x - 2y\dfrac{dy}{dx}\right)$,

$\dfrac{dy}{dx} = \dfrac{x[25 - 4(x^2 + y^2)]}{y[25 + 4(x^2 + y^2)]}$; at $(3,1)$ $\dfrac{dy}{dx} = -9/13$

31. $2x + x\dfrac{dy}{dx} + y + 2y\dfrac{dy}{dx} = 0$. Substitute $y = -2x$ to obtain $-3x\dfrac{dy}{dx} = 0$. Since $x = \pm 1$ at the indicated points, $\dfrac{dy}{dx} = 0$ there.

33. **(a)**

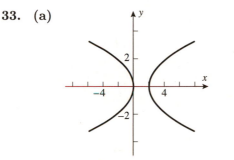

(b) Implicit differentiation of the curve yields $(4y^3 + 2y)\dfrac{dy}{dx} = 2x - 1$ so $\dfrac{dy}{dx} = 0$ only if $x = 1/2$ but $y^4 + y^2 \geq 0$ so $x = 1/2$ is impossible..

(c) $x^2 - x - (y^4 + y^2) = 0$, so by the Quadratic Formula, $x = \dfrac{-1 \pm \sqrt{(2y^2 + 1)^2}}{2} = 1 + y^2, -y^2$, and we have the two parabolas $x = -y^2, x = 1 + y^2$.

35. Solve the simultaneous equations $y = x, x^2 - xy + y^2 = 4$ to get $x^2 - x^2 + x^2 = 4, x = \pm 2, y = x = \pm 2$, so the points of intersection are $(2,2)$ and $(-2,-2)$.

By implicit differentiation, $\dfrac{dy}{dx} = \dfrac{y - 2x}{2y - x}$. When $x = y = 2$, $\dfrac{dy}{dx} = -1$; when $x = y = -2$, $\dfrac{dy}{dx} = -1$; the slopes are equal.

37. The point $(1,1)$ is on the graph, so $1 + a = b$. The slope of the tangent line at $(1,1)$ is $-4/3$; use implicit differentiation to get $\dfrac{dy}{dx} = -\dfrac{2xy}{x^2 + 2ay}$ so at $(1,1)$, $-\dfrac{2}{1 + 2a} = -\dfrac{4}{3}$, $1 + 2a = 3/2$, $a = 1/4$ and hence $b = 1 + 1/4 = 5/4$.

39. We shall find when the curves intersect and check that the slopes are negative reciprocals. For the intersection solve the simultaneous equations $x^2 + (y - c)^2 = c^2$ and $(x - k)^2 + y^2 = k^2$ to obtain $cy = kx = \dfrac{1}{2}(x^2 + y^2)$. Thus $x^2 + y^2 = cy + kx$, or $y^2 - cy = -x^2 + kx$, and $\dfrac{y - c}{x} = -\dfrac{x - k}{y}$.

Differentiating the two families yields (black) $\dfrac{dy}{dx} = -\dfrac{x}{y - c}$, and (gray) $\dfrac{dy}{dx} = -\dfrac{x - k}{y}$. But it was proven that these quantities are negative reciprocals of each other.

41. (a)

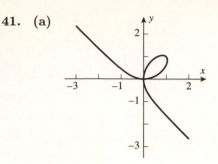

(b) $x \approx 0.84$

(c) Use implicit differentiation to get $dy/dx = (2y - 3x^2)/(3y^2 - 2x)$, so $dy/dx = 0$ if $y = (3/2)x^2$. Substitute this into $x^3 - 2xy + y^3 = 0$ to obtain $27x^6 - 16x^3 = 0, x^3 = 16/27, x = 2^{4/3}/3$ and hence $y = 2^{5/3}/3$.

43. Implicit differentiation of the equation of the curve yields $rx^{r-1} + ry^{r-1}\dfrac{dy}{dx} = 0$. At the point $(1,1)$ this becomes $r + r\dfrac{dy}{dx} = 0, \dfrac{dy}{dx} = -1$.

EXERCISE SET 2.8

1. $\dfrac{dy}{dt} = 3\dfrac{dx}{dt}$

    (a) $\dfrac{dy}{dt} = 3(2) = 6$ **(b)** $-1 = 3\dfrac{dx}{dt}, \dfrac{dx}{dt} = -\dfrac{1}{3}$

3. $8x\dfrac{dx}{dt} + 18y\dfrac{dy}{dt} = 0$

    (a) $8\dfrac{1}{2\sqrt{2}} \cdot 3 + 18\dfrac{1}{3\sqrt{2}}\dfrac{dy}{dt} = 0, \dfrac{dy}{dt} = -2$ **(b)** $8\left(\dfrac{1}{3}\right)\dfrac{dx}{dt} - 18\dfrac{\sqrt{5}}{9} \cdot 8 = 0, \dfrac{dx}{dt} = 6\sqrt{5}$

5. (b) $A = x^2$ **(c)** $\dfrac{dA}{dt} = 2x\dfrac{dx}{dt}$

    (d) Find $\dfrac{dA}{dt}\bigg|_{x=3}$ given that $\dfrac{dx}{dt}\bigg|_{x=3} = 2$. From Part (c), $\dfrac{dA}{dt}\bigg|_{x=3} = 2(3)(2) = 12 \text{ ft}^2/\text{min}.$

7. (a) $V = \pi r^2 h$, so $\dfrac{dV}{dt} = \pi\left(r^2\dfrac{dh}{dt} + 2rh\dfrac{dr}{dt}\right).$

    (b) Find $\dfrac{dV}{dt}\bigg|_{\substack{h=6,\\r=10}}$ given that $\dfrac{dh}{dt}\bigg|_{\substack{h=6,\\r=10}} = 1$ and $\dfrac{dr}{dt}\bigg|_{\substack{h=6,\\r=10}} = -1$. From Part (a),

    $\dfrac{dV}{dt}\bigg|_{\substack{h=6,\\r=10}} = \pi[10^2(1) + 2(10)(6)(-1)] = -20\pi \text{ in}^3/\text{s}$; the volume is decreasing.

9. (a) $\tan\theta = \dfrac{y}{x}$, so $\sec^2\theta\dfrac{d\theta}{dt} = \dfrac{x\dfrac{dy}{dt} - y\dfrac{dx}{dt}}{x^2}, \dfrac{d\theta}{dt} = \dfrac{\cos^2\theta}{x^2}\left(x\dfrac{dy}{dt} - y\dfrac{dx}{dt}\right)$

    (b) Find $\dfrac{d\theta}{dt}\bigg|_{\substack{x=2,\\y=2}}$ given that $\dfrac{dx}{dt}\bigg|_{\substack{x=2,\\y=2}} = 1$ and $\dfrac{dy}{dt}\bigg|_{\substack{x=2,\\y=2}} = -\dfrac{1}{4}.$

When $x = 2$ and $y = 2$, $\tan\theta = 2/2 = 1$ so $\theta = \dfrac{\pi}{4}$ and $\cos\theta = \cos\dfrac{\pi}{4} = \dfrac{1}{\sqrt{2}}$. Thus

from Part (a), $\dfrac{d\theta}{dt}\Big|_{\substack{x=2,\\y=2}} = \dfrac{(1/\sqrt{2})^2}{2^2}\left[2\left(-\dfrac{1}{4}\right) - 2(1)\right] = -\dfrac{5}{16}$ rad/s; θ is decreasing.

11. Let A be the area swept out, and θ the angle through which the minute hand has rotated. Find $\dfrac{dA}{dt}$ given that $\dfrac{d\theta}{dt} = \dfrac{\pi}{30}$ rad/min; $A = \dfrac{1}{2}r^2\theta = 8\theta$, so $\dfrac{dA}{dt} = 8\dfrac{d\theta}{dt} = \dfrac{4\pi}{15}$ in^2/min.

13. Find $\dfrac{dr}{dt}\Big|_{A=9}$ given that $\dfrac{dA}{dt} = 6$. From $A = \pi r^2$ we get $\dfrac{dA}{dt} = 2\pi r\dfrac{dr}{dt}$ so $\dfrac{dr}{dt} = \dfrac{1}{2\pi r}\dfrac{dA}{dt}$. If $A = 9$

then $\pi r^2 = 9$, $r = 3/\sqrt{\pi}$ so $\dfrac{dr}{dt}\Big|_{A=9} = \dfrac{1}{2\pi(3/\sqrt{\pi})}(6) = 1/\sqrt{\pi}$ mi/h.

15. Find $\dfrac{dV}{dt}\Big|_{r=9}$ given that $\dfrac{dr}{dt} = -15$. From $V = \dfrac{4}{3}\pi r^3$ we get $\dfrac{dV}{dt} = 4\pi r^2\dfrac{dr}{dt}$ so

$\dfrac{dV}{dt}\Big|_{r=9} = 4\pi(9)^2(-15) = -4860\pi$. Air must be removed at the rate of 4860π cm^3/min.

17. Find $\dfrac{dx}{dt}\Big|_{y=5}$ given that $\dfrac{dy}{dt} = -2$. From $x^2 + y^2 = 13^2$ we get

$2x\dfrac{dx}{dt} + 2y\dfrac{dy}{dt} = 0$ so $\dfrac{dx}{dt} = -\dfrac{y}{x}\dfrac{dy}{dt}$. Use $x^2 + y^2 = 169$ to find that

$x = 12$ when $y = 5$ so $\dfrac{dx}{dt}\Big|_{y=5} = -\dfrac{5}{12}(-2) = \dfrac{5}{6}$ ft/s.

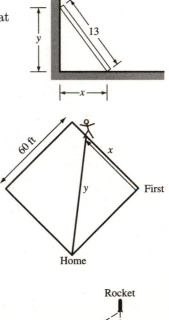

19. Let x denote the distance from first base and y the distance from home plate. Then $x^2 + 60^2 = y^2$ and $2x\dfrac{dx}{dt} = 2y\dfrac{dy}{dt}$. When $x = 50$ then $y = 10\sqrt{61}$ so

$\dfrac{dy}{dt} = \dfrac{x}{y}\dfrac{dx}{dt} = \dfrac{50}{10\sqrt{61}}(25) = \dfrac{125}{\sqrt{61}}$ ft/s.

21. Find $\dfrac{dy}{dt}\Big|_{x=4000}$ given that $\dfrac{dx}{dt}\Big|_{x=4000} = 880$. From

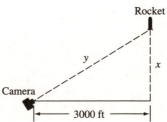

$y^2 = x^2 + 3000^2$ we get $2y\dfrac{dy}{dt} = 2x\dfrac{dx}{dt}$ so $\dfrac{dy}{dt} = \dfrac{x}{y}\dfrac{dx}{dt}$.

If $x = 4000$, then $y = 5000$ so

$\dfrac{dy}{dt}\Big|_{x=4000} = \dfrac{4000}{5000}(880) = 704$ ft/s.

23. (a) If x denotes the altitude, then $r - x = 3960$, the radius of the Earth. $\theta = 0$ at perigee, so $r = 4995/1.12 \approx 4460$; the altitude is $x = 4460 - 3960 = 500$ miles. $\theta = \pi$ at apogee, so $r = 4995/0.88 \approx 5676$; the altitude is $x = 5676 - 3960 = 1716$ miles.

(b) If $\theta = 120°$, then $r = 4995/0.94 \approx 5314$; the altitude is $5314 - 3960 = 1354$ miles. The rate of change of the altitude is given by

$$\frac{dx}{dt} = \frac{dr}{dt} = \frac{dr}{d\theta}\frac{d\theta}{dt} = \frac{4995(0.12\sin\theta)}{(1 + 0.12\cos\theta)^2}\frac{d\theta}{dt}.$$

Use $\theta = 120°$ and $d\theta/dt = 2.7°/\text{min} = (2.7)(\pi/180)$ rad/min to get $dr/dt \approx 27.7$ mi/min.

25. Find $\left.\dfrac{dh}{dt}\right|_{h=16}$ given that $\dfrac{dV}{dt} = 20$. The volume of water in the tank at a depth h is $V = \dfrac{1}{3}\pi r^2 h$. Use similar triangles (see figure)

to get $\dfrac{r}{h} = \dfrac{10}{24}$ so $r = \dfrac{5}{12}h$ thus $V = \dfrac{1}{3}\pi\left(\dfrac{5}{12}h\right)^2 h = \dfrac{25}{432}\pi h^3$,

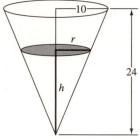

$$\frac{dV}{dt} = \frac{25}{144}\pi h^2 \frac{dh}{dt}; \quad \frac{dh}{dt} = \frac{144}{25\pi h^2}\frac{dV}{dt},$$

$$\left.\frac{dh}{dt}\right|_{h=16} = \frac{144}{25\pi(16)^2}(20) = \frac{9}{20\pi} \text{ ft/min}.$$

27. Find $\left.\dfrac{dV}{dt}\right|_{h=10}$ given that $\dfrac{dh}{dt} = 5$. $V = \dfrac{1}{3}\pi r^2 h$, but

$r = \dfrac{1}{2}h$ so $V = \dfrac{1}{3}\pi\left(\dfrac{h}{2}\right)^2 h = \dfrac{1}{12}\pi h^3$, $\dfrac{dV}{dt} = \dfrac{1}{4}\pi h^2 \dfrac{dh}{dt}$,

$$\left.\frac{dV}{dt}\right|_{h=10} = \frac{1}{4}\pi(10)^2(5) = 125\pi \text{ ft}^3/\text{min}.$$

29. With s and h as shown in the figure, we want to find $\dfrac{dh}{dt}$ given that $\dfrac{ds}{dt} = 500$. From the figure,

$h = s\sin 30° = \dfrac{1}{2}s$ so $\dfrac{dh}{dt} = \dfrac{1}{2}\dfrac{ds}{dt} = \dfrac{1}{2}(500) = 250$ mi/h.

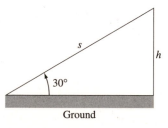

31. Find $\dfrac{dy}{dt}$ given that $\left.\dfrac{dx}{dt}\right|_{y=125} = -12$. From $x^2 + 10^2 = y^2$ we

get $2x\dfrac{dx}{dt} = 2y\dfrac{dy}{dt}$ so $\dfrac{dy}{dt} = \dfrac{x}{y}\dfrac{dx}{dt}$. Use $x^2 + 100 = y^2$ to find that

$x = \sqrt{15,525} = 15\sqrt{69}$ when $y = 125$ so $\dfrac{dy}{dt} = \dfrac{15\sqrt{69}}{125}(-12) =$

$-\dfrac{36\sqrt{69}}{25}$. The rope must be pulled at the rate of $\dfrac{36\sqrt{69}}{25}$ ft/min.

33. Find $\left.\dfrac{dx}{dt}\right|_{\theta=\pi/4}$ given that $\dfrac{d\theta}{dt} = \dfrac{2\pi}{10} = \dfrac{\pi}{5}$ rad/s.

Then $x = 4\tan\theta$ (see figure) so $\dfrac{dx}{dt} = 4\sec^2\theta\,\dfrac{d\theta}{dt}$,

$$\left.\dfrac{dx}{dt}\right|_{\theta=\pi/4} = 4\left(\sec^2\dfrac{\pi}{4}\right)\left(\dfrac{\pi}{5}\right) = 8\pi/5 \text{ km/s.}$$

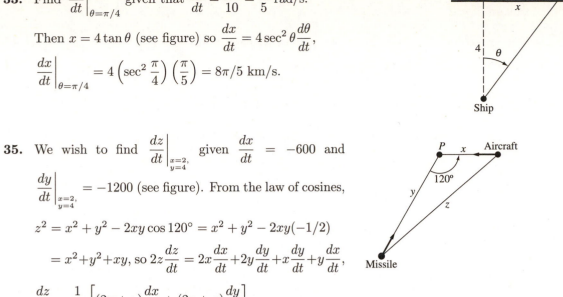

35. We wish to find $\left.\dfrac{dz}{dt}\right|_{\substack{x=2,\\y=4}}$ given $\dfrac{dx}{dt} = -600$ and

$\left.\dfrac{dy}{dt}\right|_{\substack{x=2,\\y=4}} = -1200$ (see figure). From the law of cosines,

$$z^2 = x^2 + y^2 - 2xy\cos 120° = x^2 + y^2 - 2xy(-1/2)$$

$$= x^2 + y^2 + xy, \text{ so } 2z\dfrac{dz}{dt} = 2x\dfrac{dx}{dt} + 2y\dfrac{dy}{dt} + x\dfrac{dy}{dt} + y\dfrac{dx}{dt},$$

$$\dfrac{dz}{dt} = \dfrac{1}{2z}\left[(2x+y)\dfrac{dx}{dt} + (2y+x)\dfrac{dy}{dt}\right].$$

When $x = 2$ and $y = 4$, $z^2 = 2^2 + 4^2 + (2)(4) = 28$, so $z = \sqrt{28} = 2\sqrt{7}$, thus

$$\left.\dfrac{dz}{dt}\right|_{\substack{x=2,\\y=4}} = \dfrac{1}{2(2\sqrt{7})}[(2(2)+4)(-600) + (2(4)+2)(-1200)] = -\dfrac{4200}{\sqrt{7}} = -600\sqrt{7} \text{ mi/h;}$$

the distance between missile and aircraft is decreasing at the rate of $600\sqrt{7}$ mi/h.

37. (a) We want $\left.\dfrac{dy}{dt}\right|_{\substack{x=1,\\y=2}}$ given that $\left.\dfrac{dx}{dt}\right|_{\substack{x=1,\\y=2}} = 6$. For convenience, first rewrite the equation as

$$xy^3 = \dfrac{8}{5} + \dfrac{8}{5}y^2 \text{ then } 3xy^2\dfrac{dy}{dt} + y^3\dfrac{dx}{dt} = \dfrac{16}{5}y\dfrac{dy}{dt}, \dfrac{dy}{dt} = \dfrac{y^3}{\dfrac{16}{5}y - 3xy^2}\dfrac{dx}{dt} \text{ so}$$

$$\left.\dfrac{dy}{dt}\right|_{\substack{x=1,\\y=2}} = \dfrac{2^3}{\dfrac{16}{5}(2) - 3(1)2^2}(6) = -60/7 \text{ units/s.}$$

(b) falling, because $\dfrac{dy}{dt} < 0$

39. The coordinates of P are $(x, 2x)$, so the distance between P and the point $(3, 0)$ is

$D = \sqrt{(x-3)^2 + (2x-0)^2} = \sqrt{5x^2 - 6x + 9}$. Find $\left.\dfrac{dD}{dt}\right|_{x=3}$ given that $\left.\dfrac{dx}{dt}\right|_{x=3} = -2$.

$\dfrac{dD}{dt} = \dfrac{5x-3}{\sqrt{5x^2-6x+9}}\dfrac{dx}{dt}$, so $\left.\dfrac{dD}{dt}\right|_{x=3} = \dfrac{12}{\sqrt{36}}(-2) = -4$ units/s.

41. Solve $\dfrac{dx}{dt} = 3\dfrac{dy}{dt}$ given $y = x/(x^2 + 1)$. Then $y(x^2 + 1) = x$. Differentiating with respect to x,

$(x^2+1)\dfrac{dy}{dx} + y(2x) = 1$. But $\dfrac{dy}{dx} = \dfrac{dy/dt}{dx/dt} = \dfrac{1}{3}$ so $(x^2+1)\dfrac{1}{3} + 2xy = 1$, $x^2 + 1 + 6xy = 3$,

$x^2 + 1 + 6x^2/(x^2+1) = 3$, $(x^2+1)^2 + 6x^2 - 3x^2 - 3 = 0$, $x^4 + 5x^2 - 2 = 0$. By the Quadratic Formula applied to x^2 we obtain $x^2 = (-5 \pm \sqrt{25+8})/2$. The minus sign is spurious since x^2 cannot be negative, so $x^2 = (\sqrt{33} - 5)/2$, $x \approx \pm 0.6101486075$, $y = \pm 0.4446235602$.

43. Find $\dfrac{dS}{dt}\Big|_{s=10}$ given that $\dfrac{ds}{dt}\Big|_{s=10} = -2$. From $\dfrac{1}{s} + \dfrac{1}{S} = \dfrac{1}{6}$ we get $-\dfrac{1}{s^2}\dfrac{ds}{dt} - \dfrac{1}{S^2}\dfrac{dS}{dt} = 0$, so

$\dfrac{dS}{dt} = -\dfrac{S^2}{s^2}\dfrac{ds}{dt}$. If $s = 10$, then $\dfrac{1}{10} + \dfrac{1}{S} = \dfrac{1}{6}$ which gives $S = 15$. So $\dfrac{dS}{dt}\Big|_{s=10} = -\dfrac{225}{100}(-2) = 4.5$

cm/s. The image is moving away from the lens.

45. Let r be the radius, V the volume, and A the surface area of a sphere. Show that $\dfrac{dr}{dt}$ is a constant given that $\dfrac{dV}{dt} = -kA$, where k is a positive constant. Because $V = \dfrac{4}{3}\pi r^3$,

$$\frac{dV}{dt} = 4\pi r^2 \frac{dr}{dt} \qquad (1)$$

But it is given that $\dfrac{dV}{dt} = -kA$ or, because $A = 4\pi r^2$, $\dfrac{dV}{dt} = -4\pi r^2 k$ which when substituted into

equation (1) gives $-4\pi r^2 k = 4\pi r^2 \dfrac{dr}{dt}$, $\dfrac{dr}{dt} = -k$.

47. Extend sides of cup to complete the cone and let V_0 be the volume of the portion added, then (see figure) $V = \dfrac{1}{3}\pi r^2 h - V_0$ where

$\dfrac{r}{h} = \dfrac{4}{12} = \dfrac{1}{3}$ so $r = \dfrac{1}{3}h$ and $V = \dfrac{1}{3}\pi\left(\dfrac{h}{3}\right)^2 h - V_0 = \dfrac{1}{27}\pi h^3 - V_0$,

$\dfrac{dV}{dt} = \dfrac{1}{9}\pi h^2 \dfrac{dh}{dt}$, $\dfrac{dh}{dt} = \dfrac{9}{\pi h^2}\dfrac{dV}{dt}$, $\dfrac{dh}{dt}\Big|_{h=9} = \dfrac{9}{\pi(9)^2}(20) = \dfrac{20}{9\pi}$ cm/s.

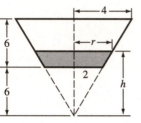

EXERCISE SET 2.9

1. (a) $f(x) \approx f(1) + f'(1)(x-1) = 1 + 3(x-1)$

(b) $f(1+\Delta x) \approx f(1) + f'(1)\Delta x = 1 + 3\Delta x$

(c) From Part (a), $(1.02)^3 \approx 1 + 3(0.02) = 1.06$. From Part (b), $(1.02)^3 \approx 1 + 3(0.02) = 1.06$.

3. (a) $f(x) \approx f(x_0) + f'(x_0)(x - x_0) = 1 + (1/(2\sqrt{1})(x-0) = 1 + (1/2)x$, so with $x_0 = 0$ and $x = -0.1$, we have $\sqrt{0.9} = f(-0.1) \approx 1 + (1/2)(-0.1) = 1 - 0.05 = 0.95$. With $x = 0.1$ we have $\sqrt{1.1} = f(0.1) \approx 1 + (1/2)(0.1) = 1.05$.

(b)

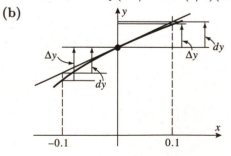

5. $f(x) = (1+x)^{15}$ and $x_0 = 0$. Thus $(1+x)^{15} \approx f(x_0) + f'(x_0)(x-x_0) = 1 + 15(1)^{14}(x-0) = 1 + 15x$.

7. $\tan x \approx \tan(0) + \sec^2(0)(x-0) = x$

9. $x^4 \approx (1)^4 + 4(1)^3(x-1)$. Set $\Delta x = x - 1$; then $x = \Delta x + 1$ and $(1+\Delta x)^4 = 1 + 4\Delta x$.

11. $\dfrac{1}{2+x} \approx \dfrac{1}{2+1} - \dfrac{1}{(2+1)^2}(x-1)$, and $2+x = 3+\Delta x$, so $\dfrac{1}{3+\Delta x} \approx \dfrac{1}{3} - \dfrac{1}{9}\Delta x$

13. $f(x) = \sqrt{x+3}$ and $x_0 = 0$, so

$$\sqrt{x+3} \approx \sqrt{3} + \dfrac{1}{2\sqrt{3}}(x-0) = \sqrt{3} + \dfrac{1}{2\sqrt{3}}x, \text{ and}$$

$$\left| f(x) - \left(\sqrt{3} + \dfrac{1}{2\sqrt{3}}x\right) \right| < 0.1 \text{ if } |x| < 1.692.$$

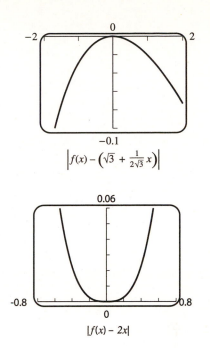

$\left| f(x) - \left(\sqrt{3} + \frac{1}{2\sqrt{3}}x\right) \right|$

15. $\tan 2x \approx \tan 0 + (\sec^2 0)(2x - 0) = 2x$,
and $|\tan 2x - 2x| < 0.1$ if $|x| < 0.3158$

$|f(x) - 2x|$

17. **(a)** The local linear approximation $\sin x \approx x$ gives $\sin 1° = \sin(\pi/180) \approx \pi/180 = 0.0174533$ and a calculator gives $\sin 1° = 0.0174524$. The relative error $|\sin(\pi/180) - (\pi/180)|/(\sin \pi/180) = 0.000051$ is very small, so for such a small value of x the approximation is very good.

(b) Use $x_0 = 45°$ (this assumes you know, or can approximate, $\sqrt{2}/2$).

(c) $44° = \dfrac{44\pi}{180}$ radians, and $45° = \dfrac{45\pi}{180} = \dfrac{\pi}{4}$ radians. With $x = \dfrac{44\pi}{180}$ and $x_0 = \dfrac{\pi}{4}$ we obtain

$$\sin 44° = \sin \dfrac{44\pi}{180} \approx \sin \dfrac{\pi}{4} + \left(\cos \dfrac{\pi}{4}\right)\left(\dfrac{44\pi}{180} - \dfrac{\pi}{4}\right) = \dfrac{\sqrt{2}}{2} + \dfrac{\sqrt{2}}{2}\left(\dfrac{-\pi}{180}\right) = 0.694765. \text{ With a}$$

calculator, $\sin 44° = 0.694658$.

19. $f(x) = x^4$, $f'(x) = 4x^3$, $x_0 = 3$, $\Delta x = 0.02$; $(3.02)^4 \approx 3^4 + (108)(0.02) = 81 + 2.16 = 83.16$

21. $f(x) = \sqrt{x}$, $f'(x) = \dfrac{1}{2\sqrt{x}}$, $x_0 = 64$, $\Delta x = 1$; $\sqrt{65} \approx \sqrt{64} + \dfrac{1}{16}(1) = 8 + \dfrac{1}{16} = 8.0625$

23. $f(x) = \sqrt{x}$, $f'(x) = \dfrac{1}{2\sqrt{x}}$, $x_0 = 81$, $\Delta x = -0.1$; $\sqrt{80.9} \approx \sqrt{81} + \dfrac{1}{18}(-0.1) \approx 8.9944$

25. $f(x) = \sin x$, $f'(x) = \cos x$, $x_0 = 0$, $\Delta x = 0.1$; $\sin 0.1 \approx \sin 0 + (\cos 0)(0.1) = 0.1$

27. $f(x) = \cos x$, $f'(x) = -\sin x$, $x_0 = \pi/6$, $\Delta x = \pi/180$;

$$\cos 31° \approx \cos 30° + \left(-\dfrac{1}{2}\right)\left(\dfrac{\pi}{180}\right) = \dfrac{\sqrt{3}}{2} - \dfrac{\pi}{360} \approx 0.8573$$

29. $\sqrt[3]{8.24} = 8^{1/3}\sqrt[3]{1.03} \approx 2(1 + \tfrac{1}{3}0.03) \approx 2.02$
$4.08^{3/2} = 4^{3/2}1.02^{3/2} = 8(1 + 0.02(3/2)) = 8.24$

31. **(a)** $dy = f'(x)dx = 2xdx = 4(1) = 4$ and
$\Delta y = (x + \Delta x)^2 - x^2 = (2+1)^2 - 2^2 = 5$

(b)

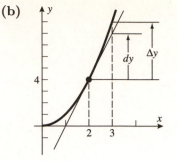

33. $dy = 3x^2 dx$; sd
$\Delta y = (x + \Delta x)^3 - x^3 = x^3 + 3x^2\Delta x + 3x(\Delta x)^2 + (\Delta x)^3 - x^3 = 3x^2\Delta x + 3x(\Delta x)^2 + (\Delta x)^3$

35. $dy = (2x - 2)dx$;
$\Delta y = [(x + \Delta x)^2 - 2(x + \Delta x) + 1] - [x^2 - 2x + 1]$
$= x^2 + 2x\,\Delta x + (\Delta x)^2 - 2x - 2\Delta x + 1 - x^2 + 2x - 1 = 2x\,\Delta x + (\Delta x)^2 - 2\Delta x$

37. **(a)** $dy = (12x^2 - 14x)dx$

(b) $dy = x\,d(\cos x) + \cos x\,dx = x(-\sin x)dx + \cos x\,dx = (-x\sin x + \cos x)dx$

39. **(a)** $dy = \left(\sqrt{1-x} - \dfrac{x}{2\sqrt{1-x}}\right)dx = \dfrac{2-3x}{2\sqrt{1-x}}dx$

(b) $dy = -17(1+x)^{-18}dx$

41. false; $dy = (dy/dx)dx$

43. false; they are equal whenever the function is linear

45. $dy = \dfrac{3}{2\sqrt{3x-2}}dx$, $x = 2$, $dx = 0.03$; $\Delta y \approx dy = \dfrac{3}{4}(0.03) = 0.0225$

47. $dy = \dfrac{1-x^2}{(x^2+1)^2}dx$, $x = 2$, $dx = -0.04$; $\Delta y \approx dy = \left(-\dfrac{3}{25}\right)(-0.04) = 0.0048$

49. **(a)** $A = x^2$ where x is the length of a side; $dA = 2x\,dx = 2(10)(\pm 0.1) = \pm 2$ ft^2.

(b) relative error in x is $\approx \dfrac{dx}{x} = \dfrac{\pm 0.1}{10} = \pm 0.01$ so percentage error in x is $\approx \pm 1\%$; relative error
in A is $\approx \dfrac{dA}{A} = \dfrac{2x\,dx}{x^2} = 2\dfrac{dx}{x} = 2(\pm 0.01) = \pm 0.02$ so percentage error in A is $\approx \pm 2\%$

51. **(a)** $x = 10\sin\theta$, $y = 10\cos\theta$ (see figure),
$$dx = 10\cos\theta\,d\theta = 10\left(\cos\dfrac{\pi}{6}\right)\left(\pm\dfrac{\pi}{180}\right) = 10\left(\dfrac{\sqrt{3}}{2}\right)\left(\pm\dfrac{\pi}{180}\right)$$
$\approx \pm 0.151$ in,
$$dy = -10(\sin\theta)d\theta = -10\left(\sin\dfrac{\pi}{6}\right)\left(\pm\dfrac{\pi}{180}\right) = -10\left(\dfrac{1}{2}\right)\left(\pm\dfrac{\pi}{180}\right)$$
$\approx \pm 0.087$ in

(b) relative error in x is $\approx \dfrac{dx}{x} = (\cot\theta)d\theta = \left(\cot\dfrac{\pi}{6}\right)\left(\pm\dfrac{\pi}{180}\right) = \sqrt{3}\left(\pm\dfrac{\pi}{180}\right) \approx \pm 0.030$
so percentage error in x is $\approx \pm 3.0\%$;
relative error in y is $\approx \dfrac{dy}{y} = -\tan\theta\,d\theta = -\left(\tan\dfrac{\pi}{6}\right)\left(\pm\dfrac{\pi}{180}\right) = -\dfrac{1}{\sqrt{3}}\left(\pm\dfrac{\pi}{180}\right) \approx \pm 0.010$
so percentage error in y is $\approx \pm 1.0\%$

53. $\dfrac{dR}{R} = \dfrac{(-2k/r^3)dr}{(k/r^2)} = -2\dfrac{dr}{r}$, but $\dfrac{dr}{r} \approx \pm 0.05$ so $\dfrac{dR}{R} \approx -2(\pm 0.05) = \pm 0.10$; percentage error in R is $\approx \pm 10\%$

55. $A = \dfrac{1}{4}(4)^2 \sin 2\theta = 4 \sin 2\theta$ thus $dA = 8 \cos 2\theta d\theta$ so, with $\theta = 30° = \pi/6$ radians and $d\theta = \pm 15' = \pm 1/4° = \pm \pi/720$ radians, $dA = 8 \cos(\pi/3)(\pm \pi/720) = \pm \pi/180 \approx \pm 0.017$ cm^2

57. $V = x^3$ where x is the length of a side; $\dfrac{dV}{V} = \dfrac{3x^2 dx}{x^3} = 3\dfrac{dx}{x}$, but $\dfrac{dx}{x} \approx \pm 0.02$ so $\dfrac{dV}{V} \approx 3(\pm 0.02) = \pm 0.06$; percentage error in V is $\approx \pm 6\%$.

59. $A = \dfrac{1}{4}\pi D^2$ where D is the diameter of the circle; $\dfrac{dA}{A} = \dfrac{(\pi D/2)dD}{\pi D^2/4} = 2\dfrac{dD}{D}$, but $\dfrac{dA}{A} \approx \pm 0.01$ so $2\dfrac{dD}{D} \approx \pm 0.01$, $\dfrac{dD}{D} \approx \pm 0.005$; maximum permissible percentage error in D is $\approx \pm 0.5\%$.

61. $V =$ volume of cylindrical rod $= \pi r^2 h = \pi r^2(15) = 15\pi r^2$; approximate ΔV by dV if $r = 2.5$ and $dr = \Delta r = 0.1$. $dV = 30\pi r\, dr = 30\pi(2.5)(0.1) \approx 23.5619$ cm^3.

63. (a) $\alpha = \Delta L/(L\Delta T) = 0.006/(40 \times 10) = 1.5 \times 10^{-5}/°C$

   (b) $\Delta L = 2.3 \times 10^{-5}(180)(25) \approx 0.1$ cm, so the pole is about 180.1 cm long.

REVIEW EXERCISES, CHAPTER 2

3. (a) $m_{\text{tan}} = \lim\limits_{w \to x} \dfrac{f(w) - f(x)}{w - x} = \lim\limits_{w \to x} \dfrac{(w^2 + 1) - (x^2 + 1)}{w - x}$

$= \lim\limits_{w \to x} \dfrac{w^2 - x^2}{w - x} = \lim\limits_{w \to x} (w + x) = 2x$

   (b) $m_{\text{tan}} = 2(2) = 4$

5. $v_{\text{inst}} = \lim\limits_{h \to 0} \dfrac{3(h+1)^{2.5} + 580h - 3}{10h} = 58 + \dfrac{1}{10}\dfrac{d}{dx}3x^{2.5}\Big|_{x=1} = 58 + \dfrac{1}{10}(2.5)(3)(1)^{1.5} = 58.75$ ft/s

7. (a) $v_{\text{ave}} = \dfrac{[3(3)^2 + 3] - [3(1)^2 + 1]}{3 - 1} = 13$ mi/h

   (b) $v_{\text{inst}} = \lim\limits_{t_1 \to 1} \dfrac{(3t_1^2 + t_1) - 4}{t_1 - 1} = \lim\limits_{t_1 \to 1} \dfrac{(3t_1 + 4)(t_1 - 1)}{t_1 - 1} = \lim\limits_{t_1 \to 1} (3t_1 + 4) = 7$ mi/h

9. (a) $\dfrac{dy}{dx} = \lim\limits_{h \to 0} \dfrac{\sqrt{9 - 4(x+h)} - \sqrt{9 - 4x}}{h} = \lim\limits_{h \to 0} \dfrac{9 - 4(x+h) - (9 - 4x)}{h(\sqrt{9 - 4(x+h)} + \sqrt{9 - 4x})}$

$= \lim\limits_{h \to 0} \dfrac{-4h}{h(\sqrt{9 - 4(x+h)} + \sqrt{9 - 4x})} = \dfrac{-4}{2\sqrt{9 - 4x}} = \dfrac{-2}{\sqrt{9 - 4x}}$

   (b) $\dfrac{dy}{dx} = \lim\limits_{h \to 0} \dfrac{\dfrac{x+h}{x+h+1} - \dfrac{x}{x+1}}{h} = \lim\limits_{h \to 0} \dfrac{(x+h)(x+1) - x(x+h+1)}{h(x+h+1)(x+1)}$

$= \lim\limits_{h \to 0} \dfrac{h}{h(x+h+1)(x+1)} = \dfrac{1}{(x+1)^2}$

11. **(a)** $x = -2, -1, 1, 3$

 (b) $(-\infty, -2),\ (-1, 1),\ (3, +\infty)$

 (c) $(-2, -1),\ (1, 3)$

 (d) $g''(x) = f''(x)\sin x + 2f'(x)\cos x - f(x)\sin x;\ g''(0) = 2f'(0)\cos 0 = 2(2)(1) = 4$

13. **(a)** The slope of the tangent line $\approx \dfrac{10 - 2.2}{2050 - 1950} = 0.078$ billion, so in 2000 the world population was increasing at the rate of about 78 million per year.

 (b) $\dfrac{dN/dt}{N} \approx \dfrac{0.078}{6} = 0.013 = 1.3\ \%/\text{year}$

15. **(a)** $f'(x) = 2x\sin x + x^2\cos x$ **(c)** $f''(x) = 4x\cos x + (2 - x^2)\sin x$

17. **(a)** $f'(x) = \dfrac{6x^2 + 8x - 17}{(3x + 2)^2}$ **(c)** $f''(x) = \dfrac{118}{(3x + 2)^3}$

19. **(a)** $\dfrac{dW}{dt} = 200(t - 15)$; at $t = 5$, $\dfrac{dW}{dt} = -2000$; the water is running out at the rate of 2000 gal/min.

 (b) $\dfrac{W(5) - W(0)}{5 - 0} = \dfrac{10000 - 22500}{5} = -2500$; the average rate of flow out is 2500 gal/min.

21. **(a)** $f'(x) = 2x,\ f'(1.8) = 3.6$

 (b) $f'(x) = (x^2 - 4x)/(x - 2)^2,\ f'(3.5) = -7/9 \approx -0.777778$

23. f is continuous at $x = 1$ because it is differentiable there, thus $\lim\limits_{h \to 0} f(1 + h) = f(1)$ and so $f(1) = 0$ because $\lim\limits_{h \to 0} \dfrac{f(1 + h)}{h}$ exists; $f'(1) = \lim\limits_{h \to 0} \dfrac{f(1 + h) - f(1)}{h} = \lim\limits_{h \to 0} \dfrac{f(1 + h)}{h} = 5$.

25. The equation of such a line has the form $y = mx$. The points (x_0, y_0) which lie on both the line and the parabola and for which the slopes of both curves are equal satisfy $y_0 = mx_0 = x_0^3 - 9x_0^2 - 16x_0$, so that $m = x_0^2 - 9x_0 - 16$. By differentiating, the slope is also given by $m = 3x_0^2 - 18x_0 - 16$. Equating, we have $x_0^2 - 9x_0 - 16 = 3x_0^2 - 18x_0 - 16$, or $2x_0^2 - 9x_0 = 0$. The root $x_0 = 0$ corresponds to $m = -16, y_0 = 0$ and the root $x_0 = 9/2$ corresponds to $m = -145/4, y_0 = -1305/8$. So the line $y = -16x$ is tangent to the curve at the point $(0, 0)$, and the line $y = -145x/4$ is tangent to the curve at the point $(9/2, -1305/8)$.

27. The slope of the tangent line is the derivative

$$y' = 2x\Big|_{x = \frac{1}{2}(a+b)} = a + b.$$ The slope of the secant is

$$\dfrac{a^2 - b^2}{a - b} = a + b,$$ so they are equal.

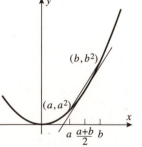

29. **(a)** $8x^7 - \dfrac{3}{2\sqrt{x}} - 15x^{-4}$

 (b) $2 \cdot 101(2x + 1)^{100}(5x^2 - 7) + 10x(2x + 1)^{101} = (2x + 1)^{100}(1030x^2 + 10x - 1414)$

31. **(a)** $2(x - 1)\sqrt{3x + 1} + \dfrac{3}{2\sqrt{3x + 1}}(x - 1)^2 = \dfrac{(x - 1)(15x + 1)}{2\sqrt{3x + 1}}$

(b) $\quad 3\left(\dfrac{3x+1}{x^2}\right)^2 \dfrac{x^2(3)-(3x+1)(2x)}{x^4} = -\dfrac{3(3x+1)^2(3x+2)}{x^7}$

33. Set $f'(x)=0$: $f'(x)=6(2)(2x+7)^5(x-2)^5+5(2x+7)^6(x-2)^4=0$, so $2x+7=0$ or $x-2=0$ or, factoring out $(2x+7)^5(x-2)^4$, $12(x-2)+5(2x+7)=0$. This reduces to $x=-7/2$, $x=2$, or $22x+11=0$, so the tangent line is horizontal at $x=-7/2, 2, -1/2$.

35. Suppose the line is tangent to $y=x^2+1$ at (x_0,y_0) and tangent to $y=-x^2-1$ at (x_1,y_1). Since it's tangent to $y=x^2+1$, its slope is $2x_0$; since it's tangent to $y=-x^2-1$, its slope is $-2x_1$. Hence $x_1=-x_0$ and $y_1=-y_0$. Since the line passes through both points, its slope is $\dfrac{y_1-y_0}{x_1-x_0}=\dfrac{-2y_0}{-2x_0}=\dfrac{y_0}{x_0}=\dfrac{x_0^2+1}{x_0}$. Thus $2x_0=\dfrac{x_0^2+1}{x_0}$, so $2x_0^2=x_0^2+1$, $x_0^2=1$, and $x_0=\pm 1$. So there are two lines which are tangent to both graphs, namely $y=2x$ and $y=-2x$.

37. The line $y-x=2$ has slope $m_1=1$ so we set $m_2=\dfrac{d}{dx}(3x-\tan x)=3-\sec^2 x=1$, or $\sec^2 x=2$, $\sec x=\pm\sqrt{2}$ so $x=n\pi\pm\pi/4$ where $n=0,\pm 1,\pm 2,\dots$.

39. $3=f(\pi/4)=(M+N)\sqrt{2}/2$ and $1=f'(\pi/4)=(M-N)\sqrt{2}/2$. Add these two equations to get $4=\sqrt{2}M$, $M=2^{3/2}$. Subtract to obtain $2=\sqrt{2}N$, $N=\sqrt{2}$. Thus $f(x)=2\sqrt{2}\sin x+\sqrt{2}\cos x$. $f'\left(\dfrac{3\pi}{4}\right)=-3$ so tangent line is $y-1=-3\left(x-\dfrac{3\pi}{4}\right)$.

41. $f'(x)=2xf(x)$, $f(2)=5$

(a) $g(x)=f(\sec x)$, $g'(x)=f'(\sec x)\sec x\tan x=2\cdot 2f(2)\cdot 2\cdot\sqrt{3}=40\sqrt{3}$.

(b) $h'(x)=4\left[\dfrac{f(x)}{x-1}\right]^3\dfrac{(x-1)f'(x)-f(x)}{(x-1)^2}$,

$h'(2)=4\dfrac{5^3}{1}\dfrac{f'(2)-f(2)}{1}=4\cdot 5^3\dfrac{2\cdot 2f(2)-f(2)}{1}=4\cdot 5^3\cdot 3\cdot 5=7500$

43. Yes, g must be differentiable (where $f'\neq 0$); this can be inferred from the graphs. Note that if $f'=0$ at a point then g' cannot exist (infinite slope).

45. (a) $3x^2+x\dfrac{dy}{dx}+y-2=0$, $\dfrac{dy}{dx}=\dfrac{2-y-3x^2}{x}$

(b) $y=(1+2x-x^3)/x=1/x+2-x^2$, $dy/dx=-1/x^2-2x$

(c) $\dfrac{dy}{dx}=\dfrac{2-(1/x+2-x^2)-3x^2}{x}=-1/x^2-2x$

47. $-\dfrac{1}{y^2}\dfrac{dy}{dx}-\dfrac{1}{x^2}=0$ so $\dfrac{dy}{dx}=-\dfrac{y^2}{x^2}$

49. $\left(x\dfrac{dy}{dx}+y\right)\sec(xy)\tan(xy)=\dfrac{dy}{dx}$, $\dfrac{dy}{dx}=\dfrac{y\sec(xy)\tan(xy)}{1-x\sec(xy)\tan(xy)}$

51. $\dfrac{dy}{dx}=\dfrac{3x}{4y}$, $\dfrac{d^2y}{dx^2}=\dfrac{(4y)(3)-(3x)(4dy/dx)}{16y^2}=\dfrac{12y-12x(3x/(4y))}{16y^2}=\dfrac{12y^2-9x^2}{16y^3}=\dfrac{-3(3x^2-4y^2)}{16y^3}$, but $3x^2-4y^2=7$ so $\dfrac{d^2y}{dx^2}=\dfrac{-3(7)}{16y^3}=-\dfrac{21}{16y^3}$

53. $\dfrac{dy}{dx}=\tan(\pi y/2)+x(\pi/2)\dfrac{dy}{dx}\sec^2(\pi y/2)$, $\dfrac{dy}{dx}\Big]_{y=1/2}=1+(\pi/4)\dfrac{dy}{dx}\Big]_{y=1/2}(2)$, $\dfrac{dy}{dx}\Big]_{y=1/2}=\dfrac{2}{2-\pi}$

55. Substitute $y = mx$ into $x^2 + xy + y^2 = 4$ to get $x^2 + mx^2 + m^2x^2 = 4$, which has distinct solutions $x = \pm 2/\sqrt{m^2 + m + 1}$. They are distinct because $m^2 + m + 1 = (m + 1/2)^2 + 3/4 \geq 3/4$, so $m^2 + m + 1$ is never zero.

Note that the points of intersection occur in pairs (x_0, y_0) and $(-x_0, -y_0)$. By implicit differentiation, the slope of the tangent line to the ellipse is given by

$dy/dx = -(2x + y)/(x + 2y)$. Since the slope is unchanged if we replace (x, y) with $(-x, -y)$, it follows that the slopes are equal at the two point of intersection.

Finally we must examine the special case $x = 0$ which cannot be written in the form $y = mx$. If $x = 0$ then $y = \pm 2$, and the formula for dy/dx gives $dy/dx = -1/2$, so the slopes are equal.

57. By implicit differentiation, $3x^2 - y - xy' + 3y^2y' = 0$, so $y' = (3x^2 - y)/(x - 3y^2)$. This derivative exists except when $x = 3y^2$. Substituting this into the original equation $x^3 - xy + y^3 = 0$, one has $27y^6 - 3y^3 + y^3 = 0$, $y^3(27y^3 - 2) = 0$. The unique solution in the first quadrant is $y = 2^{1/3}/3, x = 3y^2 = 2^{2/3}/3$

59. $A = \pi r^2$ and $\dfrac{dr}{dt} = -5$, so $\dfrac{dA}{dt} = \dfrac{dA}{dr}\dfrac{dr}{dt} = 2\pi r(-5) = -500\pi$, so the area is shrinking at a rate of $500\pi \, \text{m}^2/\text{min}$.

61. **(a)** $\Delta x = 1.5 - 2 = -0.5$; $dy = \dfrac{-1}{(x-1)^2}\Delta x = \dfrac{-1}{(2-1)^2}(-0.5) = 0.5$; and

$\Delta y = \dfrac{1}{(1.5 - 1)} - \dfrac{1}{(2-1)} = 2 - 1 = 1$.

(b) $\Delta x = 0 - (-\pi/4) = \pi/4$; $dy = \left(\sec^2(-\pi/4)\right)(\pi/4) = \pi/2$; and $\Delta y = \tan 0 - \tan(-\pi/4) = 1$.

(c) $\Delta x = 3 - 0 = 3$; $dy = \dfrac{-x}{\sqrt{25 - x^2}} = \dfrac{-0}{\sqrt{25 - (0)^2}}(3) = 0$; and

$\Delta y = \sqrt{25 - 3^2} - \sqrt{25 - 0^2} = 4 - 5 = -1$.

63. **(a)** $h = 115\tan\phi$, $dh = 115\sec^2\phi \, d\phi$; with $\phi = 51° = \dfrac{51}{180}\pi$ radians and

$d\phi = \pm 0.5° = \pm 0.5\left(\dfrac{\pi}{180}\right)$ radians, $h \pm dh = 115(1.2349) \pm 2.5340 = 142.0135 \pm 2.5340$, so the height lies between 139.48 m and 144.55 m.

(b) If $|dh| \leq 5$ then $|d\phi| \leq \dfrac{5}{115}\cos^2\dfrac{51}{180}\pi \approx 0.017$ radian, or $ld\phi| \leq 0.98°$.

MAKING CONNECTIONS, CHAPTER 2

1. **(a)** By property (ii), $f(0) = f(0 + 0) = f(0)f(0)$, so $f(0) = 0$ or 1. By property (iii), $f(0) \neq 0$, so $f(0) = 1$.

(b) By property (ii), $f(x) = f\left(\dfrac{x}{2} + \dfrac{x}{2}\right) = f\left(\dfrac{x}{2}\right)^2 \geq 0$. If $f(x) = 0$, then

$1 = f(0) = f(x + (-x)) = f(x)f(-x) = 0 \cdot f(-x) = 0$, a contradiction. Hence $f(x) > 0$.

(c) $f'(x) = \lim_{h\to 0}\dfrac{f(x+h) - f(x)}{h} = \lim_{h\to 0}\dfrac{f(x)f(h) - f(x)}{h} = \lim_{h\to 0}f(x)\dfrac{f(h) - 1}{h} = f(x)\lim_{h\to 0}\dfrac{f(h) - f(0)}{h}$

$= f(x)f'(0) = f(x)$

3. **(a)** For brevity, we omit the "(x)" throughout.

$$(f \cdot g \cdot h)' = \frac{d}{dx}[(f \cdot g) \cdot h] = (f \cdot g) \cdot \frac{dh}{dx} + h \cdot \frac{d}{dx}(f \cdot g) = f \cdot g \cdot h' + h \cdot \left(f \cdot \frac{dg}{dx} + g \cdot \frac{df}{dx}\right)$$

$$= f' \cdot g \cdot h + f \cdot g' \cdot h + f \cdot g \cdot h'$$

(b) $(f \cdot g \cdot h \cdot k)' = \dfrac{d}{dx}[(f \cdot g \cdot h) \cdot k] = (f \cdot g \cdot h) \cdot \dfrac{dk}{dx} + k \cdot \dfrac{d}{dx}(f \cdot g \cdot h)$

$$= f \cdot g \cdot h \cdot k' + k \cdot (f' \cdot g \cdot h + f \cdot g' \cdot h + f \cdot g \cdot h') = f' \cdot g \cdot h \cdot k + f \cdot g' \cdot h \cdot k + f \cdot g \cdot h' \cdot k + f \cdot g \cdot h \cdot k'$$

(c) Theorem: If $n \geq 1$ and $f_1, \cdots, f_n$ are differentiable functions of x, then

$$(f_1 \cdot f_2 \cdot \cdots \cdot f_n)' = \sum_{i=1}^{n} f_1 \cdot \cdots \cdot f_{i-1} \cdot f_i' \cdot f_{i+1} \cdot \cdots \cdot f_n.$$

Proof: For $n = 1$ the statement is obviously true: $f_1' = f_1'$. If the statement is true for $n - 1$, then

$$(f_1 \cdot f_2 \cdot \cdots \cdot f_n)' = \frac{d}{dx}[(f_1 \cdot f_2 \cdot \cdots \cdot f_{n-1}) \cdot f_n]$$

$$= (f_1 \cdot f_2 \cdot \cdots \cdot f_{n-1}) \cdot f_n' + f_n \cdot (f_1 \cdot f_2 \cdot \cdots \cdot f_{n-1})'$$

$$= f_1 \cdot f_2 \cdot \cdots \cdot f_{n-1} \cdot f_n' + f_n \cdot \sum_{i=1}^{n-1} f_1 \cdot \cdots \cdot f_{i-1} \cdot f_i' \cdot f_{i+1} \cdot \cdots \cdot f_{n-1}$$

$$= \sum_{i=1}^{n} f_1 \cdot \cdots \cdot f_{i-1} \cdot f_i' \cdot f_{i+1} \cdot \cdots \cdot f_n$$

so the statement is true for n. By induction, it's true for all n.

5. (a) By the chain rule, $\dfrac{d}{dx}\left([g(x)]^{-1}\right) = -[g(x)]^{-2}g'(x) = -\dfrac{g'(x)}{[g(x)]^2}$. By the product rule,

$$h'(x) = f(x) \cdot \frac{d}{dx}\left([g(x)]^{-1}\right) + [g(x)]^{-1} \cdot \frac{d}{dx}[f(x)] = -\frac{f(x)g'(x)}{[g(x)]^2} + \frac{f'(x)}{g(x)} = \frac{g(x)f'(x) - f(x)g'(x)}{[g(x)]^2}.$$

(b) By the product rule, $f'(x) = \dfrac{d}{dx}[h(x)g(x)] = h(x)g'(x) + g(x)h'(x)$. So

$$h'(x) = \frac{1}{g(x)}[f'(x) - h(x)g'(x)] = \frac{1}{g(x)}\left[f'(x) - \frac{f(x)}{g(x)}g'(x)\right] = \frac{g(x)f'(x) - f(x)g'(x)}{[g(x)]^2}.$$

CHAPTER 3
The Derivative in Graphing and Applications

EXERCISE SET 3.1

1. (a) $f' > 0$ and $f'' > 0$

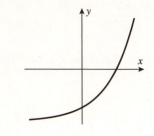

(b) $f' > 0$ and $f'' < 0$

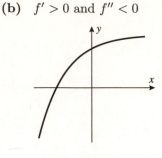

(c) $f' < 0$ and $f'' > 0$

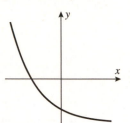

(d) $f' < 0$ and $f'' < 0$

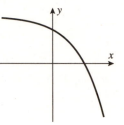

3. A: $dy/dx < 0$, $d^2y/dx^2 > 0$
B: $dy/dx > 0$, $d^2y/dx^2 < 0$
C: $dy/dx < 0$, $d^2y/dx^2 < 0$

5. An inflection point occurs when f'' changes sign: at $x = -1, 0, 1$ and 2.

7. (a) $[4, 6]$ (b) $[1, 4]$ and $[6, 7]$ (c) $(1, 2)$ and $(3, 5)$
(d) $(2, 3)$ and $(5, 7)$ (e) $x = 2, 3, 5$

9. (a) f is increasing on $[1, 3]$ (b) f is decreasing on $(-\infty, 1], [3, +\infty)$
(c) f is concave up on $(-\infty, 2), (4, +\infty)$ (d) f is concave down on $(2, 4)$
(e) points of inflection at $x = 2, 4$

11. True, by Definition 3.1.1(b).

13. False. Let $f(x) = (x - 1)^3$. Then f is increasing on $[0, 2]$, but $f'(1) = 0$.

15. $f'(x) = 2(x - 3/2)$ (a) $[3/2, +\infty)$ (b) $(-\infty, 3/2]$
$f''(x) = 2$ (c) $(-\infty, +\infty)$ (d) nowhere
    (e) none

17. $f'(x) = 6(2x + 1)^2$ (a) $(-\infty, +\infty)$ (b) nowhere
$f''(x) = 24(2x + 1)$ (c) $(-1/2, +\infty)$ (d) $(-\infty, -1/2)$
    (e) $-1/2$

19. $f'(x) = 12x^2(x - 1)$ (a) $[1, +\infty)$ (b) $(-\infty, 1]$
$f''(x) = 36x(x - 2/3)$ (c) $(-\infty, 0), (2/3, +\infty)$ (d) $(0, 2/3)$
    (e) $0, 2/3$

21. $f'(x) = -\dfrac{3(x^2 - 3x + 1)}{(x^2 - x + 1)^3}$ **(a)** $\left[\dfrac{3-\sqrt{5}}{2}, \dfrac{3+\sqrt{5}}{2}\right]$ **(b)** $\left(-\infty, \dfrac{3-\sqrt{5}}{2}\right], \left[\dfrac{3+\sqrt{5}}{2}, +\infty\right)$

$f''(x) = \dfrac{6x(2x^2 - 8x + 5)}{(x^2 - x + 1)^4}$ **(c)** $\left(0, 2 - \dfrac{\sqrt{6}}{2}\right), \left(2 + \dfrac{\sqrt{6}}{2}, +\infty\right)$

(d) $(-\infty, 0), \left(2 - \dfrac{\sqrt{6}}{2}, 2 + \dfrac{\sqrt{6}}{2}\right)$ **(e)** $0, 2 - \dfrac{\sqrt{6}}{2}, 2 + \dfrac{\sqrt{6}}{2}$

23. $f'(x) = \dfrac{2x + 1}{3(x^2 + x + 1)^{2/3}}$ **(a)** $[-1/2, +\infty)$ **(b)** $(-\infty, -1/2]$

$f''(x) = -\dfrac{2(x + 2)(x - 1)}{9(x^2 + x + 1)^{5/3}}$ **(c)** $(-2, 1)$ **(d)** $(-\infty, -2), (1, +\infty)$

(e) $-2, 1$

25. $f'(x) = \dfrac{4(x^{2/3} - 1)}{3x^{1/3}}$ **(a)** $[-1, 0], [1, +\infty)$ **(b)** $(-\infty, -1], [0, 1]$

(c) $(-\infty, 0), (0, +\infty)$ **(d)** nowhere

$f''(x) = \dfrac{4(x^{5/3} + x)}{9x^{7/3}}$ **(e)** none

27. $f'(x) = \cos x + \sin x$
$f''(x) = -\sin x + \cos x$

Increasing: $[-\pi/4, 3\pi/4]$
Decreasing: $[-\pi, -\pi/4], [3\pi/4, \pi]$
Concave up: $(-3\pi/4, \pi/4)$
Concave down: $(-\pi, -3\pi/4), (\pi/4, \pi)$
Inflection points: $-3\pi/4, \pi/4$

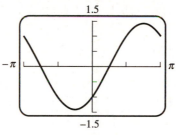

29. $f'(x) = -\dfrac{1}{2}\sec^2(x/2)$

$f''(x) = -\dfrac{1}{2}\tan(x/2)\sec^2(x/2))$
Increasing: nowhere
Decreasing: $(-\pi, \pi)$
Concave up: $(-\pi, 0)$
Concave down: $(0, \pi)$
Inflection point: 0

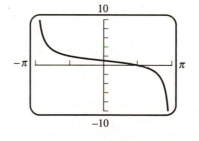

31. $f(x) = 1 + \sin 2x$
$f'(x) = 2\cos 2x$
$f''(x) = -4\sin 2x$
Increasing: $[-\pi, -3\pi/4], [-\pi/4, \pi/4], [3\pi/4, \pi]$
Decreasing: $[-3\pi/4, -\pi/4], [\pi/4, 3\pi/4]$
Concave up: $(-\pi/2, 0), (\pi/2, \pi)$
Concave down: $(-\pi, -\pi/2), (0, \pi/2)$
Inflection points: $-\pi/2, 0, \pi/2$

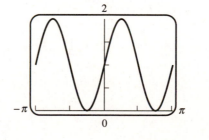

33. **(a)**

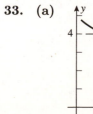

(b)

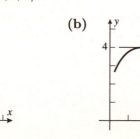

(c)

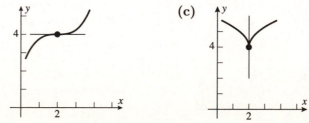

35. $f'(x) = 1/3 - 1/[3(1+x)^{2/3}]$ so f is increasing on $[0, +\infty)$
thus if $x > 0$, then $f(x) > f(0) = 0$, $1 + x/3 - \sqrt[3]{1+x} > 0$,
$\sqrt[3]{1+x} < 1 + x/3$.

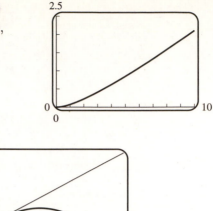

37. $x \geq \sin x$ on $[0, +\infty)$: let $f(x) = x - \sin x$.
Then $f(0) = 0$ and $f'(x) = 1 - \cos x \geq 0$,
so $f(x)$ is increasing on $[0, +\infty)$.

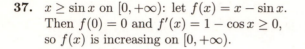

39. Points of inflection at $x = -2, +2$. Concave up on
$(-5, -2)$ and $(2, 5)$; concave down on $(-2, 2)$. Increasing on $[-3.5829, 0.2513]$ and $[3.3316, 5]$, and decreasing on
$[-5, -3.5829]$ and $[0.2513, 3.3316]$.

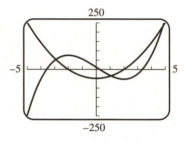

41. $f''(x) = 2\dfrac{90x^3 - 81x^2 - 585x + 397}{(3x^2 - 5x + 8)^3}$. The denominator has complex roots, so is always positive;
hence the x-coordinates of the points of inflection of $f(x)$ are the roots of the numerator (if it
changes sign). A plot of the numerator over $[-5, 5]$ shows roots lying in $[-3, -2]$, $[0, 1]$, and $[2, 3]$.
To six decimal places the roots are $x \approx -2.464202, 0.662597, 2.701605$.

43. $f(x_1) - f(x_2) = x_1^2 - x_2^2 = (x_1 + x_2)(x_1 - x_2) < 0$ if $x_1 < x_2$ for x_1, x_2 in $[0, +\infty)$, so $f(x_1) < f(x_2)$
and f is thus increasing.

45. **(a)** True. If $x_1 < x_2$ where x_1 and x_2 are in I, then $f(x_1) < f(x_2)$ and $g(x_1) < g(x_2)$, so
$f(x_1) + g(x_1) < f(x_2) + g(x_2)$, $(f + g)(x_1) < (f + g)(x_2)$. Thus $f + g$ is increasing on I.

(b) False. If $f(x) = g(x) = x$ then f and g are both increasing on $(-\infty, 0)$, but $(f \cdot g)(x) = x^2$
is decreasing there.

47. **(a)** $f(x) = x$, $g(x) = 2x$ **(b)** $f(x) = x$, $g(x) = x + 6$ **(c)** $f(x) = 2x$, $g(x) = x$

49. **(a)** $f''(x) = 6ax + 2b = 6a\left(x + \dfrac{b}{3a}\right)$, $f''(x) = 0$ when $x = -\dfrac{b}{3a}$. f changes its direction of
concavity at $x = -\dfrac{b}{3a}$ so $-\dfrac{b}{3a}$ is an inflection point.

(b) If $f(x) = ax^3 + bx^2 + cx + d$ has three x-intercepts, then it has three roots, say x_1, x_2 and
x_3, so we can write $f(x) = a(x - x_1)(x - x_2)(x - x_3) = ax^3 + bx^2 + cx + d$, from which it
follows that $b = -a(x_1 + x_2 + x_3)$. Thus $-\dfrac{b}{3a} = \dfrac{1}{3}(x_1 + x_2 + x_3)$, which is the average.

(c) $f(x) = x(x^2 - 3x + 2) = x(x - 1)(x - 2)$ so the intercepts are 0, 1, and 2 and the average is
1. $f''(x) = 6x - 6 = 6(x - 1)$ changes sign at $x = 1$. The inflection point is at (1,0). f is
concave up for $x > 1$, concave down for $x < 1$.

51. **(a)** Let $x_1 < x_2$ belong to (a, b). If both belong to $(a, c]$ or both belong to $[c, b)$ then we have $f(x_1) < f(x_2)$ by hypothesis. So assume $x_1 < c < x_2$. We know by hypothesis that $f(x_1) < f(c)$, and $f(c) < f(x_2)$. We conclude that $f(x_1) < f(x_2)$.

(b) Use the same argument as in part (a), but with inequalities reversed.

53. By Theorem 3.1.2, f is decreasing on any interval $[(2n\pi + \pi/2, 2(n+1)\pi + \pi/2]$ $(n = 0, \pm 1, \pm 2, \cdots)$, because $f'(x) = -\sin x + 1 < 0$ on $(2n\pi + \pi/2, 2(n+1)\pi + \pi/2)$. By Exercise 51(b) we can piece these intervals together to show that $f(x)$ is decreasing on $(-\infty, +\infty)$.

55.

57.

59. "Either the rate at which the temperature is falling decreases for a while and then begins to increase, or it increases for a while and then begins to decrease". If $T(t)$ is the temperature at time t, then the rate at which the temperature is falling is $-T'(t)$. If this is decreasing then $T'(t)$ is increasing, so $T(t)$ is concave up; if it's increasing then $T'(t)$ is decreasing, so $T(t)$ is concave down. When $-T'(t)$ changes from decreasing to increasing or vice versa, the direction of concavity of $T(t)$ changes, so the graph of the temperature has an inflection point.

EXERCISE SET 3.2

1. **(a)**   **(b)**

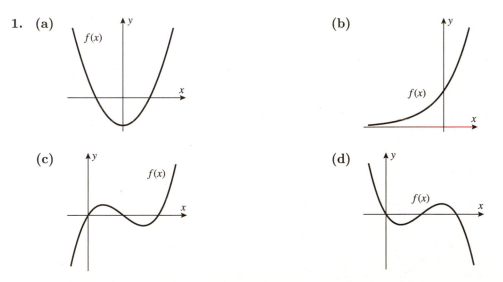

   (c)   **(d)**

3. **(a)** $f'(x) = 6x - 6$ and $f''(x) = 6$, with $f'(1) = 0$. For the first derivative test, $f' < 0$ for $x < 1$ and $f' > 0$ for $x > 1$. For the second derivative test, $f''(1) > 0$.

   (b) $f'(x) = 3x^2 - 3$ and $f''(x) = 6x$. $f'(x) = 0$ at $x = \pm 1$. First derivative test: $f' > 0$ for $x < -1$ and $x > 1$, and $f' < 0$ for $-1 < x < 1$, so there is a relative maximum at $x = -1$, and a relative minimum at $x = 1$. Second derivative test: $f'' < 0$ at $x = -1$, a relative maximum; and $f'' > 0$ at $x = 1$, a relative minimum.

5. **(a)** $f'(x) = 4(x - 1)^3$, $g'(x) = 3x^2 - 6x + 3$ so $f'(1) = g'(1) = 0$.

(b) $f''(x) = 12(x-1)^2$, $g''(x) = 6x - 6$, so $f''(1) = g''(1) = 0$, which yields no information.

(c) $f' < 0$ for $x < 1$ and $f' > 0$ for $x > 1$, so there is a relative minimum at $x = 1$; $g'(x) = 3(x-1)^2 > 0$ on both sides of $x = 1$, so there is no relative extremum at $x = 1$.

7. $f'(x) = 16x^3 - 32x = 16x(x^2 - 2)$, so $x = 0, \pm\sqrt{2}$ are stationary points.

9. $f'(x) = \dfrac{-x^2 - 2x + 3}{(x^2 + 3)^2}$, so $x = -3, 1$ are the stationary points.

11. $f'(x) = \dfrac{2x}{3(x^2 - 25)^{2/3}}$; so $x = 0$ is the stationary point; $x = \pm 5$ are critical points which are not stationary points.

13. $f(x) = |\sin x| = \begin{cases} \sin x, & \sin x \geq 0 \\ -\sin x, & \sin x < 0 \end{cases}$ so $f'(x) = \begin{cases} \cos x, & \sin x > 0 \\ -\cos x, & \sin x < 0 \end{cases}$ and $f'(x)$ does not exist when $x = n\pi$, $n = 0, \pm 1, \pm 2, \cdots$ (the points where $\sin x = 0$) because $\lim\limits_{x \to n\pi^-} f'(x) \neq \lim\limits_{x \to n\pi^+} f'(x)$ (see Theorem preceding Exercise 65, Section 2.3); these are critical points which are not stationary points. Now $f'(x) = 0$ when $\pm\cos x = 0$ provided $\sin x \neq 0$ so $x = \pi/2 + n\pi$, $n = 0, \pm 1, \pm 2, \cdots$ are stationary points.

15. False. Let $f(x) = (x-1)^2(2x-3)$. Then $f'(x) = 2(x-1)(3x-4)$; $f'(x)$ changes sign from $+$ to $-$ at $x = 1$, so f has a relative maximum at $x = 1$. But $f(2) = 1 > 0 = f(1)$.

17. False. Let $f(x) = x + (x-1)^2$. Then $f'(x) = 2x - 1$ and $f''(x) = 2$, so $f''(1) > 0$. But $f'(1) = 1 \neq 0$, so f does not have a relative extremum at $x = 1$.

19.

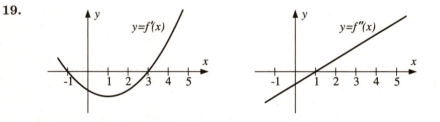

21. **(a)** none

(b) $x = 1$ because f' changes sign from $+$ to $-$ there

(c) none because $f'' = 0$ (never changes sign)

(d)

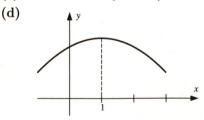

23. **(a)** $x = 2$ because $f'(x)$ changes sign from $-$ to $+$ there.

(b) $x = 0$ because $f'(x)$ changes sign from $+$ to $-$ there.

(c) $x = 1, 3$ because $f''(x)$ changes sign at these points.

(d)

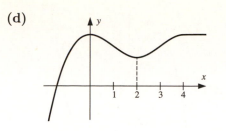

25. critical points $x = 0, 5^{1/3}$: f': $- - - 0 - - - 0 + + +$

   $x = 0$: neither $05^{1/3}$

   $x = 5^{1/3}$: relative minimum

27. critical points $x = -2, 2/3$: f': $- - - \infty + + + 0 - - -$

   $x = -2$: relative minimum $-22/3$

   $x = 2/3$: relative maximum

29. $f'(x) = 8 - 6x$: critical point $x = 4/3$

   $f''(4/3) = -6$: f has a relative maximum of $19/3$ at $x = 4/3$

31. $f'(x) = 2\cos 2x$: critical points at $x = \pi/4, 3\pi/4$

   $f''(\pi/4) = -4$: f has a relative maximum of 1 at $x = \pi/4$

   $f''(3\pi/4) = 4$: f has a relative minimum of -1 at $x = 3\pi/4$

33. $f'(x) = 4x^3 - 12x^2 + 8x$: critical points at $x = 0, 1, 2$

   relative minimum of 0 at $x = 0$

   relative maximum of 1 at $x = 1$ $- - - \ 0 \ + + + \ 0 \ - - - \ 0 \ + + +$

   relative minimum of 0 at $x = 2$ 012

35. $f'(x) = 5x^4 + 8x^3 + 3x^2$: critical points at $x = -3/5, -1, 0$

   $f''(-3/5) = 18/25$: f has a relative minimum of $-108/3125$ at $x = -3/5$

   $f''(-1) = -2$: f has a relative maximum of 0 at $x = -1$

   $f''(0) = 0$: Theorem 3.2.5 with $m = 3$: f has an inflection point at $x = 0$

37. $f'(x) = \dfrac{2(x^{1/3} + 1)}{x^{1/3}}$: critical point at $x = -1, 0$

   $f''(-1) = -\dfrac{2}{3}$: f has a relative maximum of 1 at $x = -1$

   f' does not exist at $x = 0$. By inspection it is a relative minimum of 0.

39. $f'(x) = -\dfrac{5}{(x-2)^2}$; no extrema

41. $f'(x)$ is undefined at $x = 0, 3$, so these are critical points. Elsewhere,

   $f'(x) = \begin{cases} 2x - 3 & \text{if } x < 0 \text{ or } x > 3; \\ 3 - 2x & \text{if } 0 < x < 3. \end{cases}$

   $f'(x) = 0$ for $x = 3/2$, so this is also a critical point.

   $f''(3/2) = -2$, so relative maximum of $9/4$ at $x = 3/2$

   By first derivative test, relative minimum of 0 at $x = 0$ and $x = 3$

43.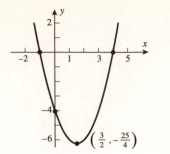

45.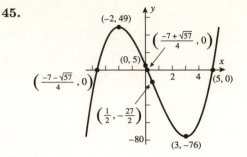

47.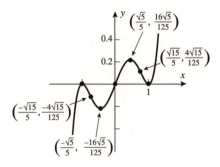

49.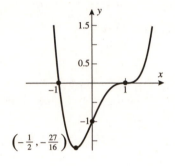

51.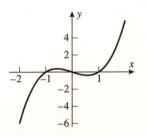

53. **(a)** $\displaystyle\lim_{x \to -\infty} y = -\infty$, $\displaystyle\lim_{x \to +\infty} y = +\infty$;
curve crosses x-axis at $x = 0, 1, -1$

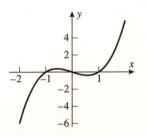

(b) $\displaystyle\lim_{x \to \pm\infty} y = +\infty$;
curve never crosses x-axis

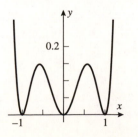

(c) $\lim\limits_{x \to -\infty} y = -\infty, \lim\limits_{x \to +\infty} y = +\infty;$
curve crosses x-axis at $x = -1$

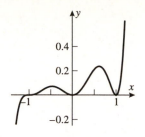

(d) $\lim\limits_{x \to \pm\infty} y = +\infty;$
curve crosses x-axis at $x = 0, 1$

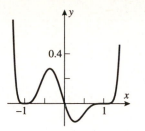

55. $f'(x) = 2 \cos 2x$ if $\sin 2x > 0$,
$f'(x) = -2 \cos 2x$ if $\sin 2x < 0$,
$f'(x)$ does not exist when $x = \pi/2, \pi, 3\pi/2$;
critical numbers $x = \pi/4, 3\pi/4, 5\pi/4, 7\pi/4, \pi/2, \pi, 3\pi/2$
relative minimum of 0 at $x = \pi/2, \pi, 3\pi/2$;
relative maximum of 1 at $x = \pi/4, 3\pi/4, 5\pi/4, 7\pi/4$

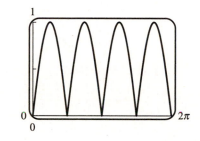

57. $f'(x) = -\sin 2x$;
critical numbers $x = \pi/2, \pi, 3\pi/2$
relative minimum of 0 at $x = \pi/2, 3\pi/2$;
relative maximum of 1 at $x = \pi$

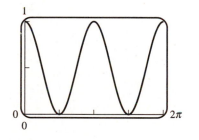

59. Relative minima at $x \approx -3.58, 3.33$;
relative maximum at $x \approx 0.25$

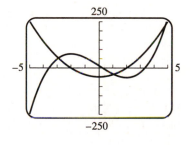

61. Relative maximum at $x \approx -0.272$,
relative minimum at $x \approx 0.224$

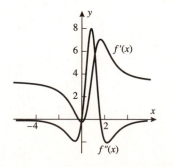

63. $f'(x) = \dfrac{4x^3 - \sin 2x}{2\sqrt{x^4 + \cos^2 x}}$,

$f''(x) = \dfrac{6x^2 - \cos 2x}{\sqrt{x^4 + \cos^2 x}} - \dfrac{(4x^3 - \sin 2x)(4x^3 - \sin 2x)}{4(x^4 + \cos^2 x)^{3/2}}$

Relative minima at $x \approx \pm 0.618$, relative maximum at $x = 0$

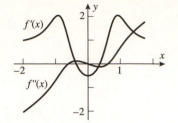

65. **(a)** Let $f(x) = x^2 + \dfrac{k}{x}$, then $f'(x) = 2x - \dfrac{k}{x^2} = \dfrac{2x^3 - k}{x^2}$. f has a relative extremum when $2x^3 - k = 0$, so $k = 2x^3 = 2(3)^3 = 54$.

   (b) Let $f(x) = \dfrac{x}{x^2 + k}$, then $f'(x) = \dfrac{k - x^2}{(x^2 + k)^2}$. f has a relative extremum when $k - x^2 = 0$, so $k = x^2 = 3^2 = 9$.

67. **(a)** Because h and g have relative maxima at x_0, $h(x) \le h(x_0)$ for all x in I_1 and $g(x) \le g(x_0)$ for all x in I_2, where I_1 and I_2 are open intervals containing x_0. If x is in both I_1 and I_2 then both inequalities are true and by addition so is $h(x) + g(x) \le h(x_0) + g(x_0)$ which shows that $h + g$ has a relative maximum at x_0.

   (b) By counterexample; both $h(x) = -x^2$ and $g(x) = -2x^2$ have relative maxima at $x = 0$ but $h(x) - g(x) = x^2$ has a relative minimum at $x = 0$ so in general $h - g$ does not necessarily have a relative maximum at x_0.

69. The first derivative test applies in many cases where the second derivative test does not. For example, it implies that $|x|$ has a relative minimum at $x = 0$, but the second derivative test does not, since $|x|$ is not differentiable there.

The second derivative test is often easier to apply, since we only need to compute $f'(x_0)$ and $f''(x_0)$, instead of analyzing $f'(x)$ at values of x near x_0. For example, let $f(x) = 10x^3 + \cos x$. Then $f'(x) = 30x^2 - \sin x$ and $f''(x) = 60x - \cos x$. Since $f'(0) = 0$ and $f''(0) = -1$, the second derivative test tells us that f has a relative maximum at $x = 0$. To prove this using the first derivative test is slightly more difficult, since we need to determine the sign of $f'(x)$ for x near, but not equal to, 0.

EXERCISE SET 3.3

1. Vertical asymptote $x = 4$
horizontal asymptote $y = -2$

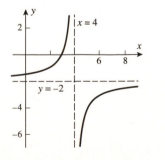

3. Vertical asymptotes $x = \pm 2$
horizontal asymptote $y = 0$

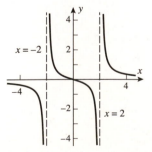

5. No vertical asymptotes
horizontal asymptote $y = 1$

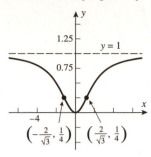

$\left(-\frac{2}{\sqrt{3}}, \frac{1}{4}\right)$ $\left(\frac{2}{\sqrt{3}}, \frac{1}{4}\right)$

7. Vertical asymptote $x = 1$
horizontal asymptote $y = 1$

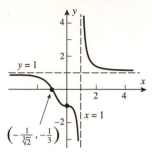

$\left(-\frac{1}{\sqrt[3]{2}}, -\frac{1}{3}\right)$

9. Vertical asymptote $x = 0$
horizontal asymptote $y = 3$

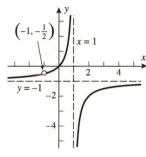

$(2, 3)$ $\left(6, \frac{25}{9}\right)$ $\left(4, \frac{11}{4}\right)$

11. Vertical asymptote $x = 1$
horizontal asymptote $y = 9$

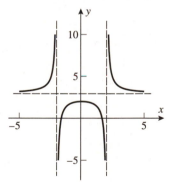

$\left(\frac{1}{3}, 9\right)$ $(-1, 1)$ $\left(-\frac{1}{3}, 0\right)$

13. Vertical asymptote $x = 1$
horizontal asymptote $y = -1$

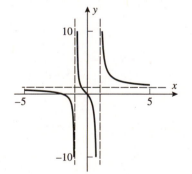

$\left(-1, -\frac{1}{2}\right)$

15. **(a)** horizontal asymptote $y = 3$ as
$x \to \pm\infty$, vertical asymptotes at $x = \pm 2$

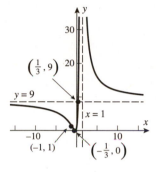

(b) horizontal asymptote of $y = 1$ as
$x \to \pm\infty$, vertical asymptotes at $x = \pm 1$

17. $\displaystyle\lim_{x \to \pm\infty} \left| \frac{x^2}{x - 3} - (x + 3) \right| = \lim_{x \to \pm\infty} \left| \frac{9}{x - 3} \right| = 0$

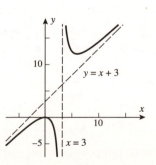

$y = x + 3$

$x = 3$

19. $y = x^2 - \dfrac{1}{x} = \dfrac{x^3 - 1}{x}$;

y-axis is a vertical asymptote;

$y' = \dfrac{2x^3 + 1}{x^2}$,

$y' = 0$ when $x = -\sqrt[3]{\dfrac{1}{2}} \approx -0.8$;

$y'' = \dfrac{2(x^3 - 1)}{x^3}$

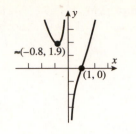

21. $y = \dfrac{(x - 2)^3}{x^2} = x - 6 + \dfrac{12x - 8}{x^2}$ so

y-axis is a vertical asymptote,

$y = x - 6$ is an oblique asymptote;

$y' = \dfrac{(x - 2)^2(x + 4)}{x^3}$,

$y'' = \dfrac{24(x - 2)}{x^4}$

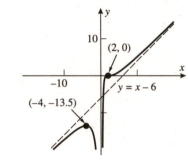

23. $y = \dfrac{x^3 - 4x - 8}{x + 2} = x^2 - 2x - \dfrac{8}{x + 2}$ so

$x = -2$ is a vertical asymptote,

$y = x^2 - 2x$ is a curvilinear asymptote as $x \to \pm\infty$

25. (a) VI   (b) I   (c) III   (d) V   (e) IV   (f) II

27. True. If the degree of P were larger than the degree of Q, then $\lim\limits_{x \to \pm\infty} f(x)$ would be infinite and the graph would not have a horizontal asymptote. If the degree of P were less than the degree of Q, then $\lim\limits_{x \to \pm\infty} f(x)$ would be zero, so the horizontal asymptote would be $y = 0$, not $y = 5$.

29. False. Let $f(x) = \sqrt[3]{x - 1}$. Then f is continuous at $x = 1$, but $\lim\limits_{x \to 1} f'(x) = \lim\limits_{x \to 1} \dfrac{1}{3}(x-1)^{-2/3} = +\infty$, so f' has a vertical asymptote at $x = 1$.

31. $y = \sqrt{4x^2 - 1}$

$$y' = \frac{4x}{\sqrt{4x^2 - 1}}$$

$$y'' = -\frac{4}{(4x^2 - 1)^{3/2}} \text{ so}$$

extrema when $x = \pm\frac{1}{2}$, no inflection points

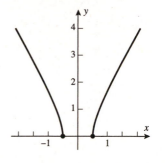

33. $y = 2x + 3x^{2/3}$;

$$y' = 2 + 2x^{-1/3};$$

$$y'' = -\frac{2}{3}x^{-4/3}$$

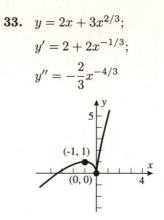

35. $y = x^{1/3}(4 - x)$;

$$y' = \frac{4(1 - x)}{3x^{2/3}};$$

$$y'' = -\frac{4(x + 2)}{9x^{5/3}}$$

37. $y = x^{2/3} - 2x^{1/3} + 4$

$$y' = \frac{2(x^{1/3} - 1)}{3x^{2/3}}$$

$$y'' = -\frac{2(x^{1/3} - 2)}{9x^{5/3}}$$

39. $y = x + \sin x$;

$y' = 1 + \cos x$, $y' = 0$ when $x = \pi + 2n\pi$;

$y'' = -\sin x$; $y'' = 0$ when $x = n\pi$

$n = 0, \pm 1, \pm 2, \ldots$

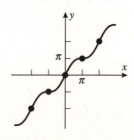

41. $y = \sqrt{3}\cos x + \sin x$;

$y' = -\sqrt{3}\sin x + \cos x$;

$y' = 0$ when $x = \pi/6 + n\pi$;

$y'' = -\sqrt{3}\cos x - \sin x$;

$y'' = 0$ when $x = 2\pi/3 + n\pi$

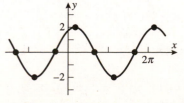

43. $y = \sin^2 x - \cos x$;

$y' = \sin x(2\cos x + 1)$;

$y' = 0$ when $x = -\pi, 0, \pi, 2\pi, 3\pi$ and when

$$x = -\frac{2}{3}\pi, \frac{2}{3}\pi, \frac{4}{3}\pi, \frac{8}{3}\pi;$$

$y'' = 4\cos^2 x + \cos x - 2$; $y'' = 0$ when

$x \approx \pm 2.57, \pm 0.94, 3.71, 5.35, 7.22, 8.86$

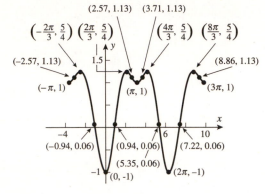

44. $y = \sqrt{\tan x}$;

$$y' = \frac{\sec^2 x}{2\sqrt{\tan x}}; \text{ so } y' > 0 \text{ always};$$

$$y'' = \sec^2 x \frac{3\tan^2 x - 1}{4(\tan x)^{3/2}},$$

$y'' = 0$ when $x = \dfrac{\pi}{6}$, $y = \dfrac{1}{\sqrt[4]{3}}$

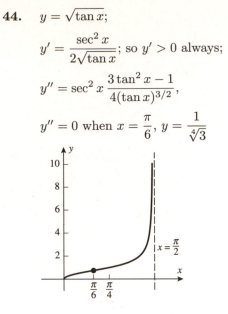

45. **(a)** $x = 1, 2.5, 4$ and $x = 3$, the latter being a cusp

(b) $(-\infty, 1], [2.5, 3)$

(c) relative maxima for $x = 1, 3$; relative minima for $x = 2.5$

(d) $x \approx 0.6, 1.9, 4$

47. Let y be the length of the other side of the rectangle, then $L = 2x + 2y$ and $xy = 400$ so $y = 400/x$ and hence $L = 2x + 800/x$. $L = 2x$ is an oblique asymptote.

$$L = 2x + \frac{800}{x} = \frac{2(x^2 + 400)}{x},$$

$$L' = 2 - \frac{800}{x^2} = \frac{2(x^2 - 400)}{x^2},$$

$$L'' = \frac{1600}{x^3},$$

$L' = 0$ when $x = 20, L = 80$

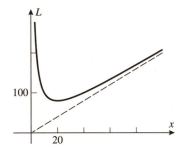

49. $y' = 0.1x^4(6x - 5)$;

critical numbers: $x = 0$, $x = 5/6$;

relative minimum at $x = 5/6$,

$y \approx -6.7 \times 10^{-3}$

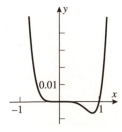

EXERCISE SET 3.4

1. relative maxima at $x = 2, 6$; absolute maximum at $x = 6$; relative minimum at $x = 4$; absolute minima at $x = 0, 4$

3. (a)

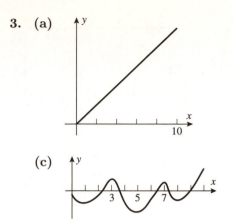

(b)

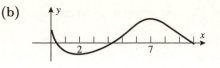

(c)

5. $x = 1$ is a point of discontinuity of f.

7. $f'(x) = 8x - 12$, $f'(x) = 0$ when $x = 3/2$; $f(1) = 2$, $f(3/2) = 1$, $f(2) = 2$ so the maximum value is 2 at $x = 1, 2$ and the minimum value is 1 at $x = 3/2$.

9. $f'(x) = 3(x - 2)^2$, $f'(x) = 0$ when $x = 2$; $f(1) = -1, f(2) = 0, f(4) = 8$ so the minimum is -1 at $x = 1$ and the maximum is 8 at $x = 4$.

11. $f'(x) = 3/(4x^2 + 1)^{3/2}$, no critical points; $f(-1) = -3/\sqrt{5}$, $f(1) = 3/\sqrt{5}$ so the maximum value is $3/\sqrt{5}$ at $x = 1$ and the minimum value is $-3/\sqrt{5}$ at $x = -1$.

13. $f'(x) = 1 - 2\cos x$, $f'(x) = 0$ when $x = \pi/3$; then $f(-\pi/4) = -\pi/4 + \sqrt{2}$; $f(\pi/3) = \pi/3 - \sqrt{3}$; $f(\pi/2) = \pi/2 - 2$, so f has a minimum of $\pi/3 - \sqrt{3}$ at $x = \pi/3$ and a maximum of $-\pi/4 + \sqrt{2}$ at $x = -\pi/4$.

15. $f(x) = 1 + |9 - x^2| = \begin{cases} 10 - x^2, & |x| \le 3 \\ -8 + x^2, & |x| > 3 \end{cases}$, $f'(x) = \begin{cases} -2x, & |x| < 3 \\ 2x, & |x| > 3 \end{cases}$ thus $f'(x) = 0$ when

$x = 0$, $f'(x)$ does not exist for x in $(-5, 1)$ when $x = -3$ because $\lim_{x \to -3^-} f'(x) \ne \lim_{x \to -3^+} f'(x)$ (see Theorem preceding Exercise 65, Section 2.3); $f(-5) = 17$, $f(-3) = 1$, $f(0) = 10$, $f(1) = 9$ so the maximum value is 17 at $x = -5$ and the minimum value is 1 at $x = -3$.

17. True, by Theorem 3.4.2.

19. True, by Theorem 3.4.3.

21. $f'(x) = 2x - 1$, $f'(x) = 0$ when $x = 1/2$; $f(1/2) = -9/4$ and $\lim_{x \to \pm\infty} f(x) = +\infty$. Thus f has a minimum of $-9/4$ at $x = 1/2$ and no maximum.

23. $f'(x) = 12x^2(1 - x)$; critical points $x = 0, 1$. Maximum value $f(1) = 1$, no minimum because $\lim_{x \to +\infty} f(x) = -\infty$.

25. No maximum or minimum because $\lim_{x \to +\infty} f(x) = +\infty$ and $\lim_{x \to -\infty} f(x) = -\infty$.

27. $\lim_{x \to -1^-} f(x) = -\infty$, so there is no absolute minimum on the interval;

$f'(x) = \dfrac{x^2 + 2x - 1}{(x + 1)^2} = 0$ at $x = -1 - \sqrt{2}$, for which $y = -2 - 2\sqrt{2} \approx -4.828$. Also $f(-5) = -13/2$, so the absolute maximum of f on the interval is $y = -2 - 2\sqrt{2}$ taken at $x = -1 - \sqrt{2}$.

29. $\lim\limits_{x\to\pm\infty} = +\infty$ so there is no absolute maximum.

$f'(x) = 4x(x-2)(x-1)$, $f'(x) = 0$ when $x = 0, 1, 2$, and $f(0) = 0, f(1) = 1, f(2) = 0$ so f has an absolute minimum of 0 at $x = 0, 2$.

31. $f'(x) = \dfrac{5(8-x)}{3x^{1/3}}$, $f'(x) = 0$ when $x = 8$ and $f'(x)$ does not exist when $x = 0$; $f(-1) = 21$, $f(0) = 0$, $f(8) = 48$, $f(20) = 0$ so the maximum value is 48 at $x = 8$ and the minimum value is 0 at $x = 0, 20$.

33. $f'(x) = -1/x^2$; no maximum or minimum because there are no critical points in $(0, +\infty)$.

35. $f'(x) = \dfrac{1 - 2\cos x}{\sin^2 x}$; $f'(x) = 0$ on $[\pi/4, 3\pi/4]$ only when $x = \pi/3$. Then $f(\pi/4) = 2\sqrt{2} - 1$, $f(\pi/3) = \sqrt{3}$ and $f(3\pi/4) = 2\sqrt{2} + 1$, so f has an absolute maximum value of $2\sqrt{2}+1$ at $x = 3\pi/4$ and an absolute minimum value of $\sqrt{3}$ at $x = \pi/3$.

37. $f'(x) = -[\cos(\cos x)]\sin x$; $f'(x) = 0$ if $\sin x = 0$ or if $\cos(\cos x) = 0$. If $\sin x = 0$, then $x = \pi$ is the critical point in $(0, 2\pi)$; $\cos(\cos x) = 0$ has no solutions because $-1 \leq \cos x \leq 1$. Thus $f(0) = \sin(1)$, $f(\pi) = \sin(-1) = -\sin(1)$, and $f(2\pi) = \sin(1)$ so the maximum value is $\sin(1) \approx 0.84147$ and the minimum value is $-\sin(1) \approx -0.84147$.

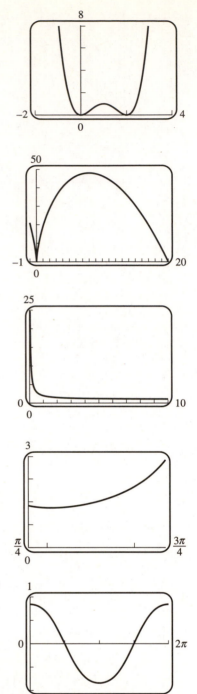

39. $f'(x) = \begin{cases} 4, & x < 1 \\ 2x - 5, & x > 1 \end{cases}$ so $f'(x) = 0$ when $x = 5/2$, and $f'(x)$ does not exist when $x = 1$

because $\lim\limits_{x\to 1^-} f'(x) \neq \lim\limits_{x\to 1^+} f'(x)$ (see Theorem preceding Exercise 65, Section 2.3); $f(1/2) = 0$, $f(1) = 2$, $f(5/2) = -1/4$, $f(7/2) = 3/4$ so the maximum value is 2 and the minimum value is $-1/4$.

41. The period of $f(x)$ is 2π, so check $f(0) = 3, f(2\pi) = 3$ and the critical points.
$f'(x) = -2\sin x - 2\sin 2x = -2\sin x(1 + 2\cos x) = 0$ on $[0, 2\pi]$ at $x = 0, \pi, 2\pi$ and $x = 2\pi/3, 4\pi/3$. Check $f(\pi) = -1, f(2\pi/3) = -3/2, f(4\pi/3) = -3/2$.

Thus f has an absolute maximum on $(-\infty, +\infty)$ of 3 at $x = 2k\pi, k = 0, \pm 1, \pm 2, \ldots$
and an absolute minimum of $-3/2$ at $x = 2k\pi \pm 2\pi/3, k = 0, \pm 1, \pm 2, \ldots$.

43. Let $f(x) = x - \sin x$, then $f'(x) = 1 - \cos x$ and so $f'(x) = 0$ when $\cos x = 1$ which has no solution
for $0 < x < 2\pi$ thus the minimum value of f must occur at 0 or 2π. $f(0) = 0$, $f(2\pi) = 2\pi$ so 0 is
the minimum value on $[0, 2\pi]$ thus $x - \sin x \geq 0$, $\sin x \leq x$ for all x in $[0, 2\pi]$.

45. Let m = slope at x, then $m = f'(x) = 3x^2 - 6x + 5$, $dm/dx = 6x - 6$; critical point for m is $x = 1$,
minimum value of m is $f'(1) = 2$

47. $\lim\limits_{x \to +\infty} f(x) = +\infty$, $\lim\limits_{x \to 8^+} f(x) = +\infty$, so there is no absolute maximum value of f for $x > 8$. By

Table 3.4.3 there must be a minimum. Since $f'(x) = \dfrac{2x(-520 + 192x - 24x^2 + x^3)}{(x - 8)^3}$, we must solve

a quartic equation to find the critical points. But it is easy to see that $x = 0$ and $x = 10$ are real
roots, and the other two are complex. Since $x = 0$ is not in the interval in question, we must have
an absolute minimum of f on $(8, +\infty)$ of 125 at $x = 10$.

49. The absolute extrema of $y(t)$ can occur at the endpoints $t = 0, 12$ or when $dy/dt = 2\sin t = 0$, i.e.
$t = 0, 12, k\pi$, $k = 1, 2, 3$; the absolute maximum is $y = 4$ at $t = \pi, 3\pi$; the absolute minimum is
$y = 0$ at $t = 0, 2\pi$.

51. $f'(x) = 2ax + b$; critical point is $x = -\dfrac{b}{2a}$

$f''(x) = 2a > 0$ so $f\left(-\dfrac{b}{2a}\right)$ is the minimum value of f, but

$f\left(-\dfrac{b}{2a}\right) = a\left(-\dfrac{b}{2a}\right)^2 + b\left(-\dfrac{b}{2a}\right) + c = \dfrac{-b^2 + 4ac}{4a}$ thus $f(x) \geq 0$ if and only if

$f\left(-\dfrac{b}{2a}\right) \geq 0$, $\dfrac{-b^2 + 4ac}{4a} \geq 0$, $-b^2 + 4ac \geq 0$, $b^2 - 4ac \leq 0$

53. If f has an absolute minimum, say at $x = a$, then, for all x, $f(x) \geq f(a) > 0$. But since
$\lim\limits_{x \to +\infty} f(x) = 0$, there is some x such that $f(x) < f(a)$. This contradiction shows that f cannot
have an absolute minimum.

On the other hand, let $f(x) = \dfrac{1}{(x^2 - 1)^2 + 1}$. Then $f(x) > 0$ for

all x. Also, $\lim\limits_{x \to +\infty} f(x) = 0$ so the x-axis is an asymptote, both

as $x \to -\infty$ and as $x \to +\infty$. But since $f(0) = \frac{1}{2} < 1 = f(1) =$
$f(-1)$, the absolute minimum of f on $[-1, 1]$ does not occur at
$x = 1$ or $x = -1$, so it is a relative minimum. (In fact it occurs at
$x = 0$.)

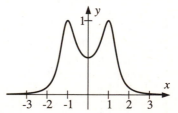

EXERCISE SET 3.5

1. If $y = x + 1/x$ for $1/2 \leq x \leq 3/2$ then $dy/dx = 1 - 1/x^2 = (x^2 - 1)/x^2$, $dy/dx = 0$ when $x = 1$. If
$x = 1/2, 1, 3/2$ then $y = 5/2, 2, 13/6$ so
   (a) y is as small as possible when $x = 1$.
   (b) y is as large as possible when $x = 1/2$.

3. $A = xy$ where $x + 2y = 1000$ so $y = 500 - x/2$ and $A = 500x - x^2/2$ for x in $[0, 1000]$; $dA/dx = 500 - x$, $dA/dx = 0$ when $x = 500$. If $x = 0$ or 1000 then $A = 0$, if $x = 500$ then $A = 125,000$ so the area is maximum when $x = 500$ ft and $y = 500 - 500/2 = 250$ ft.

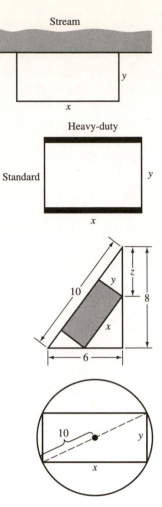

5. Let x and y be the dimensions shown in the figure and A the area, then $A = xy$ subject to the cost condition $3(2x) + 2(2y) = 6000$, or $y = 1500 - 3x/2$. Thus $A = x(1500 - 3x/2) = 1500x - 3x^2/2$ for x in $[0, 1000]$. $dA/dx = 1500 - 3x$, $dA/dx = 0$ when $x = 500$. If $x = 0$ or 1000 then $A = 0$, if $x = 500$ then $A = 375,000$ so the area is greatest when $x = 500$ ft and (from $y = 1500 - 3x/2$) when $y = 750$ ft.

7. Let x, y, and z be as shown in the figure and A the area of the rectangle, then $A = xy$ and, by similar triangles, $z/10 = y/6$, $z = 5y/3$; also $x/10 = (8 - z)/8 = (8 - 5y/3)/8$ thus $y = 24/5 - 12x/25$ so $A = x(24/5 - 12x/25) = 24x/5 - 12x^2/25$ for x in $[0, 10]$. $dA/dx = 24/5 - 24x/25$, $dA/dx = 0$ when $x = 5$. If $x = 0, 5, 10$ then $A = 0, 12, 0$ so the area is greatest when $x = 5$ in. and $y = 12/5$ in.

9. $A = xy$ where $x^2 + y^2 = 20^2 = 400$ so $y = \sqrt{400 - x^2}$ and $A = x\sqrt{400 - x^2}$ for $0 \le x \le 20$; $dA/dx = 2(200 - x^2)/\sqrt{400 - x^2}$, $dA/dx = 0$ when $x = \sqrt{200} = 10\sqrt{2}$. If $x = 0, 10\sqrt{2}, 20$ then $A = 0, 200, 0$ so the area is maximum when $x = 10\sqrt{2}$ and $y = \sqrt{400 - 200} = 10\sqrt{2}$.

11. Let $x =$ length of each side that uses the \$1 per foot fencing,
$\quad\quad y =$ length of each side that uses the \$2 per foot fencing.
The cost is $C = (1)(2x) + (2)(2y) = 2x + 4y$, but $A = xy = 3200$ thus $y = 3200/x$ so

$$C = 2x + 12800/x \text{ for } x > 0,$$
$$dC/dx = 2 - 12800/x^2, \ dC/dx = 0 \text{ when } x = 80, \ d^2C/dx^2 > 0 \text{ so}$$

C is least when $x = 80$, $y = 40$.

13. Let x and y be the dimensions of a rectangle; the perimeter is $p = 2x + 2y$. But $A = xy$ thus $y = A/x$ so $p = 2x + 2A/x$ for $x > 0$, $dp/dx = 2 - 2A/x^2 = 2(x^2 - A)/x^2$, $dp/dx = 0$ when $x = \sqrt{A}$, $d^2p/dx^2 = 4A/x^3 > 0$ if $x > 0$ so p is a minimum when $x = \sqrt{A}$ and $y = \sqrt{A}$ and thus the rectangle is a square.

15. Suppose that the lower left corner of S is at $(x, -3x)$. From the figure it's clear that the maximum area of the intersection of R and S occurs for some x in $[-4, 4]$, and the area is $A(x) = (8 - x)(12 + 3x) = 96 + 12x - 3x^2$. Since $A'(x) = 12 - 6x = 6(2 - x)$ is positive for $x < 2$ and negative for $x > 2$, $A(x)$ is increasing for x in $[-4, 2]$ and decreasing for x in $[2, 4]$. So the maximum area is $A(2) = 108$.

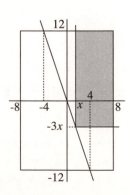

17. Suppose that the lower left corner of S is at $(x, -6x)$. From the figure it's clear that the maximum area of the intersection of R and S occurs for some x in $[-2, 2]$, and the area is $A(x) = (8-x)(12+6x) = 96+36x-6x^2$. Since $A'(x) = 36-12x = 12(3-x)$ is positive for $x < 2$, $A(x)$ is increasing for x in $[-2, 2]$. So the maximum area is $A(2) = 144$.

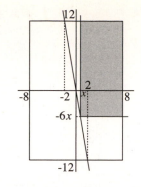

19. Let the box have dimensions x, x, y, with $y \geq x$. The constraint is $4x + y \leq 108$, and the volume $V = x^2 y$. If we take $y = 108 - 4x$ then $V = x^2(108 - 4x)$ and $dV/dx = 12x(-x + 18)$ with roots $x = 0, 18$. The maximum value of V occurs at $x = 18, y = 36$ with $V = 11,664$ in^3.

21. Let x be the length of each side of a square, then $V = x(3 - 2x)(8 - 2x) = 4x^3 - 22x^2 + 24x$ for $0 \leq x \leq 3/2$; $dV/dx = 12x^2 - 44x + 24 = 4(3x - 2)(x - 3)$, $dV/dx = 0$ when $x = 2/3$ for $0 < x < 3/2$. If $x = 0, 2/3, 3/2$ then $V = 0, 200/27, 0$ so the maximum volume is $200/27$ ft^3.

23. Let $x =$ length of each edge of base, $y =$ height, $k = \$/\text{cm}^2$ for the sides. The cost is $C = (2k)(2x^2) + (k)(4xy) = 4k(x^2 + xy)$, but $V = x^2 y = 2000$ thus $y = 2000/x^2$ so $C = 4k(x^2 + 2000/x)$ for $x > 0$, $dC/dx = 4k(2x - 2000/x^2)$, $dC/dx = 0$ when $x = \sqrt[3]{1000} = 10$, $d^2C/dx^2 > 0$ so C is least when $x = 10$, $y = 20$.

25. Let $x =$ height and width, $y =$ length. The surface area is $S = 2x^2 + 3xy$ where $x^2 y = V$, so $y = V/x^2$ and $S = 2x^2 + 3V/x$ for $x > 0$; $dS/dx = 4x - 3V/x^2$, $dS/dx = 0$ when $x = \sqrt[3]{3V/4}$, $d^2S/dx^2 > 0$ so S is minimum when $x = \sqrt[3]{\dfrac{3V}{4}}$, $y = \dfrac{4}{3}\sqrt[3]{\dfrac{3V}{4}}$.

27. Let r and h be the dimensions shown in the figure, then the volume of the inscribed cylinder is $V = \pi r^2 h$. But $r^2 + \left(\dfrac{h}{2}\right)^2 = R^2$

so $r^2 = R^2 - \dfrac{h^2}{4}$. Hence $V = \pi\left(R^2 - \dfrac{h^2}{4}\right)h = \pi\left(R^2 h - \dfrac{h^3}{4}\right)$

for $0 \leq h \leq 2R$. $\dfrac{dV}{dh} = \pi\left(R^2 - \dfrac{3}{4}h^2\right)$, $\dfrac{dV}{dh} = 0$ when

$h = 2R/\sqrt{3}$. If $h = 0, 2R/\sqrt{3}, 2R$ then $V = 0, \dfrac{4\pi}{3\sqrt{3}}R^3, 0$

so the volume is largest when $h = 2R/\sqrt{3}$ and $r = \sqrt{2/3}R$.

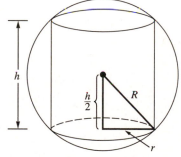

29. From (13), $S = 2\pi r^2 + 2\pi rh$. But $V = \pi r^2 h$ thus $h = V/(\pi r^2)$ and so $S = 2\pi r^2 + 2V/r$ for $r > 0$. $dS/dr = 4\pi r - 2V/r^2$, $dS/dr = 0$ if $r = \sqrt[3]{V/(2\pi)}$. Since $d^2S/dr^2 = 4\pi + 4V/r^3 > 0$, the minimum surface area is achieved when $r = \sqrt[3]{V/2\pi}$ and so $h = V/(\pi r^2) = [V/(\pi r^3)]r = 2r$.

31. The surface area is $S = \pi r^2 + 2\pi rh$ where $V = \pi r^2 h = 500$ so $h = 500/(\pi r^2)$

and $S = \pi r^2 + 1000/r$ for $r > 0$; $dS/dr = 2\pi r - 1000/r^2 = (2\pi r^3 - 1000)/r^2$,

$dS/dr = 0$ when $r = \sqrt[3]{500/\pi}$, $d^2S/dr^2 > 0$ for $r > 0$ so S is minimum when

$r = \sqrt[3]{500/\pi}$ cm and $h = \dfrac{500}{\pi r^2} = \dfrac{500}{\pi}\left(\dfrac{\pi}{500}\right)^{2/3} = \sqrt[3]{500/\pi}$ cm

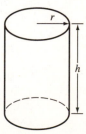

33. Let x be the length of each side of the squares and y the height of the frame, then the volume is $V = x^2 y$. The total length of the wire is L thus $8x + 4y = L$, $y = (L - 8x)/4$ so $V = x^2(L - 8x)/4 = (Lx^2 - 8x^3)/4$ for $0 \le x \le L/8$. $dV/dx = (2Lx - 24x^2)/4$, $dV/dx = 0$ for $0 < x < L/8$ when $x = L/12$. If $x = 0, L/12, L/8$ then $V = 0, L^3/1728, 0$ so the volume is greatest when $x = L/12$ and $y = L/12$.

35. Let h and r be the dimensions shown in the figure, then the volume is $V = \dfrac{1}{3}\pi r^2 h$. But $r^2 + h^2 = L^2$ thus $r^2 = L^2 - h^2$ so $V = \dfrac{1}{3}\pi(L^2 - h^2)h = \dfrac{1}{3}\pi(L^2 h - h^3)$ for $0 \le h \le L$. $\dfrac{dV}{dh} = \dfrac{1}{3}\pi(L^2 - 3h^2)$. $\dfrac{dV}{dh} = 0$ when $h = L/\sqrt{3}$.

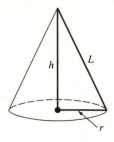

If $h = 0, L/\sqrt{3}, 0$ then $V = 0, \dfrac{2\pi}{9\sqrt{3}}L^3, 0$ so the volume is as large as possible when $h = L/\sqrt{3}$ and $r = \sqrt{2/3}L$.

37. The area of the paper is $A = \pi r L = \pi r \sqrt{r^2 + h^2}$, but $V = \dfrac{1}{3}\pi r^2 h = 100$ so $h = 300/(\pi r^2)$ and $A = \pi r\sqrt{r^2 + 90000/(\pi^2 r^4)}$.

To simplify the computations let $S = A^2$,

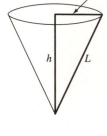

$$S = \pi^2 r^2\left(r^2 + \frac{90000}{\pi^2 r^4}\right) = \pi^2 r^4 + \frac{90000}{r^2} \text{ for } r > 0,$$

$$\frac{dS}{dr} = 4\pi^2 r^3 - \frac{180000}{r^3} = \frac{4(\pi^2 r^6 - 45000)}{r^3}, \ dS/dr = 0 \text{ when}$$

$r = \sqrt[6]{45000/\pi^2}$, $d^2 S/dr^2 > 0$, so S and hence A is least when

$r = \sqrt[6]{45000/\pi^2} = \sqrt{2}\sqrt[3]{75/\pi}$ cm, $h = \dfrac{300}{\pi}\sqrt[3]{\pi^2/45000} = 2\sqrt[3]{75/\pi}$ cm.

39. The volume of the cone is $V = \dfrac{1}{3}\pi r^2 h$. By similar tri-angles (see figure) $\dfrac{r}{h} = \dfrac{R}{\sqrt{h^2 - 2Rh}}$, $r = \dfrac{Rh}{\sqrt{h^2 - 2Rh}}$

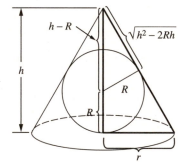

so $V = \dfrac{1}{3}\pi R^2\dfrac{h^3}{h^2 - 2Rh} = \dfrac{1}{3}\pi R^2\dfrac{h^2}{h - 2R}$ for $h > 2R$,

$\dfrac{dV}{dh} = \dfrac{1}{3}\pi R^2\dfrac{h(h - 4R)}{(h - 2R)^2}$, $\dfrac{dV}{dh} = 0$ for $h > 2R$ when

$h = 4R$, by the first derivative test V is minimum when $h = 4R$. If $h = 4R$ then $r = \sqrt{2}R$.

41. The revenue is $R(x) = x(225 - 0.25x) = 225x - 0.25x^2$. The marginal revenue is $R'(x) = 225 - 0.5x = \frac{1}{2}(450 - x)$. Since $R'(x) > 0$ for $x < 450$ and $R'(x) < 0$ for $x > 450$, the maximum revenue occurs when the company mines 450 tons of ore.

43. (a) The daily profit is

$$P = (\text{revenue}) - (\text{production cost}) = 100x - (100{,}000 + 50x + 0.0025x^2)$$
$$= -100{,}000 + 50x - 0.0025x^2$$

for $0 \le x \le 7000$, so $dP/dx = 50 - 0.005x$ and $dP/dx = 0$ when $x = 10{,}000$. Because 10,000 is not in the interval $[0, 7000]$, the maximum profit must occur at an endpoint. When $x = 0$, $P = -100{,}000$; when $x = 7000$, $P = 127{,}500$ so 7000 units should be manufactured and sold daily.

(b) Yes, because $dP/dx > 0$ when $x = 7000$ so profit is increasing at this production level.

(c) $dP/dx = 15$ when $x = 7000$, so $P(7001) - P(7000) \approx 15$, and the marginal profit is \$15.

45. The profit is

$$P = \text{(profit on nondefective)} - \text{(loss on defective)} = 100(x - y) - 20y = 100x - 120y$$

but $y = 0.01x + 0.00003x^2$ so $P = 100x - 120(0.01x + 0.00003x^2) = 98.8x - 0.0036x^2$ for $x > 0$, $dP/dx = 98.8 - 0.0072x$, $dP/dx = 0$ when $x = 98.8/0.0072 \approx 13,722$, $d^2P/dx^2 < 0$ so the profit is maximum at a production level of about 13,722 pounds.

47. The area is (see figure)
$$A = \frac{1}{2}(2\sin\theta)(4 + 4\cos\theta)$$

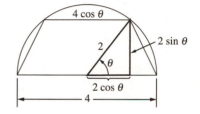

$$= 4(\sin\theta + \sin\theta\cos\theta)$$
for $0 \le \theta \le \pi/2$;
$$\begin{aligned}dA/d\theta &= 4(\cos\theta - \sin^2\theta + \cos^2\theta)\\ &= 4(\cos\theta - [1 - \cos^2\theta] + \cos^2\theta)\\ &= 4(2\cos^2\theta + \cos\theta - 1)\\ &= 4(2\cos\theta - 1)(\cos\theta + 1)\end{aligned}$$
$dA/d\theta = 0$ when $\theta = \pi/3$ for $0 < \theta < \pi/2$. If $\theta = 0, \pi/3, \pi/2$ then $A = 0, 3\sqrt{3}, 4$ so the maximum area is $3\sqrt{3}$.

49. $I = k\dfrac{\cos\phi}{\ell^2}$, k the constant of proportionality. If h is the height of the lamp above the table then $\cos\phi = h/\ell$ and $\ell = \sqrt{h^2 + r^2}$ so $I = k\dfrac{h}{\ell^3} = k\dfrac{h}{(h^2 + r^2)^{3/2}}$ for $h > 0$, $\dfrac{dI}{dh} = k\dfrac{r^2 - 2h^2}{(h^2 + r^2)^{5/2}}$, $\dfrac{dI}{dh} = 0$ when $h = r/\sqrt{2}$, by the first derivative test I is maximum when $h = r/\sqrt{2}$.

51. If $P(x_0, y_0)$ is on the curve $y = 1/x^2$, then $y_0 = 1/x_0^2$. At P the slope of the tangent line is $-2/x_0^3$ so its equation is $y - \dfrac{1}{x_0^2} = -\dfrac{2}{x_0^3}(x - x_0)$, or $y = -\dfrac{2}{x_0^3}x + \dfrac{3}{x_0^2}$. The tangent line crosses the y-axis at $\dfrac{3}{x_0^2}$, and the x-axis at $\dfrac{3}{2}x_0$. The length of the segment then is $L = \sqrt{\dfrac{9}{x_0^4} + \dfrac{9}{4}x_0^2}$ for $x_0 > 0$. For convenience, we minimize L^2 instead, so $L^2 = \dfrac{9}{x_0^4} + \dfrac{9}{4}x_0^2$, $\dfrac{dL^2}{dx_0} = -\dfrac{36}{x_0^5} + \dfrac{9}{2}x_0 = \dfrac{9(x_0^6 - 8)}{2x_0^5}$, which is 0 when $x_0^6 = 8$, $x_0 = \sqrt{2}$. $\dfrac{d^2L^2}{dx_0^2} > 0$ so L^2 and hence L is minimum when $x_0 = \sqrt{2}$, $y_0 = 1/2$.

53. At each point (x, y) on the curve the slope of the tangent line is $m = \dfrac{dy}{dx} = -\dfrac{2x}{(1 + x^2)^2}$ for any x, $\dfrac{dm}{dx} = \dfrac{2(3x^2 - 1)}{(1 + x^2)^3}$, $\dfrac{dm}{dx} = 0$ when $x = \pm 1/\sqrt{3}$, by the first derivative test the only relative maximum occurs at $x = -1/\sqrt{3}$, which is the absolute maximum because $\lim\limits_{x \to \pm\infty} m = 0$. The tangent line has greatest slope at the point $(-1/\sqrt{3}, 3/4)$.

55. Let C be the center of the circle and let θ be the angle $\angle PWE$. Then $\angle PCE = 2\theta$, so the distance along the shore from E to P is 2θ miles. Also, the distance from P to W is $2\cos\theta$ miles.

So Nancy takes $t(\theta) = \dfrac{2\theta}{8} + \dfrac{2\cos\theta}{2} = \dfrac{\theta}{4} + \cos\theta$ hours for her training routine; we wish to find the extrema of this for θ in

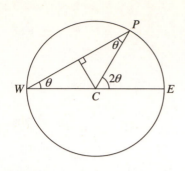

$[0, \frac{\pi}{2}]$. We have $t'(\theta) = \dfrac{1}{4} - \sin\theta$, so the only critical point in $[0, \frac{\pi}{2}]$ is $\theta = \sin^{-1}(\frac{1}{4})$. So we compute $t(0) = 1$, $t(\sin^{-1}(\frac{1}{4})) = \dfrac{1}{4}\sin^{-1}(\frac{1}{4}) + \dfrac{\sqrt{15}}{4} \approx 1.0314$, and $t(\frac{\pi}{2}) = \dfrac{\pi}{8} \approx 0.3927$.

(a) The minimum is $t(\frac{\pi}{2}) = \dfrac{\pi}{8} \approx 0.3927$. To minimize the time, Nancy should choose $P = W$; i.e. she should jog all the way from E to W, π miles.

(b) The maximum is $t(\sin^{-1}(\frac{1}{4})) = \dfrac{1}{4}\sin^{-1}(\frac{1}{4}) + \dfrac{\sqrt{15}}{4} \approx 1.0314$. To maximize the time, she should jog $2\sin^{-1}(\frac{1}{4}) \approx 0.5054$ miles.

57. With x and y as shown in the figure, the maximum length of pipe will be the smallest value of $L = x + y$. By similar triangles

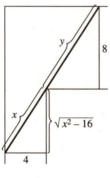

$$\frac{y}{8} = \frac{x}{\sqrt{x^2 - 16}}, \; y = \frac{8x}{\sqrt{x^2 - 16}} \text{ so}$$

$$L = x + \frac{8x}{\sqrt{x^2 - 16}} \text{ for } x > 4, \quad \frac{dL}{dx} = 1 - \frac{128}{(x^2 - 16)^{3/2}},$$

$$\frac{dL}{dx} = 0 \text{ when}$$

$$\begin{aligned}(x^2 - 16)^{3/2} &= 128 \\ x^2 - 16 &= 128^{2/3} = 16(2^{2/3}) \\ x^2 &= 16(1 + 2^{2/3}) \\ x &= 4(1 + 2^{2/3})^{1/2},\end{aligned}$$

$d^2L/dx^2 = 384x/(x^2 - 16)^{5/2} > 0$ if $x > 4$ so L is smallest when $x = 4(1 + 2^{2/3})^{1/2}$. For this value of x, $L = 4(1 + 2^{2/3})^{3/2}$ ft.

59. Let $x =$ distance from the weaker light source, $I =$ the intensity at that point, and k the constant of proportionality. Then

$$I = \frac{kS}{x^2} + \frac{8kS}{(90 - x)^2} \text{ if } 0 < x < 90;$$

$$\frac{dI}{dx} = -\frac{2kS}{x^3} + \frac{16kS}{(90 - x)^3} = \frac{2kS[8x^3 - (90 - x)^3]}{x^3(90 - x)^3} = 18\frac{kS(x - 30)(x^2 + 2700)}{x^3(x - 90)^3},$$

which is 0 when $x = 30$; $\dfrac{dI}{dx} < 0$ if $x < 30$, and $\dfrac{dI}{dx} > 0$ if $x > 30$, so the intensity is minimum at a distance of 30 cm from the weaker source.

61. $\theta = \alpha - \beta = \cot^{-1}(x/12) - \cot^{-1}(x/2)$

$$\frac{d\theta}{dx} = -\frac{12}{144 + x^2} + \frac{2}{4 + x^2} = \frac{10(24 - x^2)}{(144 + x^2)(4 + x^2)}$$

$d\theta/dx = 0$ when $x = \sqrt{24} = 2\sqrt{6}$ feet, by the first derivative test θ is maximum there.

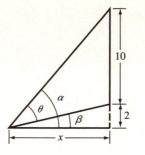

63. The total time required for the light to travel from A to P to B is

$$t = \text{(time from } A \text{ to } P) + \text{(time from } P \text{ to } B) = \frac{\sqrt{x^2 + a^2}}{v_1} + \frac{\sqrt{(c - x)^2 + b^2}}{v_2},$$

$$\frac{dt}{dx} = \frac{x}{v_1\sqrt{x^2 + a^2}} - \frac{c - x}{v_2\sqrt{(c - x)^2 + b^2}} \text{ but } x/\sqrt{x^2 + a^2} = \sin\theta_1 \text{ and}$$

$(c - x)/\sqrt{(c - x)^2 + b^2} = \sin\theta_2$ thus $\dfrac{dt}{dx} = \dfrac{\sin\theta_1}{v_1} - \dfrac{\sin\theta_2}{v_2}$ so $\dfrac{dt}{dx} = 0$ when $\dfrac{\sin\theta_1}{v_1} = \dfrac{\sin\theta_2}{v_2}$.

65. $s = (x_1 - \bar{x})^2 + (x_2 - \bar{x})^2 + \cdots + (x_n - \bar{x})^2,$
$ds/d\bar{x} = -2(x_1 - \bar{x}) - 2(x_2 - \bar{x}) - \cdots - 2(x_n - \bar{x}),$
$ds/d\bar{x} = 0$ when

$$(x_1 - \bar{x}) + (x_2 - \bar{x}) + \cdots + (x_n - \bar{x}) = 0$$
$$(x_1 + x_2 + \cdots + x_n) - n\bar{x} = 0$$
$$\bar{x} = \frac{1}{n}(x_1 + x_2 + \cdots + x_n),$$

$d^2s/d\bar{x}^2 = 2 + 2 + \cdots + 2 = 2n > 0$, so s is minimum when $\bar{x} = \dfrac{1}{n}(x_1 + x_2 + \cdots + x_n)$.

67. If we ignored the interval of possible values of the variables, we might find an extremum that is not physically meaningful, or conclude that there is no extremum. For instance, in Example 2, if we didn't restrict x to the interval $[0, 8]$, there would be no maximum value of V, since $\lim\limits_{x \to +\infty}(480x - 92x^2 + 4x^3) = +\infty$.

EXERCISE SET 3.6

1. (a) positive, negative, slowing down   **(b)** positive, positive, speeding up
(c) negative, positive, slowing down

3. (a) left because $v = ds/dt < 0$ at t_0
(b) negative because $a = d^2s/dt^2$ and the curve is concave down at $t_0 (d^2s/dt^2 < 0)$
(c) speeding up because v and a have the same sign
(d) $v < 0$ and $a > 0$ at t_1 so the particle is slowing down because v and a have opposite signs.

5.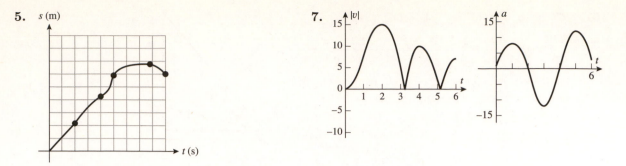

7.

9. False. A particle is speeding up when its <u>speed</u> versus time curve is increasing. When the position versus time graph is increasing, the particle is moving in the positive direction along the s-axis.

11. False. Acceleration is the <u>derivative</u> of velocity.

13. **(a)** At 60 mi/h the tangent line seems to pass through the points $(0, 20)$ and $(16, 100)$. Thus the acceleration would be $\dfrac{v_1 - v_0}{t_1 - t_0}\dfrac{88}{60} = \dfrac{100 - 20}{16 - 0}\dfrac{88}{60} \approx 7.3 \text{ ft/s}^2$.

(b) The maximum acceleration occurs at maximum slope, so when $t = 0$.

15. **(a)**

t	1	2	3	4	5
s	0.71	1.00	0.71	0.00	−0.71
v	0.56	0.00	−0.56	−0.79	−0.56
a	−0.44	−0.62	−0.44	0.00	0.44

(b) to the right at $t = 1$, stopped at $t = 2$, otherwise to the left

(c) speeding up at $t = 3$; slowing down at $t = 1, 5$; neither at $t = 2, 4$

17. **(a)** $v(t) = 3t^2 - 6t$, $a(t) = 6t - 6$

(b) $s(1) = -2$ ft, $v(1) = -3$ ft/s, speed $= 3$ ft/s, $a(1) = 0$ ft/s^2

(c) $v = 0$ at $t = 0, 2$

(d) for $t \geq 0$, $v(t)$ changes sign at $t = 2$, and $a(t)$ changes sign at $t = 1$; so the particle is speeding up for $0 < t < 1$ and $2 < t$ and is slowing down for $1 < t < 2$

(e) total distance $= |s(2) - s(0)| + |s(5) - s(2)| = |-4 - 0| + |50 - (-4)| = 58$ ft

19. **(a)** $s(t) = 9 - 9\cos(\pi t/3)$, $v(t) = 3\pi\sin(\pi t/3)$, $a(t) = \pi^2\cos(\pi t/3)$

(b) $s(1) = 9/2$ ft, $v(1) = 3\pi\sqrt{3}/2$ ft/s, speed $= 3\pi\sqrt{3}/2$ ft/s, $a(1) = \pi^2/2$ ft/s^2

(c) $v = 0$ at $t = 0, 3$

(d) for $0 < t < 5$, $v(t)$ changes sign at $t = 3$ and $a(t)$ changes sign at $t = 3/2, 9/2$; so the particle is speeding up for $0 < t < 3/2$ and $3 < t < 9/2$ and slowing down for $3/2 < t < 3$ and $9/2 < t < 5$

(e) total distance $= |s(3) - s(0)| + |s(5) - s(3)| = |18 - 0| + |9/2 - 18| = 18 + 27/2 = 63/2$ ft

21. $v(t) = \dfrac{5 - t^2}{(t^2 + 5)^2}$, $a(t) = \dfrac{2t(t^2 - 15)}{(t^2 + 5)^3}$

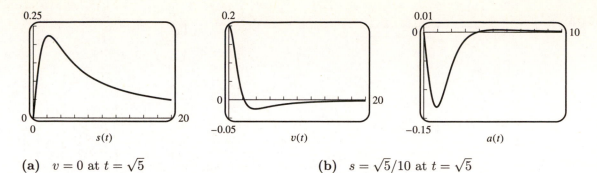

$s(t)$ $v(t)$ $a(t)$

(a) $v = 0$ at $t = \sqrt{5}$ **(b)** $s = \sqrt{5}/10$ at $t = \sqrt{5}$

(c) a changes sign at $t = \sqrt{15}$, so the particle is speeding up for $\sqrt{5} < t < \sqrt{15}$ and slowing down for $0 < t < \sqrt{5}$ and $\sqrt{15} < t$

23. $s = -4t + 3$
      $v = -4$
      $a = 0$

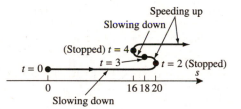

25. $s = t^3 - 9t^2 + 24t$
      $v = 3(t-2)(t-4)$
      $a = 6(t-3)$

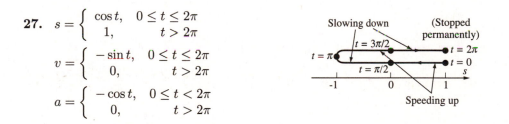

27. $s = \begin{cases} \cos t, & 0 \le t \le 2\pi \\ 1, & t > 2\pi \end{cases}$

      $v = \begin{cases} -\sin t, & 0 \le t \le 2\pi \\ 0, & t > 2\pi \end{cases}$

      $a = \begin{cases} -\cos t, & 0 \le t < 2\pi \\ 0, & t > 2\pi \end{cases}$

29.   **(a)** $v(t) = 10t - 22$, so the speed is $|v(t)| = |10t - 22|$. This decreases from 12 at $t = 1$ to 0 at $t = 2.2$ and then increases to 8 at $t = 3$. The maximum speed is 12 ft/s at $t = 1$ s.

      (b) $s(t) = t(5t - 22) < 0$ for $1 \le t \le 3$, so the maximum distance from the origin occurs when $s(t)$ is minimal. Since $v(t)$ is negative for $1 \le t < 2.2$ and positive for $2.2 < t \le 3$, this occurs when $t = 2.2$ s. The particle's position then is $s(2.2) = -24.2$ ft.

31. $s = \sin 2t$, $v = 2\cos 2t$, $a = -4\sin 2t$

      (a)   For $0 \le t \le \pi/2$, $a = 0$ when $t = 0$ or $\pi/2$. At $t = 0$, $s = 0$ and $v = 2$; at $t = \pi/2$, $s = 0$ and $v = -2$.

      (b)   For $0 \le t \le \pi/2$, $v = 0$ when $t = \pi/4$; $s = 1$ and $a = -4$.

33. **(a)**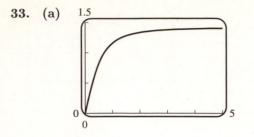

(b) $v = \dfrac{2t}{\sqrt{2t^2 + 1}}$, $\displaystyle\lim_{t \to +\infty} v = \dfrac{2}{\sqrt{2}} = \sqrt{2}$

35. **(a)** $s_1 = s_2$ if they collide, so $\frac{1}{2}t^2 - t + 3 = -\frac{1}{4}t^2 + t + 1$, $\frac{3}{4}t^2 - 2t + 2 = 0$ which has no real solution.

(b) Find the minimum value of $D = |s_1 - s_2| = \left|\frac{3}{4}t^2 - 2t + 2\right|$. From part (a), $\frac{3}{4}t^2 - 2t + 2$ is never zero, and for $t = 0$ it is positive, hence it is always positive, so $D = \frac{3}{4}t^2 - 2t + 2$.

$\dfrac{dD}{dt} = \frac{3}{2}t - 2 = 0$ when $t = \frac{4}{3}$. $\dfrac{d^2D}{dt^2} > 0$ so D is minimum when $t = \frac{4}{3}$, $D = \frac{2}{3}$.

(c) $v_1 = t - 1$, $v_2 = -\dfrac{1}{2}t + 1$. $v_1 < 0$ if $0 \le t < 1$, $v_1 > 0$ if $t > 1$; $v_2 < 0$ if $t > 2$, $v_2 > 0$ if $0 \le t < 2$. They are moving in opposite directions during the intervals $0 \le t < 1$ and $t > 2$.

37. $r(t) = \sqrt{v^2(t)}$, $r'(t) = 2v(t)v'(t)/[2\sqrt{v^2(t)}] = v(t)a(t)/|v(t)|$ so $r'(t) > 0$ (speed is increasing) if v and a have the same sign, and $r'(t) < 0$ (speed is decreasing) if v and a have opposite signs.

39. While the fuel is burning, the acceleration is positive and the rocket is speeding up. After the fuel is gone, the acceleration (due to gravity) is negative and the rocket slows down until it reaches the highest point of its flight. Then the acceleration is still negative, and the rocket speeds up as it falls, until it hits the ground. After that the acceleration is zero, and the rocket neither speeds up nor slows down.

During the powered part of the flight, the acceleration is not constant, and it's hard to say whether it will be increasing or decreasing. First, the power output of the engine may not be constant. Even if it is, the mass of the rocket decreases as the fuel is used up, which tends to increase the acceleration. But as the rocket moves faster, it encounters more air resistance, which tends to decrease the acceleration.

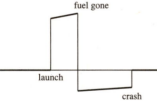

Air resistance also acts during the free-fall part of the flight. While the rocket is still rising, air resistance increases the deceleration due to gravity; while the rocket is falling, air resistance decreases the deceleration.

EXERCISE SET 3.7

1. $f(x) = x^2 - 2$, $f'(x) = 2x$, $x_{n+1} = x_n - \dfrac{x_n^2 - 2}{2x_n}$

$x_1 = 1$, $x_2 = 1.5$, $x_3 \approx 1.416666667, \ldots, x_5 \approx x_6 \approx 1.414213562$

3. $f(x) = x^3 - 6$, $f'(x) = 3x^2$, $x_{n+1} = x_n - \dfrac{x_n^3 - 6}{3x_n^2}$

$x_1 = 2$, $x_2 \approx 1.833333333$, $x_3 \approx 1.817263545, \ldots, x_5 \approx x_6 \approx 1.817120593$

5. $f(x) = x^3 - 2x - 2$, $f'(x) = 3x^2 - 2$, $x_{n+1} = x_n - \dfrac{x_n^3 - 2x_n - 2}{3x_n^2 - 2}$

$x_1 = 2$, $x_2 = 1.8$, $x_3 \approx 1.7699481865$, $x_4 \approx 1.7692926629$, $x_5 \approx x_6 \approx 1.7692923542$

7. $f(x) = x^5 + x^4 - 5$, $f'(x) = 5x^4 + 4x^3$, $x_{n+1} = x_n - \dfrac{x_n^5 + x_n^4 - 5}{5x_n^4 + 4x_n^3}$

$x_1 = 1$, $x_2 \approx 1.333333333$, $x_3 \approx 1.239420573, \ldots, x_6 \approx x_7 \approx 1.224439550$

9. There are 2 solutions.

$f(x) = x^4 + x^2 - 4$, $f'(x) = 4x^3 + 2x$

$x_{n+1} = x_n - \dfrac{x_n^4 + x_n^2 - 4}{4x_n^3 + 2x_n}$

$x_1 = -1$, $x_2 \approx -1.3333$, $x_3 \approx -1.2561$, $x_4 \approx -1.24966, \ldots,$
$x_7 \approx x_8 \approx -1.249621068.$

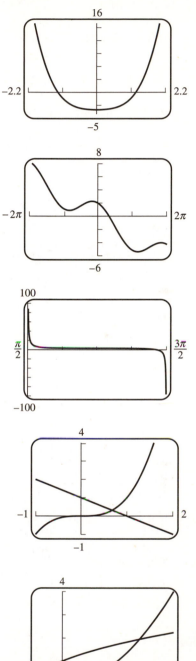

11. There is 1 solution.

$f(x) = 2\cos x - x$, $f'(x) = -2\sin x - 1$

$x_{n+1} = x_n - \dfrac{2\cos x - x}{-2\sin x - 1}$

$x_1 = 1$, $x_2 \approx 1.03004337$, $x_3 \approx 1.02986654$,
$x_4 \approx x_5 \approx 1.02986653$

13. There are infinitely many solutions.

$f(x) = x - \tan x$, $f'(x) = 1 - \sec^2 x = -\tan^2 x$

$x_{n+1} = x_n + \dfrac{x_n - \tan x_n}{\tan^2 x_n}$

$x_1 = 4.5$, $x_2 \approx 4.493613903$, $x_3 \approx 4.493409655$,
$x_4 \approx x_5 \approx 4.493409458$

15. The graphs of $y = x^3$ and $y = 1 - x$ intersect once, near $x = 0.7$.
Let $f(x) = x^3 + x - 1$, so that $f'(x) = 3x^2 + 1$, and

$x_{n+1} = x_n - \dfrac{x_n^3 + x_n - 1}{3x_n^2 + 1}.$

If $x_1 = 0.7$ then $x_2 \approx 0.68259109$, $x_3 \approx 0.68232786$,
$x_4 \approx x_5 \approx 0.68232780.$

17. The graphs of $y = x^2$ and $y = \sqrt{2x+1}$ intersect twice, near
$x = -0.5$ and $x = 1.4$. $x^2 = \sqrt{2x+1}$, $x^4 - 2x - 1 = 0$. Let
$f(x) = x^4 - 2x - 1$, then $f'(x) = 4x^3 - 2$ so

$x_{n+1} = x_n - \dfrac{x_n^4 - 2x_n - 1}{4x_n^3 - 2}.$

If $x_1 = -0.5$, then $x_2 = -0.475$, $x_3 \approx -0.474626695$,
$x_4 \approx x_5 \approx -0.474626618$;
if $x_1 = 1$, then $x_2 = 2$, $x_3 \approx 1.633333333, \ldots, x_8 \approx x_9 \approx 1.395336994.$

19. True. See the discussion before equation (1).

21. False. The function $f(x) = x^3 - x^2 - 110x$ has 3 roots: $x = -10$, $x = 0$, and $x = 11$. Newton's
method in this case gives $x_{n+1} = x_n - \dfrac{x_n^3 - x_n^2 - 110x_n}{3x_n^2 - 2x_n - 110} = \dfrac{2x_n^3 - x_n^2}{3x_n^2 - 2x_n - 110}$. Starting from $x_1 = 5$,

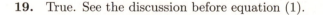

we find $x_2 = -5$, $x_3 = x_4 = x_5 = \cdots = 11$. So the method converges to the root $x = 11$, although the root closest to x_1 is $x = 0$.

23. **(a)** $f(x) = x^2 - a$, $f'(x) = 2x$, $x_{n+1} = x_n - \dfrac{x_n^2 - a}{2x_n} = \dfrac{1}{2}\left(x_n + \dfrac{a}{x_n}\right)$

(b) $a = 10$; $x_1 = 3$, $x_2 \approx 3.166666667$, $x_3 \approx 3.162280702$, $x_4 \approx x_5 \approx 3.162277660$

25. $f'(x) = x^3 + 2x - 5$; solve $f'(x) = 0$ to find the critical points. Graph $y = x^3$ and $y = -2x + 5$ to see that they intersect at a point near $x = 1.25$; $f''(x) = 3x^2 + 2$ so $x_{n+1} = x_n - \dfrac{x_n^3 + 2x_n - 5}{3x_n^2 + 2}$.

$x_1 = 1.25$, $x_2 \approx 1.3317757009$, $x_3 \approx 1.3282755613$, $x_4 \approx 1.3282688557$, $x_5 \approx 1.3282688557$ so the minimum value of $f(x)$ occurs at $x \approx 1.3282688557$ because $f''(x) > 0$; its value is approximately -4.098859132.

27. Let $f(x)$ be the square of the distance between $(1,0)$ and any point (x, x^2) on the parabola, then $f(x) = (x-1)^2 + (x^2 - 0)^2 = x^4 + x^2 - 2x + 1$ and $f'(x) = 4x^3 + 2x - 2$. Solve $f'(x) = 0$ to find the critical points; $f''(x) = 12x^2 + 2$ so $x_{n+1} = x_n - \dfrac{4x_n^3 + 2x_n - 2}{12x_n^2 + 2} = x_n - \dfrac{2x_n^3 + x_n - 1}{6x_n^2 + 1}$.

$x_1 = 1$, $x_2 \approx 0.714285714$, $x_3 \approx 0.605168701, \ldots, x_6 \approx x_7 \approx 0.589754512$; the coordinates are approximately $(0.589754512, 0.347810385)$.

29. **(a)** Let s be the arc length, and L the length of the chord, then $s = 1.5L$. But $s = r\theta$ and $L = 2r\sin(\theta/2)$ so $r\theta = 3r\sin(\theta/2)$, $\theta - 3\sin(\theta/2) = 0$.

(b) Let $f(\theta) = \theta - 3\sin(\theta/2)$, then $f'(\theta) = 1 - 1.5\cos(\theta/2)$ so $\theta_{n+1} = \theta_n - \dfrac{\theta_n - 3\sin(\theta_n/2)}{1 - 1.5\cos(\theta_n/2)}$.

$\theta_1 = 3$, $\theta_2 \approx 2.991592920$, $\theta_3 \approx 2.991563137$, $\theta_4 \approx \theta_5 \approx 2.991563136$ rad so $\theta \approx 171°$.

31. If $x = 1$, then $y^4 + y = 1$, $y^4 + y - 1 = 0$. Graph $z = y^4$ and $z = 1 - y$ to see that they intersect near $y = -1$ and $y = 1$. Let $f(y) = y^4 + y - 1$, then $f'(y) = 4y^3 + 1$ so $y_{n+1} = y_n - \dfrac{y_n^4 + y_n - 1}{4y_n^3 + 1}$.

If $y_1 = -1$, then $y_2 \approx -1.333333333$, $y_3 \approx -1.235807860, \ldots, y_6 \approx y_7 \approx -1.220744085$; if $y_1 = 1$, then $y_2 = 0.8$, $y_3 \approx 0.731233596, \ldots, y_6 \approx y_7 \approx 0.724491959$.

33. $S(25) = 250{,}000 = \dfrac{5000}{i}\left[(1+i)^{25} - 1\right]$; set $f(i) = 50i - (1+i)^{25} + 1$, $f'(i) = 50 - 25(1+i)^{24}$; solve $f(i) = 0$. Set $i_0 = .06$ and $i_{k+1} = i_k - \left[50i - (1+i)^{25} + 1\right] / \left[50 - 25(1+i)^{24}\right]$. Then $i_1 \approx 0.05430$, $i_2 \approx 0.05338$, $i_3 \approx 0.05336$, $\ldots$, $i \approx 0.053362$.

35. **(a)**

x_1	x_2	x_3	x_4	x_5	x_6	x_7	x_8	x_9	x_{10}
0.5000	−0.7500	0.2917	−1.5685	−0.4654	0.8415	−0.1734	2.7970	1.2197	0.1999

(b) The sequence x_n must diverge, since if it did converge then $f(x) = x^2 + 1 = 0$ would have a solution. It seems the x_n are oscillating back and forth in a quasi-cyclical fashion.

37. Suppose we know an interval $[a, b]$ such that $f(a)$ and $f(b)$ have opposite signs. Here are some differences between the two methods:

The Intermediate-Value method is guaranteed to converge to a root in $[a, b]$; Newton's Method starting from some x_1 in the interval might not converge, or might converge to some root outside of the interval.

If the starting approximation x_1 is close enough to the actual root, then Newton's Method converges much faster than the Intermediate-Value method.

Newton's Method can only be used if f is differentiable and we have a way to compute $f'(x)$ for any x. For the Intermediate-Value method we only need to be able to compute $f(x)$.

EXERCISE SET 3.8

1.　$f(3) = f(5) = 0; f'(x) = 2x - 8, 2c - 8 = 0, c = 4, f'(4) = 0$

3.　$f(\pi/2) = f(3\pi/2) = 0, f'(x) = -\sin x, -\sin c = 0, c = \pi$

5.　$(f(5) - f(-3))/(5 - (-3)) = 1; f'(x) = 2x - 1; 2c - 1 = 1, c = 1$

7.　$f(0) = 1, f(3) = 2, f'(x) = \dfrac{1}{2\sqrt{x+1}}, \dfrac{1}{2\sqrt{c+1}} = \dfrac{2-1}{3-0} = \dfrac{1}{3}, \sqrt{c+1} = 3/2, c+1 = 9/4, c = 5/4$

9.　(a)　$f(-2) = f(1) = 0$　　　　　　　　　　　　　　　(b)　$c \approx -1.29$
　　　　The interval is $[-2, 1]$

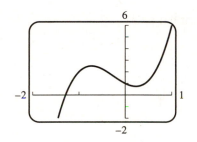

　　(c)　$x_0 = -1, x_1 = -1.5, x_2 = -1.328125,$
　　　　$x_3 \approx -1.2903686, x_4 \approx -1.2885882,$
　　　　$x_5 \approx x_6 \approx -1.2885843$

11.　False. Rolle's Theorem only applies to the case in which f is differentiable on (a, b) and the common value of $f(a)$ and $f(b)$ is zero.

13.　False. The Constant Difference Theorem states that if the <u>derivatives</u> are equal, then the <u>functions</u> differ by a constant.

15.　(a)　$f'(x) = \sec^2 x, \sec^2 c = 0$ has no solution　　　(b)　$\tan x$ is not continuous on $[0, \pi]$

17.　(a)　Two x-intercepts of f determine two solutions a and b of $f(x) = 0$; by Rolle's Theorem there exists a point c between a and b such that $f'(c) = 0$, i.e. c is an x-intercept for f'.

　　(b)　$f(x) = \sin x = 0$ at $x = n\pi$, and $f'(x) = \cos x = 0$ at $x = n\pi + \pi/2$, which lies between $n\pi$ and $(n+1)\pi$, $(n = 0, \pm1, \pm2, \ldots)$

19.　Let $s(t)$ be the position function of the automobile for $0 \le t \le 5$, then by the Mean-Value Theorem there is at least one point c in $(0, 5)$ where
　　$s'(c) = v(c) = [s(5) - s(0)]/(5 - 0) = 4/5 = 0.8$ mi/min $= 48$ mi/h.

21.　Let $f(t)$ and $g(t)$ denote the distances from the first and second runners to the starting point, and let $h(t) = f(t) - g(t)$. Since they start (at $t = 0$) and finish (at $t = t_1$) at the same time, $h(0) = h(t_1) = 0$, so by Rolle's Theorem there is a time t_2 for which $h'(t_2) = 0$, i.e. $f'(t_2) = g'(t_2)$; so they have the same velocity at time t_2.

23.　(a)　By the Constant Difference Theorem $f(x) - g(x) = k$ for some k; since $f(x_0) = g(x_0)$, $k = 0$, so $f(x) = g(x)$ for all x.

　　(b)　Set $f(x) = \sin^2 x + \cos^2 x$, $g(x) = 1$; then $f'(x) = 2\sin x \cos x - 2\cos x \sin x = 0 = g'(x)$. Since $f(0) = 1 = g(0)$, $f(x) = g(x)$ for all x.

25. **(a)** If x, y belong to I and $x < y$ then for some c in I, $\dfrac{f(y) - f(x)}{y - x} = f'(c)$,

   so $|f(x) - f(y)| = |f'(c)||x - y| \leq M|x - y|$; if $x > y$ exchange x and y; if $x = y$ the inequality also holds.

   (b) $f(x) = \sin x$, $f'(x) = \cos x$, $|f'(x)| \leq 1 = M$, so $|f(x) - f(y)| \leq |x - y|$ or $|\sin x - \sin y| \leq |x - y|$.

27. **(a)** Let $f(x) = \sqrt{x}$. By the Mean-Value Theorem there is a number c between x and y such that

   $\dfrac{\sqrt{y} - \sqrt{x}}{y - x} = \dfrac{1}{2\sqrt{c}} < \dfrac{1}{2\sqrt{x}}$ for c in (x, y), thus $\sqrt{y} - \sqrt{x} < \dfrac{y - x}{2\sqrt{x}}$

   (b) multiply through and rearrange to get $\sqrt{xy} < \dfrac{1}{2}(x + y)$.

29. **(a)** If $f(x) = x^3 + 4x - 1$ then $f'(x) = 3x^2 + 4$ is never zero, so by Exercise 28 f has at most one real root; since f is a cubic polynomial it has at least one real root, so it has exactly one real root.

   (b) Let $f(x) = ax^3 + bx^2 + cx + d$. If $f(x) = 0$ has at least two distinct real solutions r_1 and r_2, then $f(r_1) = f(r_2) = 0$ and by Rolle's Theorem there is at least one number between r_1 and r_2 where $f'(x) = 0$. But $f'(x) = 3ax^2 + 2bx + c = 0$ for $x = (-2b \pm \sqrt{4b^2 - 12ac})/(6a) = (-b \pm \sqrt{b^2 - 3ac})/(3a)$, which are not real if $b^2 - 3ac < 0$ so $f(x) = 0$ must have fewer than two distinct real solutions.

31. First we prove that $\sin x < x$ for $x > 0$. If $x > 1$, then $\sin x \leq 1 < x$. If $0 < x \leq 1$, then apply the Mean-Value Theorem with $f(x) = \sin x$ on the interval $[0, x]$: There is a number c in $(0, x)$ such that $\dfrac{\sin x}{x} = \cos c$. Since $0 < c < x \leq 1 < 2\pi$, $\cos c < 1$ so $\sin x < x$.

   Next, apply the Mean-Value Theorem to $g(x) = 6\cos x + 3x^2$ on $[0, x]$: There is some number c in $(0, x)$ such that $\dfrac{g(x) - 6}{x} = g'(c)$. Since $g'(c) = 6(c - \sin c) > 0$, $g(x) > 6$.

   Finally, apply the Mean-Value Theorem to $h(x) = x^3 + 6\sin x - 6x$ on $[0, x]$: There is some c in $(0, x)$ such that $\dfrac{h(x)}{x} = h'(c)$. Since $h'(c) = g(c) - 6 > 0$, $h(x) > 0$, which implies that $\sin x > x - \dfrac{x^3}{6}$.

33. **(a)** $\dfrac{d}{dx}[f^2(x) + g^2(x)] = 2f(x)f'(x) + 2g(x)g'(x) = 2f(x)g(x) + 2g(x)[-f(x)] = 0$,

   so $f^2(x) + g^2(x)$ is constant.

   (b) $f(x) = \sin x$ and $g(x) = \cos x$

35.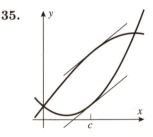

37. **(a)** similar to the proof of part (a) with $f'(c) < 0$

   (b) similar to the proof of part (a) with $f'(c) = 0$

39. If f is differentiable at $x = 1$, then f is continuous there;

   $\lim\limits_{x \to 1^+} f(x) = \lim\limits_{x \to 1^-} f(x) = f(1) = 3$, $a + b = 3$; $\lim\limits_{x \to 1^+} f'(x) = a$ and

$$\lim_{x \to 1^-} f'(x) = 6 \text{ so } a = 6 \text{ and } b = 3 - 6 = -3.$$

41. From Section 2.2 a function has a vertical tangent line at a point of its graph if the slopes of secant lines through the point approach $+\infty$ or $-\infty$. Suppose f is continuous at $x = x_0$ and $\lim_{x \to x_0^+} f(x) = +\infty$. Then a secant line through $(x_1, f(x_1))$ and $(x_0, f(x_0))$, assuming $x_1 > x_0$, will

have slope $\dfrac{f(x_1) - f(x_0)}{x_1 - x_0}$. By the Mean Value Theorem, this quotient is equal to $f'(c)$ for some

c between x_0 and x_1. But as x_1 approaches x_0, c must also approach x_0, and it is given that $\lim_{c \to x_0^+} f'(c) = +\infty$, so the slope of the secant line approaches $+\infty$. The argument can be altered

appropriately for $x_1 < x_0$, and/or for $f'(c)$ approaching $-\infty$.

43. If an object travels s miles in t hours, then at some time during the trip its instantaneous speed is exactly s/t miles per hour.

REVIEW EXERCISES, CHAPTER 3

3. $f'(x) = 2x - 5$
$f''(x) = 2$

    (a) $[5/2, +\infty)$ **(b)** $(-\infty, 5/2]$
    (c) $(-\infty, +\infty)$ **(d)** none
    (e) none

5. $f'(x) = \dfrac{4x}{(x^2 + 2)^2}$

$f''(x) = -4\dfrac{3x^2 - 2}{(x^2 + 2)^3}$

    (a) $[0, +\infty)$ **(b)** $(-\infty, 0]$
    (c) $(-\sqrt{2/3}, \sqrt{2/3})$ **(d)** $(-\infty, -\sqrt{2/3}), (\sqrt{2/3}, +\infty)$
    (e) $-\sqrt{2/3}, \sqrt{2/3}$

7. $f'(x) = \dfrac{4(x + 1)}{3x^{2/3}}$

$f''(x) = \dfrac{4(x - 2)}{9x^{5/3}}$

    (a) $[-1, +\infty)$ **(b)** $(-\infty, -1]$
    (c) $(-\infty, 0), (2, +\infty)$ **(d)** $(0, 2)$
    (e) $0, 2$

9. $f'(x) = -\sin x$
$f''(x) = -\cos x$

Increasing: $[\pi, 2\pi]$
Decreasing: $[0, \pi]$
Concave up: $(\pi/2, 3\pi/2)$
Concave down: $(0, \pi/2), (3\pi/2, 2\pi)$
Inflection points: $\pi/2, 3\pi/2$

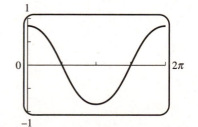

11. $f'(x) = \cos 2x$
$f''(x) = -2\sin 2x$

Increasing: $[0, \pi/4], [3\pi/4, \pi]$
Decreasing: $[\pi/4, 3\pi/4]$
Concave up: $(\pi/2, \pi)$
Concave down: $(0, \pi/2)$
Inflection point: $\pi/2$

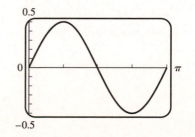

13. (a)

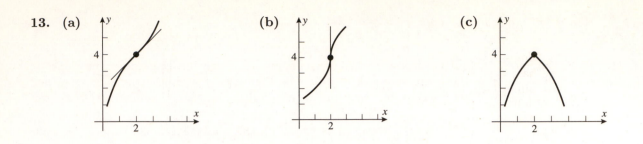

(b)

(c)

15. $f'(x) = 2ax + b$; $f'(x) > 0$ or $f'(x) < 0$ on $[0, +\infty)$ if $f'(x) = 0$ has no positive solution, so the polynomial is always increasing or always decreasing on $[0, +\infty)$ provided $-b/2a \le 0$.

21. (a) $f'(x) = (2 - x^2)/(x^2 + 2)^2$, $f'(x) = 0$ when $x = \pm\sqrt{2}$ (stationary points).

(b) $f'(x) = 8x/(x^2 + 1)^2$, $f'(x) = 0$ when $x = 0$ (stationary point).

23. (a) $f'(x) = \dfrac{7(x - 7)(x - 1)}{3x^{2/3}}$; critical numbers at $x = 0, 1, 7$;

neither at $x = 0$, relative maximum at $x = 1$, relative minimum at $x = 7$ (First Derivative Test)

(b) $f'(x) = 2\cos x(1 + 2\sin x)$; critical numbers at $x = \pi/2, 3\pi/2, 7\pi/6, 11\pi/6$; relative maximum at $x = \pi/2, 3\pi/2$, relative minimum at $x = 7\pi/6, 11\pi/6$

(c) $f'(x) = 3 - \dfrac{3\sqrt{x - 1}}{2}$; critical number at $x = 5$; relative maximum at $x = 5$

25. $\lim\limits_{x \to -\infty} f(x) = +\infty$, $\lim\limits_{x \to +\infty} f(x) = +\infty$

$f'(x) = x(4x^2 - 9x + 6)$, $f''(x) = 6(2x - 1)(x - 1)$
relative minimum at $x = 0$,
points of inflection when $x = 1/2, 1$,
no asymptotes

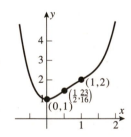

27. $\lim\limits_{x \to \pm\infty} f(x)$ doesn't exist

$f'(x) = 2x\sec^2(x^2 + 1)$,
$f''(x) = 2\sec^2(x^2 + 1)\left[1 + 4x^2\tan(x^2 + 1)\right]$
critical number at $x = 0$; relative minimum at $x = 0$
point of inflection when $1 + 4x^2\tan(x^2 + 1) = 0$

vertical asymptotes at $x = \pm\sqrt{\pi(n + \frac{1}{2}) - 1}$, $n = 0, 1, 2, \ldots$

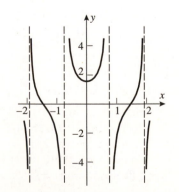

29. $f'(x) = 2\dfrac{x(x+5)}{(x^2+2x+5)^2}$, $f''(x) = -2\dfrac{2x^3+15x^2-25}{(x^2+2x+5)^3}$

critical numbers at $x = -5, 0$;
relative maximum at $x = -5$,
relative minimum at $x = 0$
points of inflection at $x \approx -7.26, -1.44, 1.20$
horizontal asymptote $y = 1$ as $x \to \pm\infty$

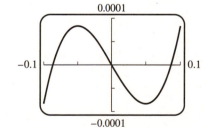

31. $\lim\limits_{x \to -\infty} f(x) = +\infty$, $\lim\limits_{x \to +\infty} f(x) = -\infty$

$$f'(x) = \begin{cases} x, & x \le 0 \\ -2x, & x > 0 \end{cases}$$

critical number at $x = 0$, no extrema

inflection point at $x = 0$ (f changes concavity)

no asymptotes

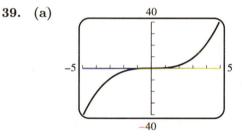

33. $f'(x) = 3x^2 + 5$; no relative extrema because there are no critical numbers.

35. $f'(x) = \frac{4}{5}x^{-1/5}$; critical number $x = 0$; relative minimum of 0 at $x = 0$ (first derivative test)

37. $f'(x) = 2x/(x^2+1)^2$; critical number $x = 0$; relative minimum of 0 at $x = 0$

39. (a)

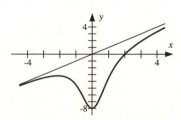

(b) $f'(x) = x^2 - \dfrac{1}{400}$, $f''(x) = 2x$

critical points at $x = \pm\dfrac{1}{20}$;

relative maximum at $x = -\dfrac{1}{20}$,

relative minimum at $x = \dfrac{1}{20}$

(c) The finer details can be seen when graphing over a much smaller x-window.

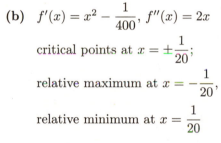

41. (a)

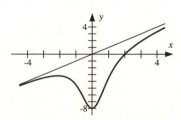

$y = x$ appears to be an asymptote for
$y = (x^3 - 8)/(x^2 + 1)$.

(b) $\dfrac{x^3-8}{x^2+1} = x - \dfrac{x+8}{x^2+1}$. Since the limit of $\dfrac{x+8}{x^2+1}$

as $x \to \pm\infty$ is 0, $y = x$ is an asymptote for

$y = \dfrac{x^3-8}{x^2+1}$.

43. $f(x) = \dfrac{(2x-1)(x^2+x-7)}{(2x-1)(3x^2+x-1)} = \dfrac{x^2+x-7}{3x^2+x-1}, \qquad x \neq 1/2$

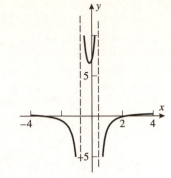

horizontal asymptote: $y = 1/3$,
vertical asymptotes: $x = (-1 \pm \sqrt{13})/6$

47. (a) True. If f has an absolute extremum at a point of (a,b) then it must, by Theorem 3.4.3, be at a critical point of f; since f is differentiable on (a,b) the critical point is a stationary point.

(b) False. It could occur at a critical point which is not a stationary point: for example, $f(x) = |x|$ on $[-1,1]$ has an absolute minimum at $x = 0$ but is not differentiable there.

49. (a) $f'(x) = 2x - 3$; critical point $x = 3/2$. Minimum value $f(3/2) = -13/4$, no maximum.

(b) No maximum or minimum because $\lim\limits_{x \to +\infty} f(x) = +\infty$ and $\lim\limits_{x \to -\infty} f(x) = -\infty$.

(c) critical point at $x = 2$; $m = -3$ at $x = 3$, $M = 0$ at $x = 2$

51. (a) $(x^2 - 1)^2$ can never be less than zero because it is the square of $x^2 - 1$; the minimum value is 0 for $x = \pm 1$, no maximum because $\lim\limits_{x \to +\infty} f(x) = +\infty$.

(b) $f'(x) = (1 - x^2)/(x^2 + 1)^2$; critical point $x = 1$. Maximum value $f(1) = 1/2$, minimum value 0 because $f(x)$ is never less than zero on $[0, +\infty)$ and $f(0) = 0$.

(c) $f'(x) = 2\sec x \tan x - \sec^2 x = (2\sin x - 1)/\cos^2 x$, $f'(x) = 0$ for x in $(0, \pi/4)$ when $x = \pi/6$; $f(0) = 2$, $f(\pi/6) = \sqrt{3}$, $f(\pi/4) = 2\sqrt{2} - 1$ so the maximum value is 2 at $x = 0$ and the minimum value is $\sqrt{3}$ at $x = \pi/6$.

53. (a)

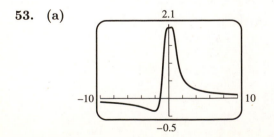

(b) minimum: $(-2.111985, -0.355116)$
maximum: $(0.372591, 2.012931)$

55. If one corner of the rectangle is at (x, y) with $x > 0$, $y > 0$, then $A = 4xy$, $y = 3\sqrt{1 - (x/4)^2}$,

$A = 12x\sqrt{1 - (x/4)^2} = 3x\sqrt{16 - x^2}$, $\dfrac{dA}{dx} = 6\dfrac{8 - x^2}{\sqrt{16 - x^2}}$, critical point at $x = 2\sqrt{2}$. Since $A = 0$

when $x = 0, 4$ and $A > 0$ otherwise, there is an absolute maximum $A = 24$ at $x = 2\sqrt{2}$. The rectangle has width $2x = 4\sqrt{2}$ and height $2y = A/(2x) = 3\sqrt{2}$.

57. $V = x(12 - 2x)^2$ for $0 \le x \le 6$;
$dV/dx = 12(x - 2)(x - 6)$, $dV/dx = 0$
when $x = 2$ for $0 < x < 6$. If $x = 0, 2, 6$
then $V = 0, 128, 0$ so the volume is largest
when $x = 2$ in.

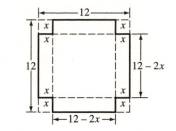

59. **(a)** Yes. If $s = 2t - t^2$ then $v = ds/dt = 2 - 2t$ and $a = dv/dt = -2$ is constant. The velocity changes sign at $t = 1$, so the particle reverses direction then.

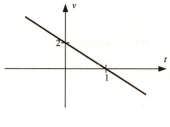

(b) Yes. If $s = t + e^{-t}$ then $v = ds/dt = 1 - e^{-t}$ and $a = dv/dt = e^{-t}$. For $t > 0$, $v > 0$ and $a > 0$, so the particle is speeding up. But $da/dt = -e^{-t} < 0$, so the acceleration is decreasing.

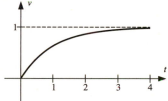

61. **(a)** $v = -2\dfrac{t(t^4 + 2t^2 - 1)}{(t^4 + 1)^2}$, $a = 2\dfrac{3t^8 + 10t^6 - 12t^4 - 6t^2 + 1}{(t^4 + 1)^3}$

(b)

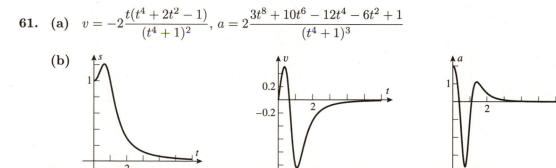

(c) It is farthest from the origin at $t \approx 0.64$ (when $v = 0$) and $s \approx 1.2$

(d) Find t so that the velocity $v = ds/dt > 0$. The particle is moving in the positive direction for $0 \le t \le 0.64$, approximately.

(e) It is speeding up when $a, v > 0$ or $a, v < 0$, so for $0 \le t < 0.36$ and $0.64 < t < 1.1$ approximately, otherwise it is slowing down.

(f) Find the maximum value of $|v|$ to obtain: maximum speed ≈ 1.05 when $t \approx 1.10$.

63. $x \approx -2.11491, 0.25410, 1.86081$

65. At a point of intersection, $x^3 = 0.5x - 1$, $x^3 - 0.5x + 1 = 0$. Let $f(x) = x^3 - 0.5x + 1$. By graphing $y = x^3$ and $y = 0.5x - 1$ it is evident that there is only one point of intersection and it occurs in the interval $[-2, -1]$; note that $f(-2) < 0$ and $f(-1) > 0$. $f'(x) = 3x^2 - 0.5$ so

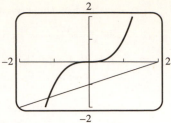

$$x_{n+1} = x_n - \frac{x_n^3 - 0.5x_n + 1}{3x_n^2 - 0.5};$$

$x_1 = -1$, $x_2 = -1.2$, $x_3 \approx -1.166492147, \ldots$, $x_5 \approx x_6 \approx -1.165373043$

67. Solve $\phi - 0.0934 \sin \phi = 2\pi(1)/1.88$ to get $\phi \approx 3.325078$ so $r = 228 \times 10^6(1 - 0.0934 \cos \phi) \approx 248.938 \times 10^6$ km.

69. **(a)** yes; $f'(0) = 0$

(b) no, f is not differentiable on $(-1, 1)$

(c) yes, $f'(\sqrt{\pi/2}) = 0$

71. $f(x) = x^6 - 2x^2 + x$ satisfies $f(0) = f(1) = 0$, so by Rolle's Theorem $f'(c) = 0$ for some c in $(0, 1)$.

MAKING CONNECTIONS, CHAPTER 3

1. **(a)** $g(x)$ has no zeros:

Since $g(x)$ is concave up for $x < 3$, its graph lies on or above the line $y = 2 - \frac{2}{3}x$, which is the tangent line at $(0, 2)$. So for $x < 3$, $g(x) \geq 2 - \frac{2}{3}x > 0$.

Since $g(x)$ is concave up for $3 \leq x < 4$, its graph lies above the line $y = 3x - 9$, which is the tangent line at $(4, 3)$. So for $3 \leq x < 4$, $g(x) > 3x - 9 \geq 0$.

Finally, if $x \geq 4$, $g(x)$ could only have a zero if $g'(a)$ were negative for some $a > 4$. But then the graph would lie below the tangent line at $(a, g(a))$, which crosses the line $y = -10$ for some $x > a$. So $g(x)$ would be less than -10 for some x.

(b) one, between 0 and 4

(c) Since $g(x)$ is concave down for $x > 4$ and $g'(4) = 3$, $g'(x) < 3$ for all $x > 4$. Hence the limit can't be 5. If it were -5 then the graph of $g(x)$ would cross the line $y = -10$ at some point. So the limit must be 0.

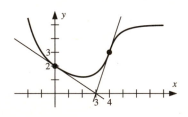

3. $g''(x) = 1 - r'(x)$, so $g(x)$ has an inflection point where the graph of $y = r'(x)$ crosses the line $y = 1$; i.e. at $x = -4$ and $x = 5$.

CHAPTER 4
Integration

EXERCISE SET 4.1

1. Endpoints $0, \frac{1}{n}, \frac{2}{n}, \ldots, \frac{n-1}{n}, 1$; using right endpoints,

$$A_n = \left[\sqrt{\frac{1}{n}} + \sqrt{\frac{2}{n}} + \cdots + \sqrt{\frac{n-1}{n}} + 1\right]\frac{1}{n}$$

n	2	5	10	50	100
A_n	0.853553	0.749739	0.710509	0.676095	0.671463

3. Endpoints $0, \frac{\pi}{n}, \frac{2\pi}{n}, \ldots, \frac{(n-1)\pi}{n}, \pi$; using right endpoints,

$$A_n = [\sin(\pi/n) + \sin(2\pi/n) + \cdots + \sin(\pi(n-1)/n) + \sin\pi]\frac{\pi}{n}$$

n	2	5	10	50	100
A_n	1.57080	1.93376	1.98352	1.99935	1.99984

5. Endpoints $1, \frac{n+1}{n}, \frac{n+2}{n}, \ldots, \frac{2n-1}{n}, 2$; using right endpoints,

$$A_n = \left[\frac{n}{n+1} + \frac{n}{n+2} + \cdots + \frac{n}{2n-1} + \frac{1}{2}\right]\frac{1}{n}$$

n	2	5	10	50	100
A_n	0.583333	0.645635	0.668771	0.688172	0.690653

7. Endpoints $0, \frac{1}{n}, \frac{2}{n}, \ldots, \frac{n-1}{n}, 1$; using right endpoints,

$$A_n = \left[\sqrt{1 - \left(\frac{1}{n}\right)^2} + \sqrt{1 - \left(\frac{2}{n}\right)^2} + \cdots + \sqrt{1 - \left(\frac{n-1}{n}\right)^2} + 0\right]\frac{1}{n}$$

n	2	5	10	50	100
A_n	0.433013	0.659262	0.726130	0.774567	0.780106

9. $3(x-1)$ **11.** $x(x+2)$ **13.** $(x+3)(x-1)$ **15.** false; the area is 4π

17. true

19. $A(6)$ represents the area between $x = 0$ and $x = 6$; $A(3)$ represents the area between $x = 0$ and $x = 3$; their difference $A(6) - A(3)$ represents the area between $x = 3$ and $x = 6$, and $A(6) - A(3) = \frac{1}{3}(6^3 - 3^3) = 63$.

21. B is also the area between the graph of $f(x) = \sqrt{x}$ and the interval $[0, 1]$ on the $y-$axis, so $A + B$ is the area of the square.

23. The area which is under the curve lies to the right of $x = 2$ (or to the left of $x = -2$). Hence $f(x) = A'(x) = 2x$; $0 = A(a) = a^2 - 4$, so take $a = 2$ (or $a = -2$ to measure the area to the left of $x = -2$).

EXERCISE SET 4.2

1. **(a)** $\displaystyle\int \frac{x}{\sqrt{1+x^2}}dx = \sqrt{1+x^2}+C$ **(b)** $\displaystyle\int x^2\cos(1+x^3)dx = \frac{1}{3}\sin(1+x^3)+C$

5. $\displaystyle\frac{d}{dx}\left[\sqrt{x^3+5}\right] = \frac{3x^2}{2\sqrt{x^3+5}}$ so $\displaystyle\int \frac{3x^2}{2\sqrt{x^3+5}}dx = \sqrt{x^3+5}+C$

7. $\displaystyle\frac{d}{dx}\left[\sin\left(2\sqrt{x}\right)\right] = \frac{\cos\left(2\sqrt{x}\right)}{\sqrt{x}}$ so $\displaystyle\int \frac{\cos\left(2\sqrt{x}\right)}{\sqrt{x}}dx = \sin\left(2\sqrt{x}\right)+C$

9. **(a)** $x^9/9+C$ **(b)** $\displaystyle\frac{7}{12}x^{12/7}+C$ **(c)** $\displaystyle\frac{2}{9}x^{9/2}+C$

11. $\displaystyle\int\left[5x+\frac{2}{3x^5}\right]dx = \int 5x\,dx + \frac{2}{3}\int \frac{1}{x^5}dx = \frac{5}{2}x^2 + \frac{2}{3}\left(\frac{-1}{4}\right)\frac{1}{x^4}C = \frac{5}{2}x^2 - \frac{1}{6x^4}+C$

13. $\displaystyle\int\left[x^{-3}-3x^{1/4}+8x^2\right]dx = \int x^{-3}dx - 3\int x^{1/4}dx + 8\int x^2\,dx = -\frac{1}{2}x^{-2} - \frac{12}{5}x^{5/4} + \frac{8}{3}x^3 + C$

15. $\displaystyle\int(x+x^4)dx = x^2/2 + x^5/5 + C$

17. $\displaystyle\int x^{1/3}(4-4x+x^2)dx = \int(4x^{1/3}-4x^{4/3}+x^{7/3})dx = 3x^{4/3} - \frac{12}{7}x^{7/3} + \frac{3}{10}x^{10/3} + C$

19. $\displaystyle\int(x+2x^{-2}-x^{-4})dx = x^2/2 - 2/x + 1/(3x^3) + C$

21. $\displaystyle\int\left[3\sin x - 2\sec^2 x\right]dx = -3\cos x - 2\tan x + C$

23. $\displaystyle\int(\sec^2 x + \sec x\tan x)dx = \tan x + \sec x + C$

25. $\displaystyle\int \frac{\sec\theta}{\cos\theta}d\theta = \int \sec^2\theta\,d\theta = \tan\theta + C$

27. $\displaystyle\int \sec x\tan x\,dx = \sec x + C$ **29.** $\displaystyle\int(1+\sin\theta)d\theta = \theta - \cos\theta + C$

31. $\displaystyle\int \frac{1-\sin x}{1-\sin^2 x}dx = \int \frac{1-\sin x}{\cos^2 x}dx = \int\left(\sec^2 x - \sec x\tan x\right)dx = \tan x - \sec x + C$

33. true   **35.** false; $y(0) = 2$

37.

39. **(a)** $y(x) = \int x^{1/3} dx = \frac{3}{4}x^{4/3} + C,\ y(1) = \frac{3}{4} + C = 2, C = \frac{5}{4};\ y(x) = \frac{3}{4}x^{4/3} + \frac{5}{4}$

(b) $y(t) = \int (\sin t + 1)\, dt = -\cos t + t + C,\ y\left(\frac{\pi}{3}\right) = -\frac{1}{2} + \frac{\pi}{3} + C = 1/2,\ C = 1 - \frac{\pi}{3};$

$y(t) = -\cos t + t + 1 - \frac{\pi}{3}$

(c) $y(x) = \int (x^{1/2} + x^{-1/2})dx = \frac{2}{3}x^{3/2} + 2x^{1/2} + C,\ y(1) = 0 = \frac{8}{3} + C,\ C = -\frac{8}{3},$

$y(x) = \frac{2}{3}x^{3/2} + 2x^{1/2} - \frac{8}{3}$

41. $s(t) = 16t^2 + C;\ s(t) = 16t^2 + 20$ **43.** $s(t) = 2t^{3/2} + C;\ s(t) = 2t^{3/2} - 15$

45. $f'(x) = \frac{2}{3}x^{3/2} + C_1;\ f(x) = \frac{4}{15}x^{5/2} + C_1 x + C_2$

47. $dy/dx = 2x + 1, y = \int (2x + 1)dx = x^2 + x + C;\ y = 0$ when $x = -3$

so $(-3)^2 + (-3) + C = 0, C = -6$ thus $y = x^2 + x - 6$

49. $f'(x) = m = -\sin x$ so $f(x) = \int (-\sin x)dx = \cos x + C;\ f(0) = 2 = 1 + C$

so $C = 1, f(x) = \cos x + 1$

51. $dy/dx = \int 6x\,dx = 3x^2 + C_1$. The slope of the tangent line is -3 so $dy/dx = -3$ when $x = 1$.

Thus $3(1)^2 + C_1 = -3, C_1 = -6$ so $dy/dx = 3x^2 - 6, y = \int (3x^2 - 6)dx = x^3 - 6x + C_2$. If $x = 1$,

then $y = 5 - 3(1) = 2$ so $(1)^2 - 6(1) + C_2 = 2, C_2 = 7$ thus $y = x^3 - 6x + 7$.

53. **(a)**

(b)

(c) $f(x) = x^2/2 - 1$

55. **(b)** because the slopes are positive for $x < 0$ and negative for $x > 0$

57. **(c)** because two changes of sign are required as well as zero slope for $x = \pm 2$

59. Theorem 4.2.2 says that every antiderivative is of the form $F(x) + C$, for some C. In particular, the antiderivative of 0 is C for some C, not for every C. (Different problems will have different values of C).

61. (a) If $x \neq 0$ then $F'(x) = \cos x, G'(x) = \cos x$

(b) For $x > 0, F(x) = \sin x, G(x) = 2 + \sin x = F(x) + 2$, but for $x < 0, F(x) = \sin x, G(x) = -1 + \sin x = F(x) - 1$

(c) No, because neither F nor G is an antiderivative on $(-\infty, +\infty)$.

63. $\displaystyle\int (\sec^2 x - 1)dx = \tan x - x + C$

65. $\displaystyle\frac{1}{2}\int (1 - \cos x)dx = \frac{1}{2}(x - \sin x) + C$

67. $v = \dfrac{1087}{2\sqrt{273}}\displaystyle\int T^{-1/2}\, dT = \dfrac{1087}{\sqrt{273}}T^{1/2} + C, v(273) = 1087 = 1087 + C$ so $C = 0, v = \dfrac{1087}{\sqrt{273}}T^{1/2}$ ft/s

EXERCISE SET 4.3

1. (a) $\displaystyle\int u^{23}du = u^{24}/24 + C = (x^2 + 1)^{24}/24 + C$

(b) $-\displaystyle\int u^3 du = -u^4/4 + C = -(\cos^4 x)/4 + C$

3. (a) $\dfrac{1}{4}\displaystyle\int \sec^2 u\, du = \dfrac{1}{4}\tan u + C = \dfrac{1}{4}\tan(4x + 1) + C$

(b) $\dfrac{1}{4}\displaystyle\int u^{1/2}du = \dfrac{1}{6}u^{3/2} + C = \dfrac{1}{6}(1 + 2y^2)^{3/2} + C$

5. (a) $-\displaystyle\int u\, du = -\dfrac{1}{2}u^2 + C = -\dfrac{1}{2}\cot^2 x + C$

(b) $\displaystyle\int u^9 du = \dfrac{1}{10}u^{10} + C = \dfrac{1}{10}(1 + \sin t)^{10} + C$

7. (a) $\displaystyle\int (u - 1)^2 u^{1/2}du = \int (u^{5/2} - 2u^{3/2} + u^{1/2})du = \dfrac{2}{7}u^{7/2} - \dfrac{4}{5}u^{5/2} + \dfrac{2}{3}u^{3/2} + C$

$$= \dfrac{2}{7}(1 + x)^{7/2} - \dfrac{4}{5}(1 + x)^{5/2} + \dfrac{2}{3}(1 + x)^{3/2} + C$$

(b) $\displaystyle\int \csc^2 u\, du = -\cot u + C = -\cot(\sin x) + C$

11. $u = 4x - 3, \quad \dfrac{1}{4}\displaystyle\int u^9\, du = \dfrac{1}{40}u^{10} + C = \dfrac{1}{40}(4x - 3)^{10} + C$

13. $u = 7x, \quad \dfrac{1}{7}\displaystyle\int \sin u\, du = -\dfrac{1}{7}\cos u + C = -\dfrac{1}{7}\cos 7x + C$

15. $u = 4x, du = 4dx; \dfrac{1}{4}\displaystyle\int \sec u \tan u\, du = \dfrac{1}{4}\sec u + C = \dfrac{1}{4}\sec 4x + C$

17. $u = 7t^2 + 12, du = 14t\, dt; \dfrac{1}{14}\displaystyle\int u^{1/2}du = \dfrac{1}{21}u^{3/2} + C = \dfrac{1}{21}(7t^2 + 12)^{3/2} + C$

19. $u = 1 - 2x, du = -2dx, -3\displaystyle\int \dfrac{1}{u^3}\, du = (-3)\left(-\dfrac{1}{2}\right)\dfrac{1}{u^2} + C = \dfrac{3}{2}\dfrac{1}{(1 - 2x)^2} + C$

21. $u = 5x^4 + 2$, $du = 20x^3\,dx$, $\dfrac{1}{20}\displaystyle\int \dfrac{du}{u^3}\,du = -\dfrac{1}{40}\dfrac{1}{u^2} + C = -\dfrac{1}{40(5x^4 + 2)^2} + C$

23. $u = 5/x$, $du = -(5/x^2)dx$; $-\dfrac{1}{5}\displaystyle\int \sin u\,du = \dfrac{1}{5}\cos u + C = \dfrac{1}{5}\cos(5/x) + C$

25. $u = \cos 3t$, $du = -3\sin 3t\,dt$, $-\dfrac{1}{3}\displaystyle\int u^4\,du = -\dfrac{1}{15}u^5 + C = -\dfrac{1}{15}\cos^5 3t + C$

27. $u = x^2$, $du = 2x\,dx$; $\dfrac{1}{2}\displaystyle\int \sec^2 u\,du = \dfrac{1}{2}\tan u + C = \dfrac{1}{2}\tan\left(x^2\right) + C$

29. $u = 2 - \sin 4\theta$, $du = -4\cos 4\theta\,d\theta$; $-\dfrac{1}{4}\displaystyle\int u^{1/2}du = -\dfrac{1}{6}u^{3/2} + C = -\dfrac{1}{6}(2 - \sin 4\theta)^{3/2} + C$

31. $u = \sec 2x$, $du = 2\sec 2x \tan 2x\,dx$; $\dfrac{1}{2}\displaystyle\int u^2\,du = \dfrac{1}{6}u^3 + C = \dfrac{1}{6}\sec^3 2x + C$

33. $u = 2y + 1$, $du = 2dy$;

$\displaystyle\int \dfrac{1}{4}(u - 1)\dfrac{1}{\sqrt{u}}\,du = \dfrac{1}{6}u^{3/2} - \dfrac{1}{2}\sqrt{u} + C = \dfrac{1}{6}(2y + 1)^{3/2} - \dfrac{1}{2}\sqrt{2y + 1} + C$

35. $\displaystyle\int \sin^2 2\theta \sin 2\theta\,d\theta = \int (1 - \cos^2 2\theta)\sin 2\theta\,d\theta$; $u = \cos 2\theta$, $du = -2\sin 2\theta\,d\theta$,

37. $u = a + bx$, $du = b\,dx$,

$\displaystyle\int (a + bx)^n\,dx = \dfrac{1}{b}\int u^n\,du = \dfrac{(a + bx)^{n+1}}{b(n + 1)} + C$

39. $u = \sin(a + bx)$, $du = b\cos(a + bx)dx$

$\dfrac{1}{b}\displaystyle\int u^n\,du = \dfrac{1}{b(n + 1)}u^{n+1} + C = \dfrac{1}{b(n + 1)}\sin^{n+1}(a + bx) + C$

41. **(a)** with $u = \sin x$, $du = \cos x\,dx$; $\displaystyle\int u\,du = \dfrac{1}{2}u^2 + C_1 = \dfrac{1}{2}\sin^2 x + C_1$;

with $u = \cos x$, $du = -\sin x\,dx$; $-\displaystyle\int u\,du = -\dfrac{1}{2}u^2 + C_2 = -\dfrac{1}{2}\cos^2 x + C_2$

(b) because they differ by a constant:

$$\left(\dfrac{1}{2}\sin^2 x + C_1\right) - \left(-\dfrac{1}{2}\cos^2 x + C_2\right) = \dfrac{1}{2}(\sin^2 x + \cos^2 x) + C_1 - C_2 = 1/2 + C_1 - C_2$$

43. $y = \displaystyle\int \sqrt{5x + 1}\,dx = \dfrac{2}{15}(5x + 1)^{3/2} + C$; $-2 = y(3) = \dfrac{2}{15}64 + C$,

so $C = -2 - \dfrac{2}{15}64 = -\dfrac{158}{15}$, and $y = \dfrac{2}{15}(5x + 1)^{3/2} - \dfrac{158}{15}$

45. **(a)** $u = x^2 + 1$, $du = 2x\,dx$; $\dfrac{1}{2}\displaystyle\int \dfrac{1}{\sqrt{u}}\,du = \sqrt{u} + C = \sqrt{x^2 + 1} + C$

(b)

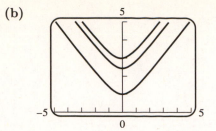

47. $f'(x) = m = \sqrt{3x+1}$, $f(x) = \int (3x+1)^{1/2}dx = \dfrac{2}{9}(3x+1)^{3/2} + C$

$f(0) = 1 = \dfrac{2}{9} + C$, $C = \dfrac{7}{9}$, so $f(x) = \dfrac{2}{9}(3x+1)^{3/2} + \dfrac{7}{9}$

EXERCISE SET 4.4

1. **(a)** $1 + 8 + 27 = 36$
 (c) $20 + 12 + 6 + 2 + 0 + 0 = 40$
 (e) $1 - 2 + 4 - 8 + 16 = 11$

 (b) $5 + 8 + 11 + 14 + 17 = 55$
 (d) $1 + 1 + 1 + 1 + 1 + 1 = 6$
 (f) $0 + 0 + 0 + 0 + 0 + 0 = 0$

3. $\displaystyle\sum_{k=1}^{10} k$

5. $\displaystyle\sum_{k=1}^{10} 2k$

7. $\displaystyle\sum_{k=1}^{6} (-1)^{k+1}(2k-1)$

9. **(a)** $\displaystyle\sum_{k=1}^{50} 2k$

 (b) $\displaystyle\sum_{k=1}^{50} (2k-1)$

11. $\dfrac{1}{2}(100)(100+1) = 5050$

13. $\dfrac{1}{6}(20)(21)(41) = 2870$

15. $\displaystyle\sum_{k=1}^{30} k(k^2 - 4) = \sum_{k=1}^{30}(k^3 - 4k) = \sum_{k=1}^{30} k^3 - 4\sum_{k=1}^{30} k = \dfrac{1}{4}(30)^2(31)^2 - 4 \cdot \dfrac{1}{2}(30)(31) = 214{,}365$

17. $\displaystyle\sum_{k=1}^{n} \dfrac{3k}{n} = \dfrac{3}{n}\sum_{k=1}^{n} k = \dfrac{3}{n} \cdot \dfrac{1}{2}n(n+1) = \dfrac{3}{2}(n+1)$

19. $\displaystyle\sum_{k=1}^{n-1} \dfrac{k^3}{n^2} = \dfrac{1}{n^2}\sum_{k=1}^{n-1} k^3 = \dfrac{1}{n^2} \cdot \dfrac{1}{4}(n-1)^2 n^2 = \dfrac{1}{4}(n-1)^2$

21. true

23. false; if $[a, b]$ consists of positive reals, true; but false on, e.g. $[-2, 1]$.

25. **(a)** $\left(2 + \dfrac{3}{n}\right)^4 \dfrac{3}{n}, \left(2 + \dfrac{6}{n}\right)^4 \dfrac{3}{n}, \left(2 + \dfrac{9}{n}\right)^4 \dfrac{3}{n}, \dots, \left(2 + \dfrac{3(n-1)}{n}\right)^4 \dfrac{3}{n}, (2+3)^4 \dfrac{3}{n}$

When $[2, 5]$ is subdivided into n equal intervals, the endpoints are $2, 2 + \dfrac{3}{n}, 2 + 2 \cdot \dfrac{3}{n}, 2 + 3 \cdot \dfrac{3}{n}, \dots, 2 + (n-1)\dfrac{3}{n}, 2 + 3 = 5$, and the right endpoint approximation to the area under the curve $y = x^4$ is given by the summands above.

(b) $\displaystyle\sum_{k=0}^{n-1}\left(2+k\cdot\frac{3}{n}\right)^4\frac{3}{n}$ gives the left endpoint approximation.

27. Endpoints $2,3,4,5,6$; $\Delta x=1$;

    (a) Left endpoints: $\displaystyle\sum_{k=1}^{4}f(x_k^*)\Delta x=7+10+13+16=46$

    (b) Midpoints: $\displaystyle\sum_{k=1}^{4}f(x_k^*)\Delta x=8.5+11.5+14.5+17.5=52$

    (c) Right endpoints: $\displaystyle\sum_{k=1}^{4}f(x_k^*)\Delta x=10+13+16+19=58$

29. Endpoints: $0,\pi/4,\pi/2,3\pi/4,\pi$; $\Delta x=\pi/4$

    (a) Left endpoints: $\displaystyle\sum_{k=1}^{4}f(x_k^*)\Delta x=\left(1+\sqrt{2}/2+0-\sqrt{2}/2\right)(\pi/4)=\pi/4$

    (b) Midpoints: $\displaystyle\sum_{k=1}^{4}f(x_k^*)\Delta x=[\cos(\pi/8)+\cos(3\pi/8)+\cos(5\pi/8)+\cos(7\pi/8)](\pi/4)$

$$=[\cos(\pi/8)+\cos(3\pi/8)-\cos(3\pi/8)-\cos(\pi/8)](\pi/4)=0$$

    (c) Right endpoints: $\displaystyle\sum_{k=1}^{4}f(x_k^*)\Delta x=\left(\sqrt{2}/2+0-\sqrt{2}/2-1\right)(\pi/4)=-\pi/4$

31. (a) $0.718771403,\ 0.705803382,\ 0.698172179$

    (b) $0.668771403,\ 0.680803382,\ 0.688172179$

    (c) $0.692835360,\ 0.693069098,\ 0.693134682$

33. (a) $4.884074734,\ 5.115572731,\ 5.248762738$

    (b) $5.684074734,\ 5.515572731,\ 5.408762738$

    (c) $5.34707029,\ 5.338362719,\ 5.334644416$

35. $\Delta x=\dfrac{3}{n}$, $x_k^*=1+\dfrac{3}{n}k$; $f(x_k^*)\Delta x=\dfrac{1}{2}x_k^*\Delta x=\dfrac{1}{2}\left(1+\dfrac{3}{n}k\right)\dfrac{3}{n}=\dfrac{3}{2}\left[\dfrac{1}{n}+\dfrac{3}{n^2}k\right]$

$$\sum_{k=1}^{n}f(x_k^*)\Delta x=\frac{3}{2}\left[\sum_{k=1}^{n}\frac{1}{n}+\sum_{k=1}^{n}\frac{3}{n^2}k\right]=\frac{3}{2}\left[1+\frac{3}{n^2}\cdot\frac{1}{2}n(n+1)\right]=\frac{3}{2}\left[1+\frac{3}{2}\frac{n+1}{n}\right]$$

$$A=\lim_{n\to+\infty}\frac{3}{2}\left[1+\frac{3}{2}\left(1+\frac{1}{n}\right)\right]=\frac{3}{2}\left(1+\frac{3}{2}\right)=\frac{15}{4}$$

37. $\Delta x=\dfrac{3}{n}$, $x_k^*=0+k\dfrac{3}{n}$; $f(x_k^*)\Delta x=\left(9-9\dfrac{k^2}{n^2}\right)\dfrac{3}{n}$

$$\sum_{k=1}^{n}f(x_k^*)\Delta x=\sum_{k=1}^{n}\left(9-9\frac{k^2}{n^2}\right)\frac{3}{n}=\frac{27}{n}\sum_{k=1}^{n}\left(1-\frac{k^2}{n^2}\right)=27-\frac{27}{n^3}\sum_{k=1}^{n}k^2$$

$$A=\lim_{n\to+\infty}\left[27-\frac{27}{n^3}\sum_{k=1}^{n}k^2\right]=27-27\left(\frac{1}{3}\right)=18$$

39. $\Delta x=\dfrac{4}{n}$, $x_k^*=2+k\dfrac{4}{n}$

$$f(x_k^*)\Delta x = (x_k^*)^3 \Delta x = \left[2 + \frac{4}{n}k\right]^3 \frac{4}{n} = \frac{32}{n}\left[1 + \frac{2}{n}k\right]^3 = \frac{32}{n}\left[1 + \frac{6}{n}k + \frac{12}{n^2}k^2 + \frac{8}{n^3}k^3\right]$$

$$\sum_{k=1}^{n} f(x_k^*)\Delta x = \frac{32}{n}\left[\sum_{k=1}^{n}1 + \frac{6}{n}\sum_{k=1}^{n}k + \frac{12}{n^2}\sum_{k=1}^{n}k^2 + \frac{8}{n^3}\sum_{k=1}^{n}k^3\right]$$

$$= \frac{32}{n}\left[n + \frac{6}{n}\cdot\frac{1}{2}n(n+1) + \frac{12}{n^2}\cdot\frac{1}{6}n(n+1)(2n+1) + \frac{8}{n^3}\cdot\frac{1}{4}n^2(n+1)^2\right]$$

$$= 32\left[1 + 3\frac{n+1}{n} + 2\frac{(n+1)(2n+1)}{n^2} + 2\frac{(n+1)^2}{n^2}\right]$$

$$A = \lim_{n\to+\infty} 32\left[1 + 3\left(1+\frac{1}{n}\right) + 2\left(1+\frac{1}{n}\right)\left(2+\frac{1}{n}\right) + 2\left(1+\frac{1}{n}\right)^2\right]$$

$$= 32[1 + 3(1) + 2(1)(2) + 2(1)^2] = 320$$

41. $\Delta x = \dfrac{3}{n},\ x_k^* = 1 + (k-1)\dfrac{3}{n}$

$$f(x_k^*)\Delta x = \frac{1}{2}x_k^*\Delta x = \frac{1}{2}\left[1 + (k-1)\frac{3}{n}\right]\frac{3}{n} = \frac{1}{2}\left[\frac{3}{n} + (k-1)\frac{9}{n^2}\right]$$

$$\sum_{k=1}^{n} f(x_k^*)\Delta x = \frac{1}{2}\left[\sum_{k=1}^{n}\frac{3}{n} + \frac{9}{n^2}\sum_{k=1}^{n}(k-1)\right] = \frac{1}{2}\left[3 + \frac{9}{n^2}\cdot\frac{1}{2}(n-1)n\right] = \frac{3}{2} + \frac{9}{4}\frac{n-1}{n}$$

$$A = \lim_{n\to+\infty}\left[\frac{3}{2} + \frac{9}{4}\left(1-\frac{1}{n}\right)\right] = \frac{3}{2} + \frac{9}{4} = \frac{15}{4}$$

43. $\Delta x = \dfrac{3}{n},\ x_k^* = 0 + (k-1)\dfrac{3}{n};\ f(x_k^*)\Delta x = \left[9 - 9\dfrac{(k-1)^2}{n^2}\right]\dfrac{3}{n}$

$$\sum_{k=1}^{n} f(x_k^*)\Delta x = \sum_{k=1}^{n}\left[9 - 9\frac{(k-1)^2}{n^2}\right]\frac{3}{n} = \frac{27}{n}\sum_{k=1}^{n}\left(1 - \frac{(k-1)^2}{n^2}\right) = 27 - \frac{27}{n^3}\sum_{k=1}^{n}k^2 + \frac{54}{n^3}\sum_{k=1}^{n}k - \frac{27}{n^2}$$

$$A = \lim_{n\to+\infty} = 27 - 27\left(\frac{1}{3}\right) + 0 + 0 = 18$$

45. Endpoints $0, \dfrac{4}{n}, \dfrac{8}{n}, \ldots, \dfrac{4(n-1)}{n}, \dfrac{4n}{n} = 4$, and midpoints $\dfrac{2}{n}, \dfrac{6}{n}, \dfrac{10}{n}, \ldots, \dfrac{4n-6}{n}, \dfrac{4n-2}{n}$. Approximate the area with the sum $\displaystyle\sum_{k=1}^{n} 2\left(\frac{4k-2}{n}\right)\frac{4}{n} = \frac{16}{n^2}\left[2\frac{n(n+1)}{2} - n\right] \to 16$ (exact) as $n\to+\infty$.

47. $\Delta x = \dfrac{1}{n},\ x_k^* = \dfrac{2k-1}{2n}$

$$f(x_k^*)\Delta x = \frac{(2k-1)^2}{(2n)^2}\frac{1}{n} = \frac{k^2}{n^3} - \frac{k}{n^3} + \frac{1}{4n^3}$$

$$\sum_{k=1}^{n} f(x_k^*)\Delta x = \frac{1}{n^3}\sum_{k=1}^{n}k^2 - \frac{1}{n^3}\sum_{k=1}^{n}k + \frac{1}{4n^3}\sum_{k=1}^{n}1$$

Using Theorem 4.4.4,

$$A = \lim_{n\to+\infty}\sum_{k=1}^{n} f(x_k^*)\Delta x = \frac{1}{3} + 0 + 0 = \frac{1}{3}$$

49. $\Delta x = \dfrac{2}{n},\ x_k^* = -1 + \dfrac{2k}{n}$

$$f(x_k^*)\Delta x = \left(-1 + \frac{2k}{n}\right)\frac{2}{n} = -\frac{2}{n} + 4\frac{k}{n^2}.$$

$$\sum_{k=1}^{n} f(x_k^*)\Delta x = -2 + \frac{4}{n^2}\sum_{k=1}^{n} k = -2 + \frac{4}{n^2}\frac{n(n+1)}{2} = -2 + 2 + \frac{2}{n}$$

$$A = \lim_{n\to+\infty}\sum_{k=1}^{n} f(x_k^*)\Delta x = 0$$

The area below the x-axis cancels the area above the x-axis.

51. $\Delta x = \dfrac{2}{n}, x_k^* = \dfrac{2k}{n}$

$$f(x_k^*) = \left[\left(\frac{2k}{n}\right)^2 - 1\right]\frac{2}{n} = \frac{8k^2}{n^3} - \frac{2}{n}$$

$$\sum_{k=1}^{n} f(x_k^*)\Delta x = \frac{8}{n^3}\sum_{k=1}^{n} k^2 - \frac{2}{n}\sum_{k=1}^{n} 1 = \frac{8}{n^3}\frac{n(n+1)(2n+1)}{6} - 2$$

$$A = \lim_{n\to+\infty}\sum_{k=1}^{n} f(x_k^*)\Delta x = \frac{16}{6} - 2 = \frac{2}{3}$$

53. (a) With x_k^* as the right endpoint, $\Delta x = \dfrac{b}{n}, x_k^* = \dfrac{b}{n}k$

$$f(x_k^*)\Delta x = (x_k^*)^3\Delta x = \frac{b^4}{n^4}k^3, \quad \sum_{k=1}^{n} f(x_k^*)\Delta x = \frac{b^4}{n^4}\sum_{k=1}^{n} k^3 = \frac{b^4}{4}\frac{(n+1)^2}{n^2}$$

$$A = \lim_{n\to+\infty}\frac{b^4}{4}\left(1 + \frac{1}{n}\right)^2 = b^4/4$$

(b) First Method (tedious)

$$\Delta x = \frac{b-a}{n}, \quad x_k^* = a + \frac{b-a}{n}k$$

$$f(x_k^*)\Delta x = (x_k^*)^3\Delta x = \left[a + \frac{b-a}{n}k\right]^3\frac{b-a}{n}$$

$$= \frac{b-a}{n}\left[a^3 + \frac{3a^2(b-a)}{n}k + \frac{3a(b-a)^2}{n^2}k^2 + \frac{(b-a)^3}{n^3}k^3\right]$$

$$\sum_{k=1}^{n} f(x_k^*)\Delta x = (b-a)\left[a^3 + \frac{3}{2}a^2(b-a)\frac{n+1}{n} + \frac{1}{2}a(b-a)^2\frac{(n+1)(2n+1)}{n^2}\right.$$

$$\left. + \frac{1}{4}(b-a)^3\frac{(n+1)^2}{n^2}\right]$$

$$A = \lim_{n\to+\infty}\sum_{k=1}^{n} f(x_k^*)\Delta x$$

$$= (b-a)\left[a^3 + \frac{3}{2}a^2(b-a) + a(b-a)^2 + \frac{1}{4}(b-a)^3\right] = \frac{1}{4}(b^4 - a^4)$$

Alternative method: Apply Part (a) of the Exercise to the interval $[0, a]$ and observe that the area under the curve and above that interval is given by $\dfrac{1}{4}a^4$. Apply Part (a) again, this time to the interval $[0, b]$ and obtain $\dfrac{1}{4}b^4$. Now subtract to obtain the correct area and the formula $A = \dfrac{1}{4}(b^4 - a^4)$.

55. If $n = 2m$ then $2m + 2(m-1) + \cdots + 2 \cdot 2 + 2 = 2 \sum_{k=1}^{m} k = 2 \cdot \frac{m(m+1)}{2} = m(m+1) = \frac{n^2 + 2n}{4}$;

if $n = 2m + 1$ then $(2m+1) + (2m-1) + \cdots + 5 + 3 + 1 = \sum_{k=1}^{m+1} (2k - 1)$

$= 2 \sum_{k=1}^{m+1} k - \sum_{k=1}^{m+1} 1 = 2 \cdot \frac{(m+1)(m+2)}{2} - (m+1) = (m+1)^2 = \frac{n^2 + 2n + 1}{4}$

57. $(3^5 - 3^4) + (3^6 - 3^5) + \cdots + (3^{17} - 3^{16}) = 3^{17} - 3^4$

59. $\left(\frac{1}{2^2} - \frac{1}{1^2} \right) + \left(\frac{1}{3^2} - \frac{1}{2^2} \right) + \cdots + \left(\frac{1}{20^2} - \frac{1}{19^2} \right) = \frac{1}{20^2} - 1 = -\frac{399}{400}$

61. (a) $\sum_{k=1}^{n} \frac{1}{(2k-1)(2k+1)} = \frac{1}{2} \sum_{k=1}^{n} \left(\frac{1}{2k-1} - \frac{1}{2k+1} \right)$

$= \frac{1}{2} \left[\left(1 - \frac{1}{3} \right) + \left(\frac{1}{3} - \frac{1}{5} \right) + \left(\frac{1}{5} - \frac{1}{7} \right) + \cdots + \left(\frac{1}{2n-1} - \frac{1}{2n+1} \right) \right]$

$= \frac{1}{2} \left[1 - \frac{1}{2n+1} \right] = \frac{n}{2n+1}$

(b) $\lim_{n \to +\infty} \frac{n}{2n+1} = \frac{1}{2}$

63. $\sum_{i=1}^{n} (x_i - \bar{x}) = \sum_{i=1}^{n} x_i - \sum_{i=1}^{n} \bar{x} = \sum_{i=1}^{n} x_i - n\bar{x}$ but $\bar{x} = \frac{1}{n} \sum_{i=1}^{n} x_i$ thus

$\sum_{i=1}^{n} x_i = n\bar{x}$ so $\sum_{i=1}^{n} (x_i - \bar{x}) = n\bar{x} - n\bar{x} = 0$

65. both are valid

67. $\sum_{k=1}^{n} (a_k - b_k) = (a_1 - b_1) + (a_2 - b_2) + \cdots + (a_n - b_n)$

$= (a_1 + a_2 + \cdots + a_n) - (b_1 + b_2 + \cdots + b_n) = \sum_{k=1}^{n} a_k - \sum_{k=1}^{n} b_k$

EXERCISE SET 4.5

1. (a) $(4/3)(1) + (5/2)(1) + (4)(2) = 71/6$ **(b)** 2

3. (a) $(-9/4)(1) + (3)(2) + (63/16)(1) + (-5)(3) = -117/16$
    (b) 3

5. $\displaystyle\int_{-1}^{2} x^2 \, dx$ **7.** $\displaystyle\int_{-3}^{3} 4x(1 - 3x) \, dx$

9. (a) $\displaystyle\lim_{\max \Delta x_k \to 0} \sum_{k=1}^{n} 2x_k^* \Delta x_k$; $a = 1$, $b = 2$ **(b)** $\displaystyle\lim_{\max \Delta x_k \to 0} \sum_{k=1}^{n} \frac{x_k^*}{x_k^* + 1} \Delta x_k$; $a = 0$, $b = 1$

11. Theorem 4.5.4(a) depends on the fact that a constant can move past an integral sign, which by Definition 4.5.1 is possible because a constant can move past a limit and/or a summation sign.

13. (a) $A = \dfrac{1}{2}(3)(3) = 9/2$

(b) $-A = -\dfrac{1}{2}(1)(1 + 2) = -3/2$

(c) $-A_1 + A_2 = -\dfrac{1}{2} + 8 = 15/2$

(d) $-A_1 + A_2 = 0$

15. (a) $A = 2(5) = 10$

(b) $0;\ A_1 = A_2$ by symmetry

(c) $A_1 + A_2 = \dfrac{1}{2}(5)(5/2) + \dfrac{1}{2}(1)(1/2)$

$= 13/2$

(d) $\dfrac{1}{2}[\pi(1)^2] = \pi/2$

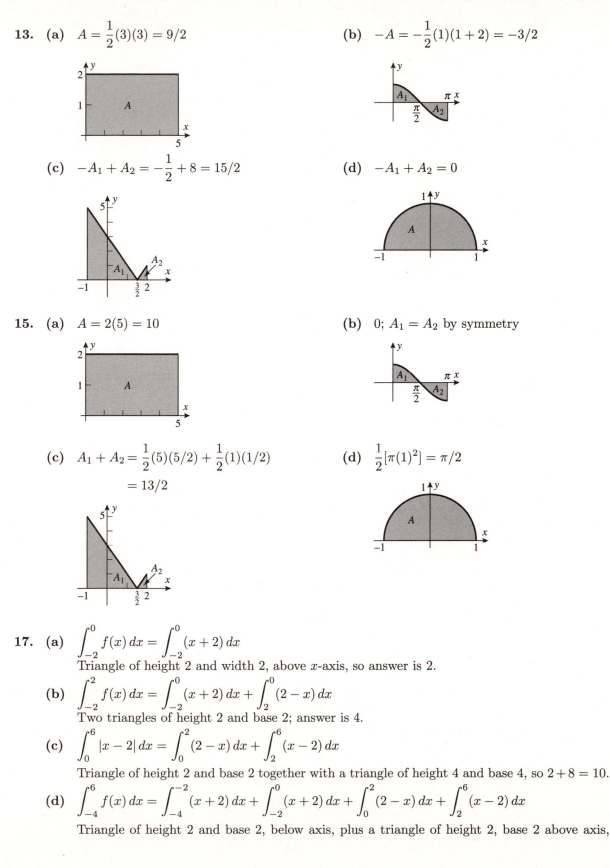

17. (a) $\displaystyle\int_{-2}^{0} f(x)\,dx = \int_{-2}^{0} (x + 2)\,dx$

Triangle of height 2 and width 2, above x-axis, so answer is 2.

(b) $\displaystyle\int_{-2}^{2} f(x)\,dx = \int_{-2}^{0} (x + 2)\,dx + \int_{2}^{0} (2 - x)\,dx$

Two triangles of height 2 and base 2; answer is 4.

(c) $\displaystyle\int_{0}^{6} |x - 2|\,dx = \int_{0}^{2} (2 - x)\,dx + \int_{2}^{6} (x - 2)\,dx$

Triangle of height 2 and base 2 together with a triangle of height 4 and base 4, so $2 + 8 = 10$.

(d) $\displaystyle\int_{-4}^{6} f(x)\,dx = \int_{-4}^{-2} (x + 2)\,dx + \int_{-2}^{0} (x + 2)\,dx + \int_{0}^{2} (2 - x)\,dx + \int_{2}^{6} (x - 2)\,dx$

Triangle of height 2 and base 2, below axis, plus a triangle of height 2, base 2 above axis, another of height 2 and base 2 above axis, and a triangle of height 4 and base 4, above axis. Thus $\int f(x) = -2 + 2 + 2 + 8 = 10$.

19. (a) 0.8   (b) −2.6   (c) −1.8   (d) −0.3

21. $\int_{-1}^{2} f(x)dx + 2\int_{-1}^{2} g(x)dx = 5 + 2(-3) = -1$

23. $\int_{1}^{5} f(x)dx = \int_{0}^{5} f(x)dx - \int_{0}^{1} f(x)dx = 1 - (-2) = 3$

25. $4\int_{-1}^{3} dx - 5\int_{-1}^{3} x dx = 4 \cdot 4 - 5(-1/2 + (3 \cdot 3)/2) = -4$

27. $\int_{0}^{1} x dx + 2\int_{0}^{1} \sqrt{1 - x^2} dx = 1/2 + 2(\pi/4) = (1 + \pi)/2$

29. false; e.g. $f(x) = 1$ if $x > 0$, $f(x) = 0$ otherwise, then f is integrable on $[-1, 1]$ but not continuous.

31. false; e.g. $f(x) = x$ on $[-2, +1]$

33. (a) $\sqrt{x} > 0$, $1 - x < 0$ on $[2, 3]$ so the integral is negative

     (b) $3 - \cos x > 0$ for all x and $x^2 \geq 0$ for all x and $x^2 > 0$ for all $x > 0$ so the integral is positive

35. If f is continuous on $[a, b]$ then f is integrable on $[a, b]$, and, considering Definition 4.5.1, for every partition and choice of $f(x^*)$ we have $\sum_{k=1}^{n} m\Delta x_k \leq \sum_{k=1}^{n} f(x_k^*)\Delta x_k \leq \sum_{k=1}^{n} M\Delta x_k$. This is equivalent to $m(b - a) \leq \sum_{k=1}^{n} f(x_k^*)\Delta x_k \leq M(b - a)$, and, taking the limit over $\max \Delta x_k \to 0$ we obtain the result.

37. $\int_{0}^{10} \sqrt{25 - (x - 5)^2} dx = \pi(5)^2/2 = 25\pi/2$ **39.** $\int_{0}^{1} (3x + 1)dx = 5/2$

41. (a) The graph of the integrand is the horizontal line $y = C$. At first, assume that $C > 0$. Then the region is a rectangle of height C whose base extends from $x = a$ to $x = b$. Thus $\int_{a}^{b} C \, dx = $ (area of rectangle) $= C(b - a)$.

If $C \leq 0$ then the rectangle lies below the axis and its integral is the negative area, i.e. $-|C|(b - a) = C(b - a)$.

     (b) Since $f(x) = C$, the Riemann sum becomes

$$\lim_{\max \Delta x_k \to 0} \sum_{k=1}^{n} f(x_k^*)\Delta x_k = \lim_{\max \Delta x_k \to 0} \sum_{k=1}^{n} C\Delta x_k = \lim_{\max \Delta x_k \to 0} C(b - a) = C(b - a).$$

By Definition 4.5.1, $\int_{a}^{b} f(x) \, dx = C(b - a)$.

43. Each subinterval of a partition of $[a, b]$ contains both rational and irrational numbers. If all x_k^* are chosen to be rational then

$$\sum_{k=1}^{n} f(x_k^*)\Delta x_k = \sum_{k=1}^{n} (1)\Delta x_k = \sum_{k=1}^{n} \Delta x_k = b - a \quad \text{so} \quad \lim_{\max \Delta x_k \to 0} \sum_{k=1}^{n} f(x_k^*)\Delta x_k = b - a.$$

If all x_k^* are irrational then $\displaystyle\lim_{\max \Delta x_k \to 0} \sum_{k=1}^{n} f(x_k^*)\Delta x_k = 0$. Thus f is not integrable on $[a, b]$ because the preceding limits are not equal.

45. (a) f is continuous on $[-1, 1]$ so f is integrable there by Theorem 4.5.2

(b) $|f(x)| \leq 1$ so f is bounded on $[-1, 1]$, and f has one point of discontinuity, so by Part (a) of Theorem 4.5.8 f is integrable on $[-1, 1]$

(c) f is not bounded on [-1,1] because $\lim_{x \to 0} f(x) = +\infty$, so f is not integrable on $[0,1]$

(d) $f(x)$ is discontinuous at the point $x = 0$ because $\lim_{x \to 0} \sin \dfrac{1}{x}$ does not exist. f is continuous elsewhere. $-1 \leq f(x) \leq 1$ for x in $[-1, 1]$ so f is bounded there. By Part (a), Theorem 4.5.8, f is integrable on $[-1, 1]$.

EXERCISE SET 4.6

1. (a) $\displaystyle\int_0^2 (2 - x)dx = (2x - x^2/2)\Big]_0^2 = 4 - 4/2 = 2$

(b) $\displaystyle\int_{-1}^1 2dx = 2x\Big]_{-1}^1 = 2(1) - 2(-1) = 4$

(c) $\displaystyle\int_1^3 (x + 1)dx = (x^2/2 + x)\Big]_1^3 = 9/2 + 3 - (1/2 + 1) = 6$

3. (a)   **(b)**   **(c)**

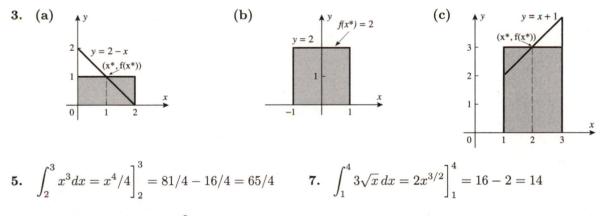

5. $\displaystyle\int_2^3 x^3 dx = x^4/4\Big]_2^3 = 81/4 - 16/4 = 65/4$ **7.** $\displaystyle\int_1^4 3\sqrt{x}\, dx = 2x^{3/2}\Big]_1^4 = 16 - 2 = 14$

9. (a) $\displaystyle\int_0^3 \sqrt{x}\, dx = \frac{2}{3}x^{3/2}\Big]_0^3 = 2\sqrt{3} = f(x^*)(3 - 0)$, so $f(x^*) = \dfrac{2}{\sqrt{3}}, x^* = \dfrac{4}{3}$

(b) $\displaystyle\int_{-12}^0 (x^2 + x)\, dx = \frac{1}{3}x^3 + \frac{1}{2}x^2\Big]_{-12}^0 = 504$, so $f(x^*)(0 - (-12)) = 504, (x^*)^2 + x^* = 42$,

$x^* = 6, -7$ but only -7 lies in the interval. $f(-7) = 49 - 7 = 42$, so the area is that of a rectangle 12 wide and 42 high.

11. $\displaystyle\int_{-2}^1 (x^2 - 6x + 12)\, dx = \left[\frac{1}{3}x^3 - 3x^2 + 12x\right]\Big]_{-2}^1 = \frac{1}{3} - 3 + 12 - \left(-\frac{8}{3} - 12 - 24\right) = 48$

13. $\displaystyle\int_1^4 \frac{4}{x^2}\, dx = -4x^{-1}\Big]_1^4 = -1 + 4 = 3$ **15.** $\dfrac{4}{5}x^{5/2}\Big]_4^9 = 844/5$

17. $-\cos\theta\Big]_{-\pi/2}^{\pi/2} = 0$ **19.** $\sin x\Big]_{-\pi/4}^{\pi/4} = \sqrt{2}$ **21.** $\left(2\sqrt{t} - 2t^{3/2}\right)\Big]_1^4 = -12$

23. (a) $\displaystyle\int_{-1}^{1} |2x-1|\,dx = \int_{-1}^{1/2}(1-2x)\,dx + \int_{1/2}^{1}(2x-1)\,dx = (x-x^2)\Big]_{-1}^{1/2} + (x^2-x)\Big]_{1/2}^{1} = \frac{5}{2}$

    (b) $\displaystyle\int_{0}^{\pi/2}\cos x\,dx + \int_{\pi/2}^{3\pi/4}(-\cos x)\,dx = \sin x\Big]_{0}^{\pi/2} - \sin x\Big]_{\pi/2}^{3\pi/4} = 2 - \sqrt{2}/2$

25. (a) $17/6$ (b) $F(x) = \begin{cases} \dfrac{1}{2}x^2, & 0 \le x \le 1 \\[2mm] \dfrac{1}{3}x^3 + \dfrac{1}{6}, & 1 < x \le 2 \end{cases}$

27. false; consider $F(x) = x^2/2$ if $x \ge 0$ and $F(x) = -x^2/2$ if $x \le 0$

29. true

31. $0.665867079;\ \displaystyle\int_{1}^{3}\frac{1}{x^2}\,dx = -\frac{1}{x}\Big]_{1}^{3} = 2/3$

33. $3.106017890;\ \displaystyle\int_{-1}^{1}\sec^2 x\,dx = \tan x\Big|_{-1}^{1} = 2\tan 1 \approx 3.114815450$

35. $A = \displaystyle\int_{0}^{3}(x^2+1)\,dx = \left(\frac{1}{3}x^3 + x\right)\Big]_{0}^{3} = 12$

37. $A = \displaystyle\int_{0}^{2\pi/3} 3\sin x\,dx = -3\cos x\Big]_{0}^{2\pi/3} = 9/2$

39. Area $= -\displaystyle\int_{0}^{1}(x^2-x)\,dx + \int_{1}^{2}(x^2-x)\,dx = 5/6 + 1/6 = 1$

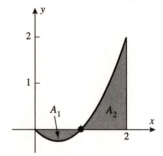

41. Area $= -\displaystyle\int_{0}^{5/4}(2\sqrt{x+1}-3)\,dx + \int_{5/4}^{3}(2\sqrt{x+1}-3)\,dx = \frac{1}{3}$

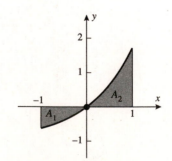

43. **(a)** $A = \int_0^{0.8} \dfrac{1}{\sqrt{1-x^2}} dx = \sin^{-1} x \Big]_0^{0.8} = \sin^{-1}(0.8)$

   (b) The calculator was in degree mode instead of radian mode; the correct answer is 0.93.

45. **(a)** the increase in height in inches, during the first ten years

   (b) the change in the radius in centimeters, during the time interval $t = 1$ to $t = 2$ seconds

   (c) the change in the speed of sound in ft/s, during an increase in temperature from $t = 32°F$ to $t = 100°F$

   (d) the displacement of the particle in cm, during the time interval $t = t_1$ to $t = t_2$ hours

47. **(a)** $F'(x) = 3x^2 - 3$

   (b) $\int_1^x (3t^2 - 3) \, dt = (t^3 - 3t) \Big]_1^x = x^3 - 3x + 2$, and $0 \dfrac{d}{dx}(x^3 - 3x + 2) = 3x^2 - 3$

49. **(a)** $\sin x^2$ **(b)** $\sqrt{1 - \cos x}$

51. $-\dfrac{x}{\cos x}$

53. $F'(x) = \sqrt{x^2 + 9}, \quad F''(x) = \dfrac{x}{\sqrt{x^2 + 9}}$

   (a) 0   **(b)** 5   **(c)** $\dfrac{4}{5}$

55. **(a)** $F'(x) = \dfrac{x-3}{x^2+7} = 0$ when $x = 3$, which is a relative minimum, and hence the absolute minimum, by the first derivative test.

   (b) increasing on $[3, +\infty)$, decreasing on $(-\infty, 3]$

   (c) $F''(x) = \dfrac{7 + 6x - x^2}{(x^2+7)^2} = \dfrac{(7-x)(1+x)}{(x^2+7)^2}$; concave up on $(-1, 7)$, concave down on $(-\infty, -1)$ and on $(7, +\infty)$

57. **(a)** $(0, +\infty)$ because f is continuous there and 1 is in $(0, +\infty)$

   (b) at $x = 1$ because $F(1) = 0$

59. **(a)** amount of water $=$ (rate of flow)(time) $= 4t$ gal, total amount $= 4(30) = 120$ gal

   (b) amount of water $= \int_0^{60} (4 + t/10) dt = 420$ gal

   (c) amount of water $= \int_0^{120} (10 + \sqrt{t}) dt = 1200 + 160\sqrt{30} \approx 2076.36$ gal

61. $\displaystyle\sum_{k=1}^{n} \dfrac{\pi}{4n} \sec^2\left(\dfrac{\pi k}{4n}\right) = \sum_{k=1}^{n} f(x_k^*)\Delta x$ where $f(x) = \sec^2 x$, $x_k^* = \dfrac{\pi k}{4n}$ and $\Delta x = \dfrac{\pi}{4n}$ for $0 \le x \le \dfrac{\pi}{4}$.

   Thus $\displaystyle\lim_{n \to +\infty} \sum_{k=1}^{n} \dfrac{\pi}{4n} \sec^2\left(\dfrac{\pi k}{4n}\right) = \lim_{n \to +\infty} \sum_{k=1}^{n} f(x_k^*)\Delta x = \int_0^{\pi/4} \sec^2 x \, dx = \tan x \Big]_0^{\pi/4} = 1$

63. Let f be continuous on a closed interval $[a, b]$ and let F be an antiderivative of f on $[a, b]$. By Theorem 4.7.2, $\dfrac{F(b) - F(a)}{b - a} = F'(x^*)$ for some x^* in (a, b). By Theorem 4.6.1,

$\displaystyle\int_a^b f(x) \, dx = F(b) - F(a)$, i.e. $\displaystyle\int_a^b f(x) \, dx = F'(x^*)(b - a) = f(x^*)(b - a)$.

EXERCISE SET 4.7

1. (a) $\text{displ} = s(3) - s(0) = \int_0^3 dt = 3$

    $\text{dist} = \int_0^3 dt = 3$

    (b) $\text{displ} = s(3) - s(0) = -\int_0^3 dt = -3$

    $\text{dist} = \int_0^3 |v(t)|\, dt = 3$

    (c) $\text{displ} = s(3) - s(0)$

    $= \int_0^3 v(t)dt = \int_0^2 (1-t)dt + \int_2^3 (t-3)dt = (t - t^2/2)\Big]_0^2 + (t^2/2 - 3t)\Big]_2^3 = -1/2;$

    $\text{dist} = \int_0^3 |v(t)|dt = (t - t^2/2)\Big]_0^1 + (t^2/2 - t)\Big]_1^2 - (t^2/2 - 3t)\Big]_2^3 = 3/2$

    (d) $\text{displ} = s(3) - s(0)$

    $= \int_0^3 v(t)dt = \int_0^1 tdt + \int_1^2 dt + \int_2^3 (5-2t)dt = t^2/2\Big]_0^1 + t\Big]_1^2 + (5t - t^2)\Big]_2^3 = 3/2;$

    $\text{dist} = \int_0^1 tdt + \int_1^2 dt + \int_2^{5/2} (5-2t)dt + \int_{5/2}^3 (2t-5)dt$

    $= t^2/2\Big]_0^1 + t\Big]_1^2 + (5t - t^2)\Big]_2^{5/2} + (t^2 - 5t)\Big]_{5/2}^3 = 2$

3. (a) $v(t) = 20 + \int_0^t a(u)du;$ add areas of the small blocks to get

    $v(4) \approx 20 + 1.4 + 3.0 + 4.7 + 6.2 = 35.3$ m/s

    (b) $v(6) = v(4) + \int_4^6 a(u)du \approx 35.3 + 7.5 + 8.6 = 51.4$ m/s

5. (a) $s(t) = t^3 - t^2 + C; 1 = s(0) = C$, so $s(t) = t^3 - t^2 + 1$

    (b) $v(t) = -\cos 3t + C_1; 3 = v(0) = -1 + C_1, C_1 = 4$, so $v(t) = -\cos 3t + 4$. Then

    $s(t) = -\frac{1}{3}\sin 3t + 4t + C_2; 3 = s(0) = C_2$, so $s(t) = -\frac{1}{3}\sin 3t + 4t + 3$

7. (a) $s(t) = \frac{3}{2}t^2 + t + C; 4 = s(2) = 6 + 2 + C, C = -4$ and $s(t) = \frac{3}{2}t^2 + t - 4$

    (b) $v(t) = -t^{-1} + C_1, 0 = v(1) = -1 + C_1, C_1 = 1$ and
    $v(t) = -t^{-1} + 1$ so $s(t) = -\ln t + t + C_2, 2 = s(1) = 1 + C_2,$
    $C_2 = 1$ and $s(t) = -\ln t + t + 1$

9. (a) $\text{displacement} = s(\pi/2) - s(0) = \int_0^{\pi/2} \sin tdt = -\cos t\Big]_0^{\pi/2} = 1$ m

    $\text{distance} = \int_0^{\pi/2} |\sin t|dt = 1$ m

    (b) $\text{displacement} = s(2\pi) - s(\pi/2) = \int_{\pi/2}^{2\pi} \cos tdt = \sin t\Big]_{\pi/2}^{2\pi} = -1$ m

    $\text{distance} = \int_{\pi/2}^{2\pi} |\cos t|dt = -\int_{\pi/2}^{3\pi/2} \cos tdt + \int_{3\pi/2}^{2\pi} \cos tdt = 3$ m

11. (a) $v(t) = t^3 - 3t^2 + 2t = t(t-1)(t-2)$

$$\text{displacement} = \int_0^3 (t^3 - 3t^2 + 2t)\,dt = 9/4 \text{ m}$$

$$\text{distance} = \int_0^3 |v(t)|\,dt = \int_0^1 v(t)\,dt + \int_1^2 -v(t)\,dt + \int_2^3 v(t)\,dt = 11/4 \text{ m}$$

(b) $\displaystyle\text{displacement} = \int_0^3 (\sqrt{t} - 2)\,dt = 2\sqrt{3} - 6 \text{ m}$

$$\text{distance} = \int_0^3 |v(t)|\,dt = -\int_0^3 v(t)\,dt = 6 - 2\sqrt{3} \text{ m}$$

13. $v = 3t - 1$

$$\text{displacement} = \int_0^2 (3t-1)\,dt = 4 \text{ m}$$

$$\text{distance} = \int_0^2 |3t-1|\,dt = \frac{13}{3} \text{ m}$$

15. $v = \displaystyle\int (1/\sqrt{3t+1}\,dt = \frac{2}{3}\sqrt{3t+1} + C; v(0) = 4/3$ so $C = 2/3, v = \frac{2}{3}\sqrt{3t+1} + 2/3$

$$\text{displacement} = \int_1^5 \left(\frac{2}{3}\sqrt{3t+1} + \frac{2}{3}\right) dt = \frac{296}{27} \text{ m}$$

$$\text{distance} = \int_1^5 \left(\frac{2}{3}\sqrt{3t+1} + \frac{2}{3}\right) dt = \frac{296}{27} \text{ m}$$

17. (a) $s = \displaystyle\int \sin\frac{1}{2}\pi t\,dt = -\frac{2}{\pi}\cos\frac{1}{2}\pi t + C$

$s = 0$ when $t = 0$ which gives $C = \dfrac{2}{\pi}$ so $s = -\dfrac{2}{\pi}\cos\dfrac{1}{2}\pi t + \dfrac{2}{\pi}$.

$a = \dfrac{dv}{dt} = \dfrac{\pi}{2}\cos\dfrac{1}{2}\pi t$. When $t = 1: s = 2/\pi,\ v = 1,\ |v| = 1,\ a = 0$.

(b) $v = -3\displaystyle\int t\,dt = -\frac{3}{2}t^2 + C_1,\ v = 0$ when $t = 0$ which gives $C_1 = 0$ so $v = -\dfrac{3}{2}t^2$

$s = -\dfrac{3}{2}\displaystyle\int t^2\,dt = -\frac{1}{2}t^3 + C_2,\ s = 1$ when $t = 0$ which gives $C_2 = 1$ so $s = -\dfrac{1}{2}t^3 + 1$.

When $t = 1: s = 1/2,\ v = -3/2,\ |v| = 3/2,\ a = -3$.

19. By inspection the velocity is positive for $t > 0$, and during the first second the particle is at most $5/2$ cm from the starting position. For $T > 1$ the displacement of the particle during the time interval $[0, T]$ is given by

$$\int_0^T v(t)\,dt = 5/2 + \int_1^T (6\sqrt{t} - 1)\,dt = 5/2 + (4t^{3/2} - t)\Big]_1^T = -1/2 + 4T^{3/2} - T,$$

and the displacement equals 4 cm if $T \approx 1.277818837$

21. $s(t) = \displaystyle\int (20t^2 - 110t + 120)\,dt = \frac{20}{3}t^3 - 55t^2 + 120t + C$. But $s = 0$ when $t = 0$, so $C = 0$ and

$s = \dfrac{20}{3}t^3 - 55t^2 + 120t$. Moreover, $a(t) = \dfrac{d}{dt}v(t) = 40t - 110$.

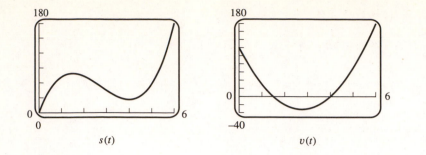

$s(t)$ $v(t)$ $a(t)$

23. true; if $a(t) = a_0$ then $v(t) = a_0 t + v_0$ **25.** false; consider $v(t) = t$ on $[-1, 1]$

27. **(a)** positive on $(0, 0.74)$ and $(2.97, 5)$, negative on $(0.75, 2.97)$

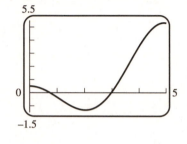

  (b) For $0 < T < 5$ the displacement is
$$\text{disp} = T/2 - \sin(T) + T\cos(T)$$

29. **(a)** $a(t) = \begin{cases} 0, & t < 4 \\ -10, & t > 4 \end{cases}$ **(b)** $v(t) = \begin{cases} 25, & t < 4 \\ 65 - 10t, & t > 4 \end{cases}$

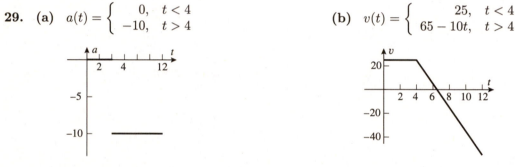

  (c) $x(t) = \begin{cases} 25t, & t < 4 \\ 65t - 5t^2 - 80, & t > 4 \end{cases}$, so $x(8) = 120$, $x(12) = -20$

  (d) $x(6.5) = 131.25$

31. $a = a_0$ ft/s^2, $v = a_0 t + v_0 = a_0 t + 132$ ft/s, $s = a_0 t^2/2 + 132t + s_0 = a_0 t^2/2 + 132t$ ft; $s = 200$ ft
when $v = 88$ ft/s. Solve $88 = a_0 t + 132$ and $200 = a_0 t^2/2 + 132t$ to get $a_0 = -\dfrac{121}{5}$ when $t = \dfrac{20}{11}$,
so $s = -12.1t^2 + 132t$, $v = -\dfrac{121}{5}t + 132$.

  (a) $a_0 = -\dfrac{121}{5}$ ft/s^2 **(b)** $v = 55$ mi/h $= \dfrac{242}{3}$ ft/s when $t = \dfrac{70}{33}$ s

  (c) $v = 0$ when $t = \dfrac{60}{11}$ s

33. The truck's velocity is $v_T = 50$ and its position is $s_T = 50t + 2500$. The car's acceleration is
$a_C = 4$ ft/s^2, so $v_C = 4t$, $s_C = 2t^2$ (initial position and initial velocity of the car are both
zero). $s_T = s_C$ when $50t + 2500 = 2t^2$, $2t^2 - 50t - 2500 = 2(t + 25)(t - 50) = 0$, $t = 50$ s and
$s_C = s_T = 2t^2 = 5000$ ft.

35. $s = 0$ and $v = 112$ when $t = 0$ so $v(t) = -32t + 112$, $s(t) = -16t^2 + 112t$

(a) $v(3) = 16$ ft/s, $v(5) = -48$ ft/s

(b) $v = 0$ when the projectile is at its maximum height so $-32t + 112 = 0$, $t = 7/2$ s, $s(7/2) = -16(7/2)^2 + 112(7/2) = 196$ ft.

(c) $s = 0$ when it reaches the ground so $-16t^2 + 112t = 0$, $-16t(t - 7) = 0$, $t = 0, 7$ of which $t = 7$ is when it is at ground level on its way down. $v(7) = -112$, $|v| = 112$ ft/s.

37. (a) $s(t) = 0$ when it hits the ground, $s(t) = -16t^2 + 16t = -16t(t - 1) = 0$ when $t = 1$ s.

(b) The projectile moves upward until it gets to its highest point where $v(t) = 0$, $v(t) = -32t + 16 = 0$ when $t = 1/2$ s.

39. $s(t) = s_0 + v_0 t - \frac{1}{2}gt^2 = 60t - 4.9t^2$ m and $v(t) = v_0 - gt = 60 - 9.8t$ m/s

(a) $v(t) = 0$ when $t = 60/9.8 \approx 6.12$ s

(b) $s(60/9.8) \approx 183.67$ m

(c) another 6.12 s; solve for t in $s(t) = 0$ to get this result, or use the symmetry of the parabola $s = 60t - 4.9t^2$ about the line $t = 6.12$ in the t-s plane

(d) also 60 m/s, as seen from the symmetry of the parabola (or compute $v(6.12)$)

41. $s(t) = -4.9t^2 + 49t + 150$ and $v(t) = -9.8t + 49$

(a) the projectile reaches its maximum height when $v(t) = 0$, $-9.8t + 49 = 0$, $t = 5$ s

(b) $s(5) = -4.9(5)^2 + 49(5) + 150 = 272.5$ m

(c) the projectile reaches its starting point when $s(t) = 150$, $-4.9t^2 + 49t + 150 = 150$, $-4.9t(t - 10) = 0$, $t = 10$ s

(d) $v(10) = -9.8(10) + 49 = -49$ m/s

(e) $s(t) = 0$ when the projectile hits the ground, $-4.9t^2 + 49t + 150 = 0$ when (use the quadratic formula) $t \approx 12.46$ s

(f) $v(12.46) = -9.8(12.46) + 49 \approx -73.1$, the speed at impact is about 73.1 m/s

43. If $g = 32$ ft/s^2, $s_0 = 7$ and v_0 is unknown, then $s(t) = 7 + v_0 t - 16t^2$ and $v(t) = v_0 - 32t$; $s = s_{max}$ when $v = 0$, or $t = v_0/32$; and $s_{max} = 208$ yields
$208 = s(v_0/32) = 7 + v_0(v_0/32) - 16(v_0/32)^2 = 7 + v_0^2/64$, so $v_0 = 8\sqrt{201} \approx 113.42$ ft/s.

EXERCISE SET 4.8

1. (a) $f_{ave} = \dfrac{1}{4 - 0} \displaystyle\int_0^4 2x\, dx = 4$ (b) $2x^* = 4$, $x^* = 2$

(c)

3. $f_{ave} = \dfrac{1}{3 - 1} \displaystyle\int_1^3 3x\, dx = \dfrac{3}{4}x^2 \Big]_1^3 = 6$ **5.** $f_{ave} = \dfrac{1}{\pi} \displaystyle\int_0^\pi \sin x\, dx = -\dfrac{1}{\pi}\cos x \Big]_0^\pi = \dfrac{2}{\pi}$

7. $f_{\text{ave}} = \dfrac{1}{2}\displaystyle\int_0^2 \dfrac{x}{(5x^2+1)^2}\,dx = -\dfrac{1}{10}(5x^2+1)^{-1}\Big]_0^2 = \dfrac{1}{21}$

9. **(a)** $\frac{1}{5}[f(0.4)+f(0.8)+f(1.2)+f(1.6)+f(2.0)] = \frac{1}{5}[0.48+1.92+4.32+7.68+12.00] = 5.28$

(b) $\dfrac{1}{20}3[(0.1)^2+(0.2)^2+\ldots+(1.9)^2+(2.0)^2] = \dfrac{861}{200} = 4.305$

(c) $f_{\text{ave}} = \dfrac{1}{2}\displaystyle\int_0^2 3x^2\,dx = \dfrac{1}{2}x^3\Big]_0^2 = 4$

(d) Parts (a) and (b) can be interpreted as being two Riemann sums (n = 5, n = 20) for the average, using right endpoints. Since f is increasing, these sums overestimate the integral.

11. **(a)** $\displaystyle\int_0^3 v(t)\,dt = \int_0^2(1-t)\,dt + \int_2^3(t-3)\,dt = -\dfrac{1}{2}$, so $v_{\text{ave}} = -\dfrac{1}{6}$

(b) $\displaystyle\int_0^3 v(t)\,dt = \int_0^1 t\,dt + \int_1^2 dt + \int_2^3(-2t+5)\,dt = \dfrac{1}{2}+1+0 = \dfrac{3}{2}$, so $v_{\text{ave}} = \dfrac{1}{2}$

13. Linear means $f(\alpha x_1 + \beta x_2) = \alpha f(x_1) + \beta f(x_2)$, so $f\left(\dfrac{a+b}{2}\right) = \dfrac{1}{2}f(a) + \dfrac{1}{2}f(b) = \dfrac{f(a)+f(b)}{2}$.

15. false; $f(x) = x, g(x) = 0$ on $[-1,2]$ **17.** true; Theorem 4.5.4(b)

19. **(a)** $v_{\text{ave}} = \dfrac{1}{4-1}\displaystyle\int_1^4 (3t^3+2)\,dt = \dfrac{1}{3}\dfrac{789}{4} = \dfrac{263}{4}$

(b) $v_{\text{ave}} = \dfrac{s(4)-s(1)}{4-1} = \dfrac{100-7}{3} = 31$

21. time to fill tank = (volume of tank)/(rate of filling) = $[\pi(3)^2 5]/(1) = 45\pi$, weight of water in tank at time $t = (62.4)$ (rate of filling)(time) = $62.4t$,

$\text{weight}_{\text{ave}} = \dfrac{1}{45\pi}\displaystyle\int_0^{45\pi} 62.4t\,dt = 1404\pi = 4410.8$ lb

23. $\displaystyle\int_0^{30} 100(1-0.0001t^2)\,dt = 2910$ cars, so an average of $\dfrac{2910}{30} = 97$ cars/min.

25. From the chart we read

$\dfrac{dV}{dt} = f(t) = \begin{cases} 40t, & 0 \le t \le 1 \\ 40, & 1 \le t \le 3 \\ -20t+100, & 3 \le t \le 5 \end{cases}$

It follows that (constants of integration are chosen to ensure that $V(0) = 0$ and that $V(t)$ is continuous)

$V(t) = \begin{cases} 20t^2, & 0 \le t \le 1 \\ 40t-20, & 1 \le t \le 3 \\ -10t^2+100t-110, & 3 \le t \le 5 \end{cases}$

Now the average rate of change of the volume of juice in the glass during these 5 seconds refers to the quantity $\frac{1}{5}(V(5)-V(0)) = \frac{1}{5}140 = 28$, and the average value of the flow rate is

$f_{\text{ave}} = \frac{1}{5}\displaystyle\int_0^1 f(t)\,dt = \frac{1}{5}\left[\int_0^1 40t\,dt + \int_1^3 40\,dt + \int_3^5(-20t+100)\,dt\right]$

$= \dfrac{1}{5}[20+80-160+200] = 28.$

27. Solve for k : $\int_0^k \sqrt{3x}\,dx = 6k$, so $\sqrt{3}\dfrac{2}{3}x^{3/2}\Big]_0^k = \dfrac{2}{3}\sqrt{3}k^{3/2} = 6k$, $k = (3\sqrt{3})^2 = 27$

EXERCISE SET 4.9

1. **(a)** $\dfrac{1}{2}\displaystyle\int_1^5 u^3\,du$ **(b)** $\dfrac{3}{2}\displaystyle\int_9^{25} \sqrt{u}\,du$

    (c) $\dfrac{1}{\pi}\displaystyle\int_{-\pi/2}^{\pi/2} \cos u\,du$ **(d)** $\displaystyle\int_1^2 (u+1)u^5\,du$

3. $u = 2x + 1$, $\dfrac{1}{2}\displaystyle\int_1^3 u^3\,du = \dfrac{1}{8}u^4\Big]_1^3 = 10$ or $\dfrac{1}{8}(2x+1)^4\Big]_0^1 = 10$

5. $u = 2x - 1$, $\dfrac{1}{2}\displaystyle\int_{-1}^1 u^3\,du = 0$, because u^3 is odd on $[-1, 1]$.

7. $u = 1 + x$, $\displaystyle\int_1^9 (u-1)u^{1/2}\,du = \int_1^9 (u^{3/2} - u^{1/2})\,du = \dfrac{2}{5}u^{5/2} - \dfrac{2}{3}u^{3/2}\Big]_1^9 = 1192/15$,

    or $\dfrac{2}{5}(1+x)^{5/2} - \dfrac{2}{3}(1+x)^{3/2}\Big]_0^8 = 1192/15$

9. $u = x/2$, $8\displaystyle\int_0^{\pi/4} \sin u\,du = -8\cos u\Big]_0^{\pi/4} = 8 - 4\sqrt{2}$, or $-8\cos(x/2)\Big]_0^{\pi/2} = 8 - 4\sqrt{2}$

11. $u = x^2 + 2$, $\dfrac{1}{2}\displaystyle\int_6^3 u^{-3}\,du = -\dfrac{1}{4u^2}\Big]_6^3 = -1/48$, or $-\dfrac{1}{4}\dfrac{1}{(x^2+2)^2}\Big]_{-2}^{-1} = -1/48$

13. $\dfrac{1}{3}\displaystyle\int_{-5}^5 \sqrt{25 - u^2}\,du = \dfrac{1}{3}\left[\dfrac{1}{2}\pi(5)^2\right] = \dfrac{25}{6}\pi$

15. $-\dfrac{1}{2}\displaystyle\int_1^0 \sqrt{1 - u^2}\,du = \dfrac{1}{2}\int_0^1 \sqrt{1 - u^2}\,du = \dfrac{1}{2}\cdot\dfrac{1}{4}[\pi(1)^2] = \pi/8$

17. $\displaystyle\int_0^1 \sin\pi t\,dt = -\dfrac{1}{\pi}\cos\pi t\Big]_0^1 = -\dfrac{1}{\pi}(-1 - 1) = 2/\pi$ m

19. $\displaystyle\int_{-1}^1 \dfrac{9}{(x+2)^2}\,dx = -9(x+2)^{-1}\Big]_{-1}^1 = -9\left[\dfrac{1}{3} - 1\right] = 6$

21. $u = 2x - 1$, $\dfrac{1}{2}\displaystyle\int_1^9 \dfrac{1}{\sqrt{u}}\,du = \sqrt{u}\Big]_1^9 = 2$ **23.** $\dfrac{2}{3}(x^3 + 9)^{1/2}\Big]_{-1}^1 = \dfrac{2}{3}(\sqrt{10} - 2\sqrt{2})$

25. $u = x^2 + 4x + 7$, $\dfrac{1}{2}\displaystyle\int_{12}^{28} u^{-1/2}\,du = u^{1/2}\Big]_{12}^{28} = \sqrt{28} - \sqrt{12} = 2(\sqrt{7} - \sqrt{3})$

27. $2\sin^2 x\big]_0^{\pi/4} = 1$

29. $\dfrac{5}{2}\sin(x^2)\Big]_0^{\sqrt{\pi}} = 0$

31. $u = 3\theta,\ \dfrac{1}{3}\displaystyle\int_{\pi/4}^{\pi/3}\sec^2 u\,du = \dfrac{1}{3}\tan u\Big]_{\pi/4}^{\pi/3} = (\sqrt{3}-1)/3$

33. $u = 4 - 3y,\ y = \dfrac{1}{3}(4-u),\ dy = -\dfrac{1}{3}du$

$$-\dfrac{1}{27}\int_4^1 \dfrac{16 - 8u + u^2}{u^{1/2}}du = \dfrac{1}{27}\int_1^4 (16u^{-1/2} - 8u^{1/2} + u^{3/2})du$$

$$= \dfrac{1}{27}\left[32u^{1/2} - \dfrac{16}{3}u^{3/2} + \dfrac{2}{5}u^{5/2}\right]_1^4 = 106/405$$

35. (b) $\displaystyle\int_0^{\pi/6}\sin^4 x(1 - \sin^2 x)\cos x\,dx = \left(\dfrac{1}{5}\sin^5 x - \dfrac{1}{7}\sin^7 x\right)\Big|_0^{\pi/6} = \dfrac{1}{160} - \dfrac{1}{896} = \dfrac{23}{4480}$

37. (a) $u = 3x + 1,\ \dfrac{1}{3}\displaystyle\int_1^4 f(u)du = 5/3$ **(b)** $u = 3x,\ \dfrac{1}{3}\displaystyle\int_0^9 f(u)du = 5/3$

  (c) $u = x^2,\ 1/2\displaystyle\int_4^0 f(u)du = -1/2\displaystyle\int_0^4 f(u)du = -1/2$

39. $\sin x = \cos(\pi/2 - x)$,

$$\int_0^{\pi/2}\sin^n x\,dx = \int_0^{\pi/2}\cos^n(\pi/2 - x)dx = -\int_{\pi/2}^0 \cos^n u\,du \quad (u = \pi/2 - x)$$

$$= \int_0^{\pi/2}\cos^n u\,du = \int_0^{\pi/2}\cos^n x\,dx \quad \text{(by replacing } u \text{ by } x)$$

41. (a) $\dfrac{1}{7}[0.74 + 0.65 + 0.56 + 0.45 + 0.35 + 0.25 + 0.16] = 0.4514285714$

  (b) $\dfrac{1}{7}\displaystyle\int_0^7 [0.5 + 0.5\sin(0.213x + 2.481)\,dx = 0.4614$

43. (a) $\displaystyle\int_0^1 \sin\pi x\,dx = 2/\pi$

45. (a) Let $u = -x$ then

$$\int_{-a}^a f(x)dx = -\int_a^{-a}f(-u)du = \int_{-a}^a f(-u)du = -\int_{-a}^a f(u)du$$

so, replacing u by x in the latter integral,

$$\int_{-a}^a f(x)dx = -\int_{-a}^a f(x)dx,\ 2\int_{-a}^a f(x)dx = 0,\ \int_{-a}^a f(x)dx = 0$$

The graph of f is symmetric about the origin so $\displaystyle\int_{-a}^0 f(x)dx$ is the negative of $\displaystyle\int_0^a f(x)dx$

thus $\displaystyle\int_{-a}^a f(x)dx = \int_{-a}^0 f(x)\,dx + \int_0^a f(x)dx = 0$

(b) $\displaystyle\int_{-a}^{a} f(x)dx = \int_{-a}^{0} f(x)dx + \int_{0}^{a} f(x)dx$, let $u = -x$ in $\displaystyle\int_{-a}^{0} f(x)dx$ to get

$$\int_{-a}^{0} f(x)dx = -\int_{a}^{0} f(-u)du = \int_{0}^{a} f(-u)du = \int_{0}^{a} f(u)du = \int_{0}^{a} f(x)dx$$

so $\displaystyle\int_{-a}^{a} f(x)dx = \int_{0}^{a} f(x)dx + \int_{0}^{a} f(x)dx = 2\int_{0}^{a} f(x)dx$

The graph of $f(x)$ is symmetric about the y-axis so there is as much signed area to the left of the y-axis as there is to the right.

47. (a) $\displaystyle I = -\int_{a}^{0} \frac{f(a-u)}{f(a-u)+f(u)} du = \int_{0}^{a} \frac{f(a-u)+f(u)-f(u)}{f(a-u)+f(u)} du$

$\displaystyle = \int_{0}^{a} du - \int_{0}^{a} \frac{f(u)}{f(a-u)+f(u)} du, I = a - I$ so $2I = a, I = a/2$

(b) $3/2$ **(c)** $\pi/4$

REVIEW EXERCISES, CHAPTER 4

1. $\displaystyle -\frac{1}{4x^2} + \frac{8}{3}x^{3/2} + C$ **3.** $-4\cos x + 2\sin x + C$

5. (a) $\displaystyle y(x) = 2\sqrt{x} - \frac{2}{3}x^{3/2} + C; y(1) = 0$, so $C = -\frac{4}{3}, y(x) = 2\sqrt{x} - \frac{2}{3}x^{3/2} - \frac{4}{3}$

(b) $\displaystyle y(x) = \sin x - \frac{5}{2}x^2 + C, y(0) = 1 = C, y(x) = \sin x - \frac{5}{2}x^2 + 1$

(c) $\displaystyle y(x) = 2 + \int_{1}^{x} t^{1/3} dt = 2 + \frac{3}{4}t^{4/3}\Big]_{1}^{x} = \frac{5}{4} + \frac{3}{4}x^{4/3}$

(d) $\displaystyle y(x) = \int_{0}^{x} te^{t^2} dt = \frac{1}{2}e^{x^2} - \frac{1}{2}$

7. (a) If $u = \sec x, du = \sec x \tan x dx$, $\displaystyle\int \sec^2 x \tan x dx = \int u du = u^2/2 + C_1 = (\sec^2 x)/2 + C_1$;

if $u = \tan x, du = \sec^2 x dx$, $\displaystyle\int \sec^2 x \tan x dx = \int u du = u^2/2 + C_2 = (\tan^2 x)/2 + C_2$.

(b) They are equal only if $\sec^2 x$ and $\tan^2 x$ differ by a constant, which is true.

9. $\displaystyle u = x^4 + 2, \frac{1}{4}\int\left(\sqrt{u} - \frac{2}{\sqrt{u}}\right), du = \frac{1}{4}\frac{2}{3}u^{3/2} - u^{1/2} + C = \frac{1}{6}(x^4+2)^{3/2} - \sqrt{x^4+2} + C$

11. $\displaystyle u = 5 + 2\sin 3x, du = 6\cos 3x dx; \int \frac{1}{6\sqrt{u}} du = \frac{1}{3}u^{1/2} + C = \frac{1}{3}\sqrt{5 + 2\sin 3x} + C$

13. $\displaystyle u = ax^3 + b, du = 3ax^2 dx; \int \frac{1}{3au^2} du = -\frac{1}{3au} + C = -\frac{1}{3a^2x^3 + 3ab} + C$

15. (a) $\displaystyle\sum_{k=0}^{14}(k+4)(k+1)$ **(b)** $\displaystyle\sum_{k=5}^{19}(k-1)(k-4)$

17. $\displaystyle\lim_{n\to+\infty}\sum_{k=1}^{n}\left[4\frac{4k}{n}-\left(\frac{4k}{n}\right)^2\right]\frac{4}{n}=\lim_{n\to+\infty}\frac{64}{n^3}\sum_{k=1}^{n}(kn-k^2)$

$\displaystyle=\lim_{n\to+\infty}\frac{64}{n^3}\left[\frac{n^2(n+1)}{2}-\frac{n(n+1)(2n+1)}{6}\right]=\lim_{n\to+\infty}\frac{64}{6n^3}[n^3-n]=\frac{32}{3}$

19. $0.7187714032, 0.6687714032, 0.6928353604$

23. (a) $\dfrac{1}{2}+\dfrac{1}{4}=\dfrac{3}{4}$
(b) $-1-\dfrac{1}{2}=-\dfrac{3}{2}$

(c) $5\left(-1-\dfrac{3}{4}\right)=-\dfrac{35}{4}$
(d) -2

(e) not enough information
(f) not enough information

25. (a) $\displaystyle\int_{-1}^{1}dx+\int_{-1}^{1}\sqrt{1-x^2}\,dx=2(1)+\pi(1)^2/2=2+\pi/2$

(b) $\dfrac{1}{3}(x^2+1)^{3/2}\Big]_{0}^{3}-\pi(3)^2/4=\dfrac{1}{3}(10^{3/2}-1)-9\pi/4$

(c) $u=x^2,\,du=2xdx;\ \dfrac{1}{2}\displaystyle\int_{0}^{1}\sqrt{1-u^2}\,du=\dfrac{1}{2}\pi(1)^2/4=\pi/8$

27. $\left(\dfrac{1}{3}x^3-2x^2+7x\right)\Big]_{-3}^{0}=48$
29. $\displaystyle\int_{1}^{3}x^{-2}dx=-\dfrac{1}{x}\Big]_{1}^{3}=2/3$

31. $\left(\dfrac{1}{2}x^2-\sec x\right)\Big]_{0}^{1}=3/2-\sec(1)$

33. $\displaystyle\int_{0}^{3/2}(3-2x)dx+\int_{3/2}^{2}(2x-3)dx=(3x-x^2)\Big]_{0}^{3/2}+(x^2-3x)\Big]_{3/2}^{2}=9/4+1/4=5/2$

35. $\displaystyle\int_{1}^{9}\sqrt{x}\,dx=\dfrac{2}{3}x^{3/2}\Big]_{1}^{9}=\dfrac{2}{3}(27-1)=52/3$

37. $A=\displaystyle\int_{1}^{2}(-x^2+3x-2)dx=\left(-\dfrac{1}{3}x^3+\dfrac{3}{2}x^2-2x\right)\Big]_{1}^{2}=1/6$

39. $A=A_1+A_2=\displaystyle\int_{0}^{1}(1-x^2)dx+\int_{1}^{3}(x^2-1)dx=2/3+20/3=22/3$

41. (a) x^3+1
(b) $F(x)=\left(\dfrac{1}{4}t^4+t\right)\Big]_{1}^{x}=\dfrac{1}{4}x^4+x-\dfrac{5}{4};\ F'(x)=x^3+1$

43. $\dfrac{1}{x^4+5}$
45. $|x-1|$

49. $F'(x)=\dfrac{1}{1+x^2}+\dfrac{1}{1+(1/x)^2}(-1/x^2)=0$ so F is constant on $(0,+\infty)$.

51. (a) The domain is $(-\infty, +\infty)$; $F(x)$ is 0 if $x = 1$, positive if $x > 1$, and negative if $x < 1$, because the integrand is positive, so the sign of the integral depends on the orientation (forwards or backwards).

(b) The domain is $[-2, 2]$; $F(x)$ is 0 if $x = -1$, positive if $-1 < x \leq 2$, and negative if $-2 \leq x < -1$; same reasons as in Part (a).

53. (a) $f_{ave} = \dfrac{1}{3}\displaystyle\int_0^3 x^{1/2}dx = 2\sqrt{3}/3; \sqrt{x^*} = 2\sqrt{3}/3, x^* = \dfrac{4}{3}$

(b) $f_{ave} = \dfrac{1}{2}\displaystyle\int_0^2 (2x - x^2) = \dfrac{1}{2}\left(x^2 - \dfrac{1}{3}x^3\right)\Big]_0^2 = \dfrac{2}{3}$

55. For $0 < x < 3$ the area between the curve and the x-axis consists of two triangles of equal area but of opposite signs, hence 0. For $3 < x < 5$ the area is a rectangle of width 2 and height 3. For $5 < x < 7$ the area consists of two triangles of equal area but opposite sign, hence 0; and for $7 < x < 10$ the curve is given by $y = (4t - 37)/3$ and $\displaystyle\int_7^{10}(4t - 37)/3\, dt = -3$. Thus the desired average is $\dfrac{1}{10}(0 + 6 + 0 - 3) = 0.3$.

57. If the acceleration $a = $ const, then $v(t) = at + v_0$, $s(t) = \frac{1}{2}at^2 + v_0 t + s_0$.

59. $s(t) = \displaystyle\int(t^3 - 2t^2 + 1)dt = \dfrac{1}{4}t^4 - \dfrac{2}{3}t^3 + t + C,$

$s(0) = \dfrac{1}{4}(0)^4 - \dfrac{2}{3}(0)^3 + 0 + C = 1, C = 1, s(t) = \dfrac{1}{4}t^4 - \dfrac{2}{3}t^3 + t + 1$

61. $s(t) = \displaystyle\int(2t - 3)dt = t^2 - 3t + C, s(1) = (1)^2 - 3(1) + C = 5, C = 7, s(t) = t^2 - 3t + 7$

63. displacement $= s(6) - s(0) = \displaystyle\int_0^6 (2t - 4)dt = (t^2 - 4t)\Big]_0^6 = 12$ m

distance $= \displaystyle\int_0^6 |2t - 4|dt = \int_0^2 (4 - 2t)dt + \int_2^6 (2t - 4)dt = (4t - t^2)\Big]_0^2 + (t^2 - 4t)\Big]_2^6 = 20$ m

65. displacement $= \displaystyle\int_1^3 \left(\dfrac{1}{2} - \dfrac{1}{t^2}\right)dt = 1/3$ m

distance $= \displaystyle\int_1^3 |v(t)|dt = -\int_1^{\sqrt{2}} v(t)dt + \int_{\sqrt{2}}^3 v(t)dt = 10/3 - 2\sqrt{2}$ m

67. $v(t) = -2t + 3$

displacement $= \displaystyle\int_1^4 (-2t + 3)dt = -6$ m

distance $= \displaystyle\int_1^4 |-2t + 3|dt = \int_1^{3/2}(-2t + 3)dt + \int_{3/2}^4 (2t - 3)dt = 13/2$ m

69. Take $t = 0$ when deceleration begins, then $a = -10$ so $v = -10t + C_1$, but $v = 88$ when $t = 0$ which gives $C_1 = 88$ thus $v = -10t + 88, t \geq 0$

(a) $v = 45$ mi/h $= 66$ ft/s, $66 = -10t + 88, t = 2.2$ s

(b) $v = 0$ (the car is stopped) when $t = 8.8$ s

$$s = \int v\, dt = \int (-10t + 88)dt = -5t^2 + 88t + C_2, \text{ and taking } s = 0 \text{ when } t = 0, C_2 = 0 \text{ so}$$

$s = -5t^2 + 88t$. At $t = 8.8$, $s = 387.2$. The car travels 387.2 ft before coming to a stop.

71. From the free-fall model $s = -\frac{1}{2}gt^2 + v_0t + s_0$ the ball is caught when $s_0 = -\frac{1}{2}gt_1^2 + v_0t_1 + s_0$ with the positive root $t_1 = 2v_0/g$ so the average speed of the ball while it is up in the air is average

$$\text{speed} = \frac{1}{t_1}\int_0^{t_1}|v_0 - gt|\, dt = \frac{g}{2v_0}\left[\int_0^{v_0/g}(v_0 - gt)\,gt + \int_{v_0/g}^{2v_0/g}(gt - v_0)\, dt\right] = v_0/2.$$

73. $u = 2x + 1$, $\dfrac{1}{2}\displaystyle\int_1^3 u^4 du = \dfrac{1}{10}u^5\Big]_1^3 = 121/5$, or $\dfrac{1}{10}(2x+1)^5\Big]_0^1 = 121/5$

75. $\dfrac{2}{3}(3x+1)^{1/2}\Big]_0^1 = 2/3$ **77.** $\dfrac{1}{3\pi}\sin^3 \pi x\Big]_0^1 = 0$

MAKING CONNECTIONS, CHAPTER 4

1. (a) $\displaystyle\sum_{k=1}^{n} 2x_k^*\Delta x_k = \sum_{k=1}^{n}(x_k + x_{k-1})(x_k - x_{k-1}) = \sum_{k=1}^{n}(x_k^2 - x_{k-1}^2) = \sum_{k=1}^{n}x_k^2 - \sum_{k=0}^{n-1}x_k^2 = b^2 - a^2$

(b) By Theorem 4.5.2, f is integrable on $[a,b]$. Using Part (a) of Definition 4.5.1, in which we choose any partition and use the midpoints $x_k^* = (x_k + x_{k-1})/2$, we see from Part (a) of this exercise that the Riemann sum is equal to $x_n^2 - x_0^2 = b^2 - a^2$. Since the right side of this equation does not depend on partitions, the limit of the Riemann sums as $\max(\Delta x_k) \to 0$ is equal to $b^2 - a^2$.

3. Use the partition $0 < 8(1)^3/n^3 < 8(2)^3/n^3 < \ldots < 8(n-1)^3/n^3 < 8$ with x_k^* as the right endpoint of the k-th interval, $x_k^* = 8k^3/n^3$. Then

$$\sum_{k=1}^{n} f(x_k^*)\Delta x_k = \sum_{k=1}^{n}\sqrt[3]{8k^3/n^3}\left(\frac{8k^3}{n^3} - \frac{8(k-1)^3}{n^3}\right) = \sum_{k=1}^{n}\frac{16}{n^4}(k^4 - k(k-1)^3)$$

$$= \frac{16}{n^4}\frac{3n^4 + 2n^3 - n^2}{4} \to 16\frac{3}{4} = 12 \text{ as } n \to \infty.$$

5. (a) $\displaystyle\sum_{k=1}^{n} g(x_k^*)\Delta x_k = \sum_{k=1}^{n} 2x_k^* f((x_k^*)^2)\Delta x_k = \sum_{k=1}^{n}(x_k + x_{k-1})f((x_k^*)^2)(x_k - x_{k-1})$

$$= \sum_{k=1}^{n} f((x_k^*)^2)(x_k^2 - x_{k-1}^2) = \sum_{k=1}^{n} f(u_k^*)\Delta u_k. \text{ The two Riemann sums are equal.}$$

(b) In Part (a) note that $\Delta u_k = \Delta x_k^2 = x_k^2 - x_{k-1}^2 = (x_k + x_{k-1})\Delta x_k$, and since $2 \le x_k \le 3$, $4\Delta x_k \le \Delta u_k$ and $\Delta u_k \le 6\Delta x_k$, so that $\max\{u_k\}$ tends to zero iff $\max\{x_k\}$ tends to zero.

$$\int_2^3 g(x)\, dx = \lim_{\max(\Delta x_k) \to 0}\sum_{k=1}^{n} g(x_k^*)\Delta x_k = \lim_{\max(\Delta u_k) \to 0}\sum_{k=1}^{n} f(u_k^*)\Delta u_k = \int_4^9 f(u)\, du$$

(c) Since the symbol g is already in use, we shall use γ to denote the mapping $u = \gamma(x) = x^2$ of Theorem 4.9.1. Applying the Theorem,

$$\int_4^9 f(u)\, du = \int_2^3 f(\gamma(x))\gamma'(x)\, dx = \int_2^3 f(x^2)2x\, dx = \int_2^3 g(x)\, dx.$$

CHAPTER 5
Applications of the Definite Integral in Geometry, Science, and Engineering

EXERCISE SET 5.1

1. $A = \int_{-1}^{2} (x^2 + 1 - x)\, dx = (x^3/3 + x - x^2/2) \Big]_{-1}^{2} = 9/2$

3. $A = \int_{1}^{2} (y - 1/y^2)\, dy = (y^2/2 + 1/y) \Big]_{1}^{2} = 1$

5. (a) $A = \int_{0}^{2} (2x - x^2)\, dx = 4/3$

    (b) $A = \int_{0}^{4} (\sqrt{y} - y/2)\, dy = 4/3$

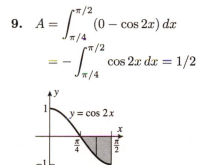

7. $A = \int_{1/4}^{1} (\sqrt{x} - x^2)\, dx = 49/192$

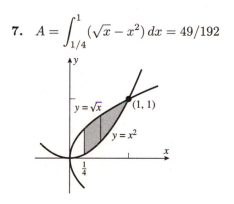

9. $A = \int_{\pi/4}^{\pi/2} (0 - \cos 2x)\, dx$

$\qquad = -\int_{\pi/4}^{\pi/2} \cos 2x\, dx = 1/2$

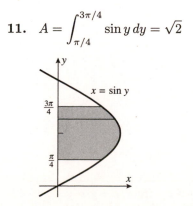

11. $A = \int_{\pi/4}^{3\pi/4} \sin y\, dy = \sqrt{2}$

119

13. $y = 2 + |x - 1| = \begin{cases} 3 - x, & x \le 1 \\ 1 + x, & x \ge 1 \end{cases}$,

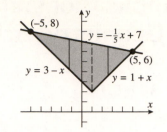

$$A = \int_{-5}^{1} \left[\left(-\frac{1}{5}x + 7 \right) - (3 - x) \right] dx$$

$$+ \int_{1}^{5} \left[\left(-\frac{1}{5}x + 7 \right) - (1 + x) \right] dx$$

$$= \int_{-5}^{1} \left(\frac{4}{5}x + 4 \right) dx + \int_{1}^{5} \left(6 - \frac{6}{5}x \right) dx$$

$$= 72/5 + 48/5 = 24$$

15. $A = \int_{0}^{1} (x^3 - 4x^2 + 3x)\, dx$

$$+ \int_{1}^{3} [-(x^3 - 4x^2 + 3x)]\, dx$$

$$= 5/12 + 32/12 = 37/12$$

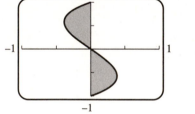

17. From the symmetry of the region

$$A = 2 \int_{\pi/4}^{5\pi/4} (\sin x - \cos x)\, dx = 4\sqrt{2}$$

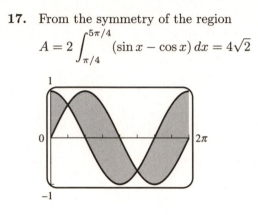

19. $A = \int_{-1}^{0} (y^3 - y)\, dy + \int_{0}^{1} -(y^3 - y)\, dy$

$$= 1/2$$

21. True. If $f(x) - g(x) = c > 0$ then $f(x) > g(x)$ so Formula (1) implies that

$$A = \int_{a}^{b} [f(x) - g(x)]\, dx = \int_{a}^{b} c\, dx = c(b - a).$$

If $g(x) - f(x) = c > 0$ then $g(x) > f(x)$ so

$$A = \int_{a}^{b} [g(x) - f(x)]\, dx = \int_{a}^{b} c\, dx = c(b - a).$$

23. True. Since f and g are distinct, there is some point c in $[a, b]$ for which $f(c) \ne g(c)$. Suppose $f(c) > g(c)$. (The case $f(c) < g(c)$ is similar.) Let $p = f(c) - g(c) > 0$. Since $f - g$ is continuous, there is an interval $[d, e]$ containing c such that $f(x) - g(x) > p/2$ for all x in $[d, e]$. So

$\int_d^e [f(x) - g(x)]\, dx \geq \dfrac{p}{2}(e - d) > 0.$ Hence

$$0 = \int_a^b [f(x) - g(x)]\, dx = \int_a^d [f(x) - g(x)]\, dx + \int_d^e [f(x) - g(x)]\, dx + \int_e^b [f(x) - g(x)]\, dx$$

$$> \int_a^d [f(x) - g(x)]\, dx + \int_b^e [f(x) - g(x)]\, dx,$$

so at least one of $\int_a^d [f(x) - g(x)]\, dx$ and $\int_b^e [f(x) - g(x)]\, dx$ is negative. Therefore $f(t) - g(t) < 0$ for some point t in one of the intervals $[a, d]$ and $[b, e]$. So the graph of f is above the graph of g at $x = c$ and below it at $x = t$; by the Intermediate-Value Theorem, the curves cross somewhere between c and t.

(Note: It is not necessarily true that the curves cross at a point. For example, let

$$f(x) = \begin{cases} x & \text{if } x < 0; \\ 0 & \text{if } 0 \leq x \leq 1; \\ x - 1 & \text{if } x > 1, \end{cases}$$

and $g(x) = 0$. Then $\int_{-1}^2 [f(x) - g(x)]\, dx = 0$, and the curves cross between -1 and 2, but there's no single point at which they cross; they coincide for x in $[0, 1]$.)

25. Solve $3 - 2x = x^6 + 2x^5 - 3x^4 + x^2$ to find the real roots $x = -3, 1$; from a plot it is seen that the line is above the polynomial when $-3 < x < 1$, so $A = \int_{-3}^1 (3 - 2x - (x^6 + 2x^5 - 3x^4 + x^2))\, dx = 9152/105$

27. $\int_0^k 2\sqrt{y}\, dy = \int_k^9 2\sqrt{y}\, dy$

$\int_0^k y^{1/2}\, dy = \int_k^9 y^{1/2}\, dy$

$\dfrac{2}{3} k^{3/2} = \dfrac{2}{3}(27 - k^{3/2})$

$k^{3/2} = 27/2$

$k = (27/2)^{2/3} = 9/\sqrt[3]{4}$

29. (a) $A = \int_0^2 (2x - x^2)\, dx = 4/3$

(b) $y = mx$ intersects $y = 2x - x^2$ where $mx = 2x - x^2$, $x^2 + (m - 2)x = 0$, $x(x + m - 2) = 0$ so $x = 0$ or $x = 2 - m$. The area below the curve and above the line is

$$\int_0^{2-m} (2x - x^2 - mx)\, dx = \int_0^{2-m} [(2 - m)x - x^2]\, dx = \left[\dfrac{1}{2}(2 - m)x^2 - \dfrac{1}{3}x^3\right]_0^{2-m} = \dfrac{1}{6}(2 - m)^3$$

so $(2 - m)^3/6 = (1/2)(4/3) = 2/3$, $(2 - m)^3 = 4$, $m = 2 - \sqrt[3]{4}$.

31. The curves intersect at $x = 0$ and, by Newton's Method, at $x \approx 2.595739080 = b$, so

$$A \approx \int_0^b (\sin x - 0.2x)\, dx = -\left[\cos x + 0.1x^2\right]_0^b \approx 1.180898334$$

33. The x-coordinates of the points of intersection are $a \approx -0.423028$ and $b \approx 1.725171$; the area is

$$\int_a^b (2\sin x - x^2 + 1)\, dx \approx 2.542696.$$

35. $\int_0^{60} [v_2(t) - v_1(t)]\,dt = s_2(60) - s_2(0) - [s_1(60) - s_1(0)]$, but they are even at time $t = 60$, so $s_2(60) = s_1(60)$. Consequently the integral gives the difference $s_1(0) - s_2(0)$ of their starting points in meters.

37. **(a)** It gives the area of the region that is between f and g when $f(x) > g(x)$ <u>minus</u> the area of the region between f and g when $f(x) < g(x)$, for $a \le x \le b$.

(b) It gives the area of the region that is between f and g for $a \le x \le b$.

39. Solve $x^{1/2} + y^{1/2} = a^{1/2}$ for y to get

$$y = (a^{1/2} - x^{1/2})^2 = a - 2a^{1/2}x^{1/2} + x$$

$$A = \int_0^a (a - 2a^{1/2}x^{1/2} + x)\,dx = a^2/6$$

41. First find all solutions of the equation $f(x) = g(x)$ in the interval $[a, b]$; call them $c_1, \cdots, c_n$. Let $c_0 = a$ and $c_{n+1} = b$. For $i = 0, 1, \cdots, n$, $f(x) - g(x)$ has constant sign on $[c_i, c_{i+1}]$, so the area bounded by $x = c_i$ and $x = c_{i+1}$ is either $\int_{c_i}^{c_{i+1}} [f(x) - g(x)]\,dx$ or $\int_{c_i}^{c_{i+1}} [g(x) - f(x)]\,dx$. Compute each of these $n + 1$ areas and add them to get the area bounded by $x = a$ and $x = b$.

EXERCISE SET 5.2

1. $V = \pi \int_{-1}^3 (3 - x)\,dx = 8\pi$

3. $V = \pi \int_0^2 \frac{1}{4}(3 - y)^2\,dy = 13\pi/6$

5. $V = \pi \int_{\pi/4}^{\pi/2} \cos x\,dx = (1 - \sqrt{2}/2)\pi$

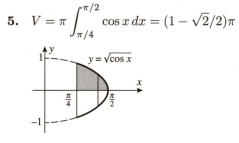

7. $V = \pi \int_{-1}^3 (1 + y)\,dy = 8\pi$

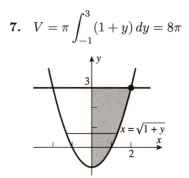

9. $V = \int_0^2 x^4\,dx = 32/5$

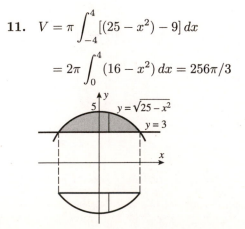

11. $V = \pi \int_{-4}^4 [(25 - x^2) - 9]\,dx$

$$= 2\pi \int_0^4 (16 - x^2)\,dx = 256\pi/3$$

13. $V = \pi \int_0^4 [(4x)^2 - (x^2)^2] \, dx$

$= \pi \int_0^4 (16x^2 - x^4) \, dx = 2048\pi/15$

15. $V = \int_0^1 \left(y^{1/3} \right)^2 dy = \dfrac{3}{5}$

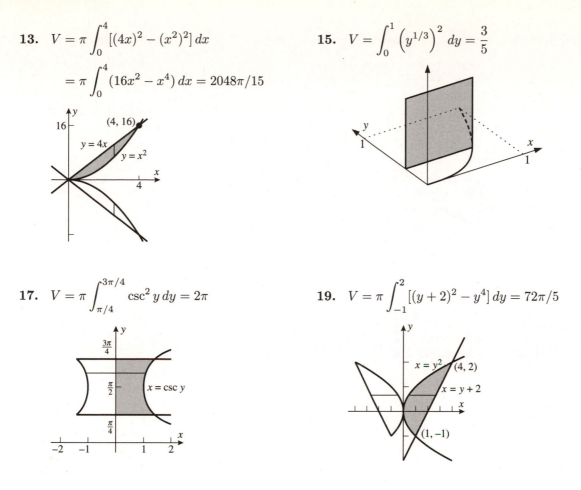

17. $V = \pi \int_{\pi/4}^{3\pi/4} \csc^2 y \, dy = 2\pi$

19. $V = \pi \int_{-1}^2 [(y+2)^2 - y^4] \, dy = 72\pi/5$

21. False. For example, consider the pyramid in Example 1, with the roles of the x- and y-axes interchanged.

23. False. For example, let S be the solid generated by rotating the region under $y = x^2$ over the interval $[0, 1]$. Then $A(x) = \pi x^4$.

25. $V = \pi \int_{-a}^a \dfrac{b^2}{a^2}(a^2 - x^2) \, dx = 4\pi ab^2/3$

27. $V = \pi \int_{-1}^0 (x + 1) \, dx$

$+ \pi \int_0^1 [(x+1) - 2x] \, dx$

$= \pi/2 + \pi/2 = \pi$

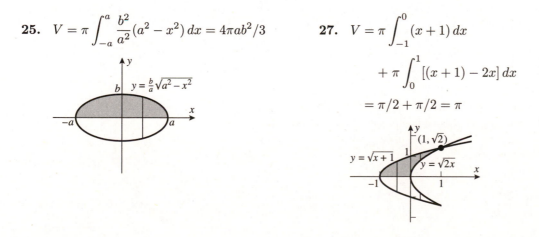

29. Partition the interval $[a, b]$ with $a = x_0 < x_1 < x_2 < \ldots < x_{n-1} < x_n = b$. Let x_k^* be an arbitrary point of $[x_{k-1}, x_k]$. The disk in question is obtained by revolving about the line $y = k$ the rectangle for which $x_{k-1} < x < x_k$, and y lies between $y = k$ and $y = f(x)$; the volume of this disk is

$\Delta V_k = \pi (f(x_k^*) - k)^2 \Delta x_k$, and the total volume is given by $V = \pi \displaystyle\int_a^b (f(x) - k)^2 \, dx$.

31. **(a)** Intuitively, it seems that a line segment which is revolved about a line which is perpendicular to the line segment will generate a larger area, the farther it is from the line. This is because the average point on the line segment will be revolved through a circle with a greater radius, and thus sweeps out a larger circle.

Consider the line segment which connects a point (x, y) on the curve $y = \sqrt{3 - x}$ to the point $(x, 0)$ beneath it. If this line segment is revolved around the x-axis we generate an area πy^2. If on the other hand the segment is revolved around the line $y = 2$ then the area of the resulting (infinitely thin) washer is $\pi[2^2 - (2 - y)^2]$. So the question can be reduced to asking whether $y^2 \geq [2^2 - (2 - y)^2]$, $y^2 \geq 4y - y^2$, or $y \geq 2$. In the present case the curve $y = \sqrt{3 - x}$ always satisfies $y \leq 2$, so V_2 has the larger volume.

(b) The volume of the solid generated by revolving the area around the x-axis is

$$V_1 = \pi \int_{-1}^{3} (3 - x) \, dx = 8\pi, \text{ and the volume generated by revolving the area around the line}$$

$y = 2$ is $V_2 = \pi \displaystyle\int_{-1}^{3} [2^2 - (2 - \sqrt{3 - x})^2] \, dx = \dfrac{40}{3}\pi$

33. $V = \pi \displaystyle\int_0^3 (9 - y^2)^2 \, dy$

$= \pi \displaystyle\int_0^3 (81 - 18y^2 + y^4) \, dy$

$= 648\pi / 5$

35. $V = \pi \displaystyle\int_0^1 [(\sqrt{x} + 1)^2 - (x + 1)^2] \, dx$

$= \pi \displaystyle\int_0^1 (2\sqrt{x} - x - x^2) \, dx = \pi / 2$

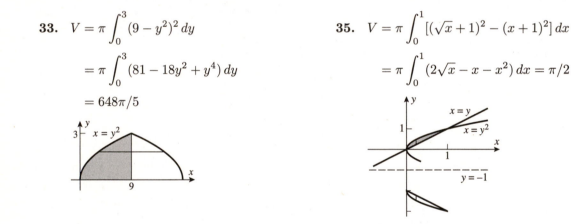

37. The region is given by the inequalities $0 \leq y \leq 1$, $\sqrt{y} \leq x \leq \sqrt[3]{y}$. For each y in the interval $[0, 1]$ the cross-section of the solid perpendicular to the axis $x = 1$ is a washer with outer radius $1 - \sqrt{y}$ and inner radius $1 - \sqrt[3]{y}$. The area of this washer is

$A(y) = \pi[(1 - \sqrt{y})^2 - (1 - \sqrt[3]{y})^2]$

$\quad = \pi(-2y^{1/2} + y + 2y^{1/3} - y^{2/3})$,

so the volume is

$V = \displaystyle\int_0^1 A(y) \, dy = \pi \displaystyle\int_0^1 (-2y^{1/2} + y + 2y^{1/3} - y^{2/3}) \, dy$

$= \pi \left[-\dfrac{4}{3}y^{3/2} + \dfrac{1}{2}y^2 + \dfrac{3}{2}y^{4/3} - \dfrac{3}{5}y^{5/3} \right]_0^1 = \dfrac{\pi}{15}$

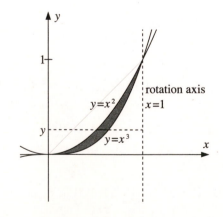

39. $A(x) = \pi(x^2/4)^2 = \pi x^4/16,$

$V = \int_0^{20} (\pi x^4/16)\, dx = 40{,}000\pi \ \text{ft}^3$

41. $V = \int_0^1 (x - x^2)^2\, dx$

$= \int_0^1 (x^2 - 2x^3 + x^4)\, dx = 1/30$

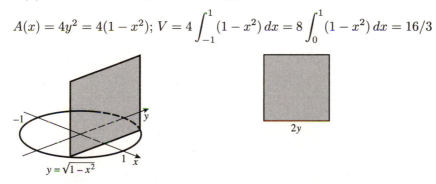

43. On the upper half of the circle, $y = \sqrt{1 - x^2}$, so:

(a) $A(x)$ is the area of a semicircle of radius y, so

$A(x) = \pi y^2/2 = \pi(1 - x^2)/2; \ V = \dfrac{\pi}{2} \int_{-1}^1 (1 - x^2)\, dx = \pi \int_0^1 (1 - x^2)\, dx = 2\pi/3$

(b) $A(x)$ is the area of a square of side $2y$, so

$A(x) = 4y^2 = 4(1 - x^2); \ V = 4 \int_{-1}^1 (1 - x^2)\, dx = 8 \int_0^1 (1 - x^2)\, dx = 16/3$

(c) $A(x)$ is the area of an equilateral triangle with sides $2y$, so

$A(x) = \dfrac{\sqrt{3}}{4}(2y)^2 = \sqrt{3}y^2 = \sqrt{3}(1 - x^2);$

$V = \int_{-1}^1 \sqrt{3}(1 - x^2)\, dx = 2\sqrt{3} \int_0^1 (1 - x^2)\, dx = 4\sqrt{3}/3$

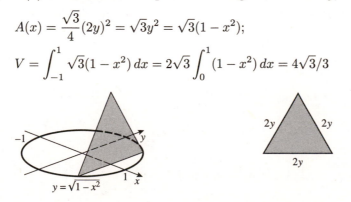

45. The two curves cross at $x = b \approx 1.403288534$, so

$V = \pi \int_0^b ((2x/\pi)^2 - \sin^{16} x)\, dx + \pi \int_b^{\pi/2} (\sin^{16} x - (2x/\pi)^2)\, dx \approx 0.710172176.$

47. (a) $V = \pi \int_{r-h}^{r} (r^2 - y^2)\, dy = \pi(rh^2 - h^3/3) = \frac{1}{3}\pi h^2(3r - h)$

(b) By the Pythagorean Theorem,

$r^2 = (r - h)^2 + \rho^2,\ 2hr = h^2 + \rho^2$; from part (a),

$V = \dfrac{\pi h}{3}(3hr - h^2) = \dfrac{\pi h}{3}\left(\dfrac{3}{2}(h^2 + \rho^2) - h^2\right)$

$\quad = \dfrac{1}{6}\pi h(3\rho^2 + h^2)$

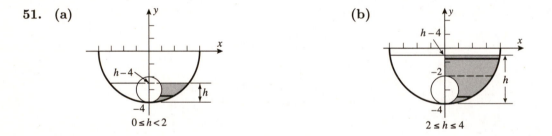

49. (a) The bulb is approximately a sphere of radius 1.25 cm attached to a cylinder of radius 0.625 cm and length 2.5 cm, so its volume is roughly $\frac{4}{3}\pi(1.25)^3 + \pi(0.625)^2 \cdot 2.5 \approx 11.25$ cm. (Other answers are possible, depending on how we approximate the light bulb using familiar shapes.)

(b) $\Delta x = \dfrac{5}{10} = 0.5;\ \{y_0, y_1, \cdots, y_{10}\} = \{0, 2.00, 2.45, 2.45, 2.00, 1.46, 1.26, 1.25, 1.25, 1.25, 1.25\};$

$\text{left} = \pi \sum_{i=0}^{9} \left(\dfrac{y_i}{2}\right)^2 \Delta x \approx 11.157;$

$\text{right} = \pi \sum_{i=1}^{10} \left(\dfrac{y_i}{2}\right)^2 \Delta x \approx 11.771;\ V \approx \text{average} = 11.464\ \text{cm}^3$

51. (a)   **(b)**

If the cherry is partially submerged then $0 \le h < 2$ as shown in Figure (a); if it is totally submerged then $2 \le h \le 4$ as shown in Figure (b). The radius of the glass is 4 cm and that of the cherry is 1 cm so points on the sections shown in the figures satisfy the equations $x^2 + y^2 = 16$ and $x^2 + (y + 3)^2 = 1$. We will find the volumes of the solids that are generated when the shaded regions are revolved about the y-axis.

For $0 \le h < 2$,

$$V = \pi \int_{-4}^{h-4} [(16 - y^2) - (1 - (y + 3)^2)]\, dy = 6\pi \int_{-4}^{h-4} (y + 4)\, dy = 3\pi h^2;$$

for $2 \le h \le 4$,

$$V = \pi \int_{-4}^{-2} [(16 - y^2) - (1 - (y + 3)^2)]\, dy + \pi \int_{-2}^{h-4} (16 - y^2)\, dy$$

$$= 6\pi \int_{-4}^{-2} (y + 4)\, dy + \pi \int_{-2}^{h-4} (16 - y^2)\, dy = 12\pi + \frac{1}{3}\pi(12h^2 - h^3 - 40)$$

$$= \frac{1}{3}\pi(12h^2 - h^3 - 4)$$

so

$$V = \begin{cases} 3\pi h^2 & \text{if } 0 \le h < 2 \\ \dfrac{1}{3}\pi(12h^2 - h^3 - 4) & \text{if } 2 \le h \le 4 \end{cases}$$

53. $\tan\theta = h/x$ so $h = x\tan\theta$,

$$A(y) = \frac{1}{2}hx = \frac{1}{2}x^2\tan\theta = \frac{1}{2}(r^2 - y^2)\tan\theta$$

because $x^2 = r^2 - y^2$,

$$V = \frac{1}{2}\tan\theta \int_{-r}^{r}(r^2 - y^2)\,dy$$

$$= \tan\theta\int_{0}^{r}(r^2 - y^2)\,dy = \frac{2}{3}r^3\tan\theta\cdot$$

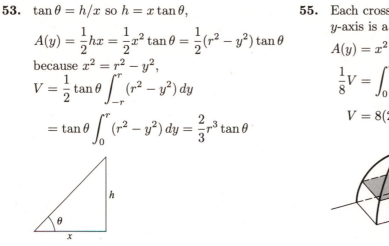

55. Each cross section perpendicular to the y-axis is a square so

$$A(y) = x^2 = r^2 - y^2,$$

$$\frac{1}{8}V = \int_{0}^{r}(r^2 - y^2)\,dy$$

$$V = 8(2r^3/3) = 16r^3/3$$

57. Position an x-axis perpendicular to the bases of the solids. Let a be the smallest x-coordinate of any point in either solid, and let b be the largest. Let $A(x)$ be the common area of the cross-sections of the solids at x-coordinate x. By equation (3), each solid has volume $V = \displaystyle\int_{a}^{b} A(x)\,dx$, so they are equal.

EXERCISE SET 5.3

1. $V = \displaystyle\int_{1}^{2} 2\pi x(x^2)\,dx = 2\pi\int_{1}^{2} x^3\,dx = 15\pi/2$

3. $V = \displaystyle\int_{0}^{1} 2\pi y(2y - 2y^2)\,dy = 4\pi\int_{0}^{1}(y^2 - y^3)\,dy = \pi/3$

5. $V = \displaystyle\int_{0}^{1} 2\pi(x)(x^3)\,dx$

$$= 2\pi\int_{0}^{1} x^4\,dx = 2\pi/5$$

7. $V = \displaystyle\int_{1}^{3} 2\pi x(1/x)\,dx = 2\pi\int_{1}^{3} dx = 4\pi$

9. $V = \int_1^2 2\pi x[(2x-1) - (-2x+3)]\,dx$

$= 8\pi \int_1^2 (x^2 - x)\,dx = 20\pi/3$

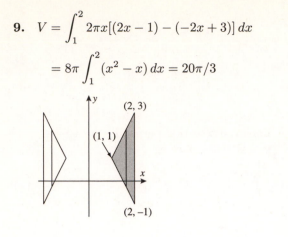

11. $V = \int_0^1 2\pi y^3\,dy = \pi/2$

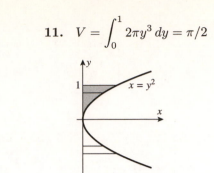

13. $V = \int_0^1 2\pi y(1 - \sqrt{y})\,dy$

$= 2\pi \int_0^1 (y - y^{3/2})\,dy = \pi/5$

15. True. The surface area of the cylinder is $2\pi \cdot$ [average radius] $\cdot$ [height], so by equation (1) the volume equals the thickness times the surface area.

17. True. In 5.3.2 we integrate over an interval on the x-axis, which is perpendicular to the y-axis, which is the axis of revolution.

19. $V = 2\pi \int_0^\pi x \sin x\,dx = 2\pi(\sin x - x \cos x)\Big]_0^\pi = 2\pi^2$
20. $V = 2\pi \int_0^{\pi/2} x \cos x\,dx = \pi^2 - 2\pi$

21. The volume is given by $2\pi \int_0^k x \sin x\,dx = 2\pi(\sin k - k \cos k) = 8$; solve for k to get $k \approx 1.736796$.

23. **(a)** $V = \int_0^1 2\pi x(x^3 - 3x^2 + 2x)\,dx = 7\pi/30$

(b) much easier; the method of slicing would require that x be expressed in terms of y.

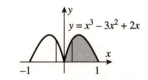

25. **(a)** For x in $[0,1]$, the cross-section with x-coordinate x has length x, and its distance from the axis of revolution is $1 - x$, so the volume is $\int_0^1 2\pi(1 - x)x\,dx$.

(b) For y in $[0,1]$, the cross-section with y-coordinate y has length $1 - y$, and its distance from the axis of revolution is $1 + y$, so the volume is $\int_0^1 2\pi(1 + y)(1 - y)\,dy$.

27. $V = \int_1^2 2\pi(x+1)(1/x^3)\,dx$

$$= 2\pi \int_1^2 (x^{-2} + x^{-3})\,dx = 7\pi/4$$

29. $x = \dfrac{h}{r}(r-y)$ is an equation of the line through $(0, r)$ and $(h, 0)$ so

$$V = \int_0^r 2\pi y \left[\frac{h}{r}(r-y)\right] dy$$

$$= \frac{2\pi h}{r} \int_0^r (ry - y^2)\,dy = \pi r^2 h/3$$

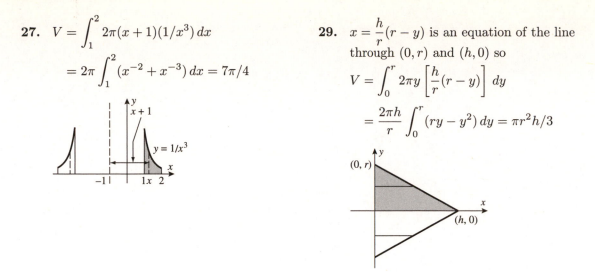

31. Let the sphere have radius R, the hole radius r. By the Pythagorean Theorem, $r^2 + (L/2)^2 = R^2$. Use cylindrical shells to calculate the volume of the solid obtained by rotating about the y-axis the region $r < x < R$, $-\sqrt{R^2 - x^2} < y < \sqrt{R^2 - x^2}$:

$$V = \int_r^R (2\pi x)2\sqrt{R^2 - x^2}\,dx = -\frac{4}{3}\pi(R^2 - x^2)^{3/2}\Big]_r^R = \frac{4}{3}\pi(L/2)^3,$$

so the volume is independent of R.

33. $V_x = \pi \int_{1/2}^b \dfrac{1}{x^2}\,dx = \pi(2 - 1/b)$, $V_y = 2\pi \int_{1/2}^b dx = \pi(2b - 1)$;

$V_x = V_y$ if $2 - 1/b = 2b - 1$, $2b^2 - 3b + 1 = 0$, solve to get $b = 1/2$ (reject) or $b = 1$.

35. In the method of disks/washers, we integrate the area of a flat surface, perpendicular to the axis of revolution. The variable of integration measures distance along the axis of revolution.

In the method of cylindrical shells, we integrate the area of a curved surface surrounding the axis of revolution. The variable of integration measures distance perpendicular to the axis of revolution.

EXERCISE SET 5.4

1. By the Theorem of Pythagoras, the length is $\sqrt{(2-1)^2 + (4-2)^2} = \sqrt{1+4} = \sqrt{5}$.

    (a) $\dfrac{dy}{dx} = 2$, $L = \int_1^2 \sqrt{1+4}\,dx = \sqrt{5}$

    (b) $\dfrac{dx}{dy} = \dfrac{1}{2}$, $L = \int_2^4 \sqrt{1 + 1/4}\,dy = 2\sqrt{5/4} = \sqrt{5}$

3. $f'(x) = \dfrac{9}{2}x^{1/2}$, $1 + [f'(x)]^2 = 1 + \dfrac{81}{4}x$,

$$L = \int_0^1 \sqrt{1 + 81x/4}\,dx = \frac{8}{243}\left(1 + \frac{81}{4}x\right)^{3/2}\Big]_0^1 = (85\sqrt{85} - 8)/243$$

5. $\dfrac{dy}{dx} = \dfrac{2}{3}x^{-1/3}$, $1 + \left(\dfrac{dy}{dx}\right)^2 = 1 + \dfrac{4}{9}x^{-2/3} = \dfrac{9x^{2/3} + 4}{9x^{2/3}}$,

$$L = \int_1^8 \dfrac{\sqrt{9x^{2/3} + 4}}{3x^{1/3}}\,dx = \dfrac{1}{18}\int_{13}^{40} u^{1/2}\,du, \quad u = 9x^{2/3} + 4$$

$$= \dfrac{1}{27}u^{3/2}\Big]_{13}^{40} = \dfrac{1}{27}(40\sqrt{40} - 13\sqrt{13}) = \dfrac{1}{27}(80\sqrt{10} - 13\sqrt{13})$$

or (alternate solution)

$$x = y^{3/2}, \dfrac{dx}{dy} = \dfrac{3}{2}y^{1/2}, 1 + \left(\dfrac{dx}{dy}\right)^2 = 1 + \dfrac{9}{4}y = \dfrac{4 + 9y}{4},$$

$$L = \dfrac{1}{2}\int_1^4 \sqrt{4 + 9y}\,dy = \dfrac{1}{18}\int_{13}^{40} u^{1/2}\,du = \dfrac{1}{27}(80\sqrt{10} - 13\sqrt{13})$$

7. $x = g(y) = \dfrac{1}{24}y^3 + 2y^{-1}$, $g'(y) = \dfrac{1}{8}y^2 - 2y^{-2}$,

$$1 + [g'(y)]^2 = 1 + \left(\dfrac{1}{64}y^4 - \dfrac{1}{2} + 4y^{-4}\right) = \dfrac{1}{64}y^4 + \dfrac{1}{2} + 4y^{-4} = \left(\dfrac{1}{8}y^2 + 2y^{-2}\right)^2,$$

$$L = \int_2^4 \left(\dfrac{1}{8}y^2 + 2y^{-2}\right)\,dy = 17/6$$

9. False. The derivative $\dfrac{dy}{dx} = -\dfrac{x}{\sqrt{1 - x^2}}$ is not defined at $x = \pm 1$, so it is not continuous on $[-1, 1]$.

11. True. If $f(x) = mx + c$ then the approximation equals

$$\sum_{k=1}^{n} \sqrt{1 + m^2}\,\Delta x_k = \sum_{k=1}^{n} \sqrt{1 + m^2}\,(x_k - x_{k-1}) = \sqrt{1 + m^2}\,(x_n - x_0) = (b - a)\sqrt{1 + m^2}$$

and the arc length is the distance from $(a, ma + c)$ to $(b, mb + c)$, which equals

$$\sqrt{(b - a)^2 + [(mb + c) - (ma + c)]^2} = \sqrt{(b - a)^2 + [m(b - a)]^2} = (b - a)\sqrt{1 + m^2}.$$

So each approximation equals the arc length.

13. (a)   **(b)** dy/dx does not exist at $x = 0$.

(c) $x = g(y) = y^{3/2}$, $g'(y) = \dfrac{3}{2}y^{1/2}$,

$$L = \int_0^1 \sqrt{1 + 9y/4}\,dy \quad \text{(portion for } -1 \le x \le 0\text{)}$$

$$+ \int_0^4 \sqrt{1 + 9y/4}\,dy \quad \text{(portion for } 0 \le x \le 8\text{)}$$

$$= \dfrac{8}{27}\left(\dfrac{13}{8}\sqrt{13} - 1\right) + \dfrac{8}{27}(10\sqrt{10} - 1) = (13\sqrt{13} + 80\sqrt{10} - 16)/27$$

15. (a) The function $y = f(x) = x^2$ is inverse to the function $x = g(y) = \sqrt{y}$: $f(g(y)) = y$ for $1/4 \le y \le 4$, and $g(f(x)) = x$ for $1/2 \le x \le 2$. Geometrically this means that the graphs of $y = f(x)$ and $x = g(y)$ are symmetric to each other with respect to the line $y = x$ and hence have the same arc length.

(b) $L_1 = \displaystyle\int_{1/2}^{2} \sqrt{1 + (2x)^2}\, dx$ and $L_2 = \displaystyle\int_{1/4}^{4} \sqrt{1 + \left(\frac{1}{2\sqrt{x}}\right)^2}\, dx$

Make the change of variables $x = \sqrt{y}$ in the first integral to obtain

$$L_1 = \int_{1/4}^{4} \sqrt{1 + (2\sqrt{y})^2}\,\frac{1}{2\sqrt{y}}\, dy = \int_{1/4}^{4} \sqrt{\left(\frac{1}{2\sqrt{y}}\right)^2 + 1}\, dy = L_2$$

(c) $L_1 = \displaystyle\int_{1/4}^{4} \sqrt{1 + \left(\frac{1}{2\sqrt{y}}\right)^2}\, dy,$ $L_2 = \displaystyle\int_{1/2}^{2} \sqrt{1 + (2y)^2}\, dy$

(d) For L_1, $\Delta x = \dfrac{3}{20}$, $x_k = \dfrac{1}{2} + k\dfrac{3}{20} = \dfrac{3k + 10}{20}$, and thus

$$L_1 \approx \sum_{k=1}^{10} \sqrt{(\Delta x)^2 + [f(x_k) - f(x_{k-1})]^2}$$

$$= \sum_{k=1}^{10} \sqrt{\left(\frac{3}{20}\right)^2 + \left(\frac{(3k+10)^2 - (3k+7)^2}{400}\right)^2} \approx 4.072396336$$

For L_2, $\Delta x = \dfrac{15}{40} = \dfrac{3}{8}$, $x_k = \dfrac{1}{4} + \dfrac{3k}{8} = \dfrac{3k + 2}{8}$, and thus

$$L_2 \approx \sum_{k=1}^{10} \sqrt{\left(\frac{3}{8}\right)^2 + \left[\sqrt{\frac{3k+2}{8}} - \sqrt{\frac{3k-1}{8}}\right]^2} \approx 4.071626502$$

(e) Each polygonal path is shorter than the curve segment, so both approximations in (d) are smaller than the actual length. Hence the larger one, the approximation for L_1, is better.

(f) For L_1, $\Delta x = \dfrac{3}{20}$, the midpoint is $x_k^* = \dfrac{1}{2} + \left(k - \dfrac{1}{2}\right)\dfrac{3}{20} = \dfrac{6k + 17}{40}$, and thus

$$L_1 \approx \sum_{k=1}^{10} \frac{3}{20} \sqrt{1 + \left(2\frac{6k+17}{40}\right)^2} \approx 4.072396336.$$

For L_2, $\Delta x = \dfrac{15}{40}$, and the midpoint is $x_k^* = \dfrac{1}{4} + \left(k - \dfrac{1}{2}\right)\dfrac{15}{40} = \dfrac{6k + 1}{16}$, and thus

$$L_2 \approx \sum_{k=1}^{10} \frac{15}{40} \sqrt{1 + \left(4\frac{6k+1}{16}\right)^{-1}} \approx 4.066160149$$

(g) $L_1 = \displaystyle\int_{1/2}^{2} \sqrt{1 + (2x)^2}\, dx \approx 4.0729,$ $L_2 = \displaystyle\int_{1/4}^{4} \sqrt{1 + \left(\frac{1}{2\sqrt{x}}\right)^2}\, dx \approx 4.0729$

17. (a) The function $y = f(x) = 1 + 1/x$ is inverse to the function $x = g(y) = 1/(y-1)$: $f(g(y)) = y$ for $4/3 \leq y \leq 2$, and $g(f(x)) = x$ for $1 \leq x \leq 3$. Geometrically this means that the graphs of $y = f(x)$ and $x = g(y)$ are symmetric to each other with respect to the line $y = x$.

(b) $L_1 = \displaystyle\int_1^3 \sqrt{1 + x^{-4}}\, dx, \qquad L_2 = \int_{4/3}^2 \sqrt{1 + (x-1)^{-4}}\, dx$

In the expression for L_1 make the change of variable $x = (y-1)^{-1}$ to obtain

$$L_1 = \int_2^{4/3} \sqrt{1 + (y-1)^4}\,\left(-(y-1)^{-2}\right) dy = \int_{4/3}^2 \sqrt{1 + (y-1)^{-4}}\, dy = L_2$$

(c) $L_1 = \displaystyle\int_{4/3}^2 \sqrt{1 + (y-1)^{-4}}\, dy, \qquad L_2 = \int_1^3 \sqrt{1 + y^{-4}}\, dy$

(d) For L_1, $\Delta x = 1/5$, $x_k = 1 + k/5$, and thus

$$L_1 \approx \sum_{k=1}^{10} \sqrt{(\Delta x)^2 + [f(x_k) - f(x_{k-1})]^2}$$

$$= \sum_{k=1}^{10} \sqrt{\left(\frac{1}{5}\right)^2 + \left[\left(1 + \frac{1}{1 + k/5}\right) - \left(1 + \frac{1}{1 + (k-1)/5}\right)\right]^2}$$

$$= \sum_{k=1}^{10} \sqrt{\frac{1}{25} + \left[\frac{1}{1 + k/5} - \frac{1}{1 + (k-1)/5}\right]^2} \approx 2.145900021$$

For L_2, $\Delta x = \dfrac{1}{10}\left(2 - \dfrac{4}{3}\right) = \dfrac{1}{15}$, $x_k = \dfrac{4}{3} + \dfrac{k}{15} = \dfrac{k + 20}{15}$, and thus

$$L_2 \approx \sum_{k=1}^{10} \sqrt{\frac{1}{225} + \left[\left(\frac{1}{3} + \frac{k}{15}\right)^{-1} - \left(\frac{1}{3} + \frac{k-1}{15}\right)^{-1}\right]^2} \approx 2.146262783$$

(e) Each polygonal path is shorter than the curve segment, so both approximations in (d) are smaller than the actual length. Hence the larger one, the approximation for L_2, is better.

(f) For L_1, $\Delta x = 1/5$, the midpoint is $x_k^* = 1 + \dfrac{1}{5}\left(k - \dfrac{1}{2}\right) = \dfrac{9 + 2k}{10}$, and thus

$$L_1 \approx \sum_{k=1}^{10} \frac{1}{5}\sqrt{1 + \left(\frac{9 + 2k}{10}\right)^{-4}} \approx 2.144326862.$$

For L_2, $\Delta x = 1/15$, the midpoint is $x_k^* = \dfrac{4}{3} + \dfrac{1}{15}\left(k - \dfrac{1}{2}\right) = \dfrac{39 + 2k}{30}$, and thus

$$L_2 \approx \sum_{k=1}^{10} \frac{1}{15}\sqrt{1 + \left(\frac{9 + 2k}{30}\right)^{-4}} \approx 2.137080472$$

(g) $L_1 = \displaystyle\int_1^3 \sqrt{1 + x^{-4}}\, dx \approx 2.1466$

$$L_2 = \int_{4/3}^2 \sqrt{1 + (x-1)^{-4}}\, dx \approx 2.1466$$

19. $f'(x) = \sec x \tan x$, $0 \le \sec x \tan x \le 2\sqrt{3}$ for $0 \le x \le \pi/3$ so $\dfrac{\pi}{3} \le L \le \dfrac{\pi}{3}\sqrt{13}$.

21. $L = \displaystyle\int_0^\pi \sqrt{1 + (k\cos x)^2}\, dx$

k	1	2	1.84	1.83	1.832
L	3.8202	5.2704	5.0135	4.9977	5.0008

Experimentation yields the values in the table, which by the Intermediate-Value Theorem show that the true solution k to $L = 5$ lies between $k = 1.83$ and $k = 1.832$, so $k = 1.83$ to two decimal places.

23. $y = 0$ at $x = b = 12.54/0.41 \approx 30.585$; distance $= \displaystyle\int_0^b \sqrt{1 + (12.54 - 0.82x)^2}\, dx \approx 196.31$ yd

25. $(dx/dt)^2 + (dy/dt)^2 = (t^2)^2 + (t)^2 = t^2(t^2 + 1)$, $L = \displaystyle\int_0^1 t(t^2+1)^{1/2} dt = (2\sqrt{2} - 1)/3$

27. $(dx/dt)^2 + (dy/dt)^2 = (-2\sin 2t)^2 + (2\cos 2t)^2 = 4$, $L = \displaystyle\int_0^{\pi/2} 2\, dt = \pi$

29. (a) $(dx/dt)^2 + (dy/dt)^2 = 4\sin^2 t + \cos^2 t = 4\sin^2 t + (1 - \sin^2 t) = 1 + 3\sin^2 t$,

$$L = \int_0^{2\pi} \sqrt{1 + 3\sin^2 t}\, dt = 4\int_0^{\pi/2} \sqrt{1 + 3\sin^2 t}\, dt$$

(b) 9.69

(c) distance traveled $= \displaystyle\int_{1.5}^{4.8} \sqrt{1 + 3\sin^2 t}\, dt \approx 5.16$ cm

31. The length of the curve is approximated by the length of a polygon whose vertices lie on the graph of $y = f(x)$. Each term in the sum is the length of one edge of the approximating polygon. By the distance formula, the length of the k'th edge is $\sqrt{(\Delta x_k)^2 + (\Delta y_k)^2}$, where Δx_k is the change in x along the edge and Δy_k is the change in y along the edge. We use the Mean-Value Theorem to express Δy_k as $f'(x_k^*)\Delta x_k$. Factoring the Δx_k out of the square root yields the k'th term in the sum.

EXERCISE SET 5.5

1. $S = \displaystyle\int_0^1 2\pi(7x)\sqrt{1 + 49}\, dx = 70\pi\sqrt{2}\int_0^1 x\, dx = 35\pi\sqrt{2}$

3. $f'(x) = -x/\sqrt{4 - x^2}$, $1 + [f'(x)]^2 = 1 + \dfrac{x^2}{4 - x^2} = \dfrac{4}{4 - x^2}$,

$S = \displaystyle\int_{-1}^1 2\pi\sqrt{4 - x^2}(2/\sqrt{4 - x^2})\, dx = 4\pi\int_{-1}^1 dx = 8\pi$

5. $S = \displaystyle\int_0^2 2\pi(9y + 1)\sqrt{82}\, dy = 2\pi\sqrt{82}\int_0^2 (9y + 1)\, dy = 40\pi\sqrt{82}$

7. $g'(y) = -y/\sqrt{9 - y^2}$, $1 + [g'(y)]^2 = \dfrac{9}{9 - y^2}$, $S = \displaystyle\int_{-2}^2 2\pi\sqrt{9 - y^2} \cdot \dfrac{3}{\sqrt{9 - y^2}}\, dy = 6\pi\int_{-2}^2 dy = 24\pi$

9. $f'(x) = \dfrac{1}{2}x^{-1/2} - \dfrac{1}{2}x^{1/2}$, $1 + [f'(x)]^2 = 1 + \dfrac{1}{4}x^{-1} - \dfrac{1}{2} + \dfrac{1}{4}x = \left(\dfrac{1}{2}x^{-1/2} + \dfrac{1}{2}x^{1/2}\right)^2$,

$$S = \int_1^3 2\pi\left(x^{1/2} - \dfrac{1}{3}x^{3/2}\right)\left(\dfrac{1}{2}x^{-1/2} + \dfrac{1}{2}x^{1/2}\right)dx = \dfrac{\pi}{3}\int_1^3 (3 + 2x - x^2)\,dx = 16\pi/9$$

11. $x = g(y) = \dfrac{1}{4}y^4 + \dfrac{1}{8}y^{-2}$, $g'(y) = y^3 - \dfrac{1}{4}y^{-3}$,

$$1 + [g'(y)]^2 = 1 + \left(y^6 - \dfrac{1}{2} + \dfrac{1}{16}y^{-6}\right) = \left(y^3 + \dfrac{1}{4}y^{-3}\right)^2,$$

$$S = \int_1^2 2\pi\left(\dfrac{1}{4}y^4 + \dfrac{1}{8}y^{-2}\right)\left(y^3 + \dfrac{1}{4}y^{-3}\right)dy = \dfrac{\pi}{16}\int_1^2 (8y^7 + 6y + y^{-5})\,dy = 16{,}911\pi/1024$$

13. $f'(x) = \cos x$, $1 + [f'(x)]^2 = 1 + \cos^2 x$,

$$S = \int_0^\pi 2\pi \sin x\sqrt{1 + \cos^2 x}\,dx \approx 14.42$$

15. True, by equation (1) with $r_1 = 0$, $r_2 = r$, and $l = \sqrt{r^2 + h^2}$.

17. True. If $f(x) = c$ for all x then $f'(x) = 0$ so the approximation is $\displaystyle\sum_{k=1}^n 2\pi c\,\Delta x_k = 2\pi c(b - a)$. Since the surface is the lateral surface of a cylinder of length $b - a$ and radius c, its area is also $2\pi c(b - a)$.

19. $n = 20$, $a = 0$, $b = \pi$, $\Delta x = (b - a)/20 = \pi/20$, $x_k = k\pi/20$,

$$S \approx \pi\sum_{k=1}^{20}[\sin(k - 1)\pi/20 + \sin k\pi/20]\sqrt{(\pi/20)^2 + [\sin(k-1)\pi/20 - \sin k\pi/20]^2} \approx 14.39$$

21. $S = \displaystyle\int_a^b 2\pi[f(x) + k]\sqrt{1 + [f'(x)]^2}\,dx$

23. $f(x) = \sqrt{r^2 - x^2}$, $f'(x) = -x/\sqrt{r^2 - x^2}$, $1 + [f'(x)]^2 = r^2/(r^2 - x^2)$,

$$S = \int_{-r}^r 2\pi\sqrt{r^2 - x^2}\,(r/\sqrt{r^2 - x^2})\,dx = 2\pi r\int_{-r}^r dx = 4\pi r^2$$

25. Suppose the two planes are $y = y_1$ and $y = y_2$, where $-r \le y_1 \le y_2 \le r$. Then the area of the zone equals the area of a spherical cap of height $r - y_1$ minus the area of a spherical cap of height $r - y_2$. By Exercise 24, this is $2\pi r(r - y_1) - 2\pi r(r - y_2) = 2\pi r(y_2 - y_1)$, which only depends on the radius r and the distance $y_2 - y_1$ between the planes.

27. Note that $1 \le \sec x \le 2$ for $0 \le x \le \pi/3$. Let L be the arc length of the curve $y = \tan x$ for $0 < x < \pi/3$. Then $L = \displaystyle\int_0^{\pi/3}\sqrt{1 + \sec^2 x}\,dx$, and by Exercise 26 and the inequalities above, $2\pi L \le S \le 4\pi L$. But from the inequalities for $\sec x$ above, we can show that $\sqrt{2}\pi/3 \le L \le \sqrt{5}\pi/3$. Hence, combining the two sets of inequalities, $2\pi(\sqrt{2}\pi/3) \le 2\pi L \le S \le 4\pi L \le 4\pi\sqrt{5}\pi/3$. To obtain the inequalities in the text, observe that

$$\dfrac{2\pi^2}{3} < 2\pi\dfrac{\sqrt{2}\pi}{3} \le 2\pi L \le S \le 4\pi L \le 4\pi\dfrac{\sqrt{5}\pi}{3} < \dfrac{4\pi^2}{3}\sqrt{13}.$$

29. Let $a = t_0 < t_1 < \ldots < t_{n-1} < t_n = b$ be a partition of $[a, b]$. Then the lateral area of the frustum of slant height $\ell = \sqrt{\Delta x_k^2 + \Delta y_k^2}$ and radii $y(t_1)$ and $y(t_2)$ is $\pi(y(t_k) + y(t_{k-1}))\ell$. Thus the area of the frustum S_k is given by $S_k = \pi(y(t_{k-1}) + y(t_k))\sqrt{[x(t_k) - x(t_{k-1})]^2 + [y(t_k) - y(t_{k-1})]^2}$ with the limit as $\max \Delta t_k \to 0$ of $S = \int_a^b 2\pi y(t)\sqrt{x'(t)^2 + y'(t)^2}\, dt$

31. $x' = 2t, y' = 2, (x')^2 + (y')^2 = 4t^2 + 4$

$$S = 2\pi \int_0^4 (2t)\sqrt{4t^2 + 4}\, dt = 8\pi \int_0^4 t\sqrt{t^2 + 1}\, dt = \frac{8\pi}{3}(17\sqrt{17} - 1)$$

33. $x' = 1, y' = 4t, (x')^2 + (y')^2 = 1 + 16t^2$, $S = 2\pi \int_0^1 t\sqrt{1 + 16t^2}\, dt = \frac{\pi}{24}(17\sqrt{17} - 1)$

35. $x' = -r \sin t, y' = r \cos t, (x')^2 + (y')^2 = r^2$,

$$S = 2\pi \int_0^\pi r \sin t \sqrt{r^2}\, dt = 2\pi r^2 \int_0^\pi \sin t\, dt = 4\pi r^2$$

37. Suppose we approximate the k'th frustum by the lateral surface of a cylinder of width Δx_k and radius $f(x_k^*)$, where x_k^* is between x_{k-1} and x_k. The area of this surface is $2\pi f(x_k^*)\, \Delta x_k$. Proceeding as before, we would conclude that $S = \int_a^b 2\pi f(x)\, dx$, which is too small. Basically, when $|f'(x)| > 0$, the area of the frustum is larger than the area of the cylinder, and ignoring this results in an incorrect formula.

EXERCISE SET 5.6

1. $W = \int_0^3 F(x)\, dx = \int_0^3 (x + 1)\, dx = \left[\frac{1}{2}x^2 + x\right]_0^3 = 7.5$ ft·lb

3. Since $W = \int_a^b F(x)\, dx =$ the area under the curve, it follows that $d < 2.5$ since the area increases faster under the left part of the curve. In fact, $W_d = \int_0^d F(x)\, dx = 40d$, and

$W = \int_0^5 F(x)\, dx = 140$, so $d = 7/4$.

5. distance traveled $= \int_0^5 v(t)\, dt = \int_0^5 \frac{4t}{5}\, dt = \frac{2}{5}t^2 \Big]_0^5 = 10$ ft. The force is a constant 10 lb, so the work done is $10 \cdot 10 = 100$ ft·lb.

7. $F(x) = kx, F(0.2) = 0.2k = 100, k = 500$ N/m, $W = \int_0^{0.8} 500x\, dx = 160$ J

9. $W = \int_0^1 kx\, dx = k/2 = 10, k = 20$ lb/ft

11. False. The work depends on the force and the distance, not on the elapsed time.

13. True. By equation (6), work and energy have the same units in any system of units.

15. $W = \displaystyle\int_0^{9/2} (9 - x)62.4(25\pi)\,dx$

$\qquad = 1560\pi \displaystyle\int_0^{9/2} (9 - x)\,dx = 47{,}385\pi$ ft·lb

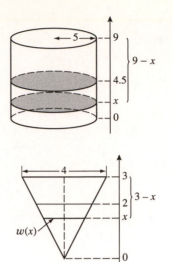

17. $w/4 = x/3, w = 4x/3,$

$\qquad W = \displaystyle\int_0^2 (3 - x)(9810)(4x/3)(6)\,dx$

$\qquad = 78480 \displaystyle\int_0^2 (3x - x^2)\,dx$

$\qquad = 261{,}600\,\text{J}$

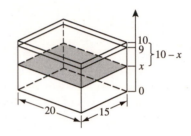

19. (a) $W = \displaystyle\int_0^9 (10 - x)62.4(300)\,dx$

$\qquad = 18{,}720 \displaystyle\int_0^9 (10 - x)\,dx$

$\qquad = 926{,}640$ ft·lb

(b) To empty the pool in one hour would require
926,640/3600 = 257.4 ft·lb of work per second
so hp of motor = 257.4/550 = 0.468.

21. $W = \displaystyle\int_0^{100} 15(100 - x)\,dx$

$\qquad = 75{,}000$ ft·lb

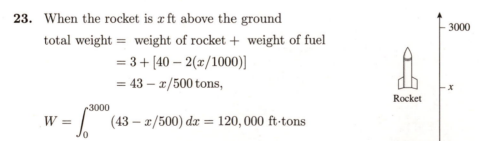

23. When the rocket is x ft above the ground

$\qquad$ total weight = weight of rocket + weight of fuel

$\qquad\qquad = 3 + [40 - 2(x/1000)]$

$\qquad\qquad = 43 - x/500$ tons,

$W = \displaystyle\int_0^{3000} (43 - x/500)\,dx = 120{,}000$ ft·tons

25. (a) $150 = k/(4000)^2, k = 2.4 \times 10^9, w(x) = k/x^2 = 2{,}400{,}000{,}000/x^2$ lb

(b) $6000 = k/(4000)^2, k = 9.6 \times 10^{10}, w(x) = \left(9.6 \times 10^{10}\right)/(x + 4000)^2$ lb

(c) $W = \displaystyle\int_{4000}^{5000} 9.6(10^{10})x^{-2}\,dx = 4{,}800{,}000$ mi·lb $= 2.5344 \times 10^{10}$ ft·lb

27. $W = \frac{1}{2}mv_f^2 - \frac{1}{2}mv_i^2 = \frac{1}{2}4.00 \times 10^5(v_f^2 - 20^2)$. But $W = F \cdot d = (6.40 \times 10^5) \cdot (3.00 \times 10^3)$, so $19.2 \times 10^8 = 2.00 \times 10^5 v_f^2 - 8.00 \times 10^7, 19200 = 2v_f^2 - 800, v_f = 100$ m/s.

29. **(a)** The kinetic energy would have decreased by $\frac{1}{2}mv^2 = \frac{1}{2}4 \cdot 10^6(15000)^2 = 4.5 \times 10^{14}$ J

(b) $(4.5 \times 10^{14})/(4.2 \times 10^{15}) \approx 0.107$ **(c)** $\frac{1000}{13}(0.107) \approx 8.24$ bombs

31. The work-energy relationship involves 4 quantities, the work W, the mass m, and the initial and final velocities v_i and v_f. In any problem in which 3 of these are given, the work-energy relationship can be used to compute the fourth.

In cases where the force is constant, we may combine equation (1) with the work-energy relationship to get $Fd = \frac{1}{2}mv_f^2 - \frac{1}{2}mv_i^2$. In this form there are 5 quantities, the force F, the distance d, the mass m, and the initial and final velocities v_i and v_f. So if any 4 of these are given, the work-energy relationship can be used to compute the fifth.

EXERCISE SET 5.7

1. **(a)** m_1 and m_3 are equidistant from $x = 5$, but m_3 has a greater mass, so the sum is positive.

(b) Let a be the unknown coordinate of the fulcrum; then the total moment about the fulcrum is $5(0 - a) + 10(5 - a) + 20(10 - a) = 0$ for equilibrium, so $250 - 35a = 0$, $a = 50/7$. The fulcrum should be placed $50/7$ units to the right of m_1.

3. By symmetry, the centroid is $(1/2, 1/2)$. We confirm this using Formulas (8) and (9) with $a = 0$, $b = 1$, $f(x) = 1$: The area is 1, so $\bar{x} = \int_0^1 x\, dx = \frac{1}{2}$ and $\bar{y} = \int_0^1 \frac{1}{2}\, dx = \frac{1}{2}$, as expected.

5. By symmetry, the centroid is $(1, 1/2)$. We confirm this using Formulas (8) and (9) with $a = 0$, $b = 2$, $f(x) = 1$: The area is 2, so $\bar{x} = \frac{1}{2}\int_0^2 x\, dx = 1$ and $\bar{y} = \frac{1}{2}\int_0^2 \frac{1}{2}\, dx = \frac{1}{2}$, as expected.

7. By symmetry, the centroid lies on the line $y = 1 - x$. To find $\bar{x}$ we use Formula (8) with $a = 0$, $b = 1$, $f(x) = x$: The area is $\frac{1}{2}$, so $\bar{x} = 2\int_0^1 x^2\, dx = \frac{2}{3}$. Hence $\bar{y} = 1 - \frac{2}{3} = \frac{1}{3}$ and the centroid is $\left(\frac{2}{3}, \frac{1}{3}\right)$.

9. We use Formulas (10) and (11) with $a = 0$, $b = 1$, $f(x) = 2 - x^2$, $g(x) = x$: The area is

$$\int_0^1 (2 - x^2 - x)\, dx = \left[2x - \frac{1}{3}x^3 - \frac{1}{2}x^2\right]_0^1 = \frac{7}{6}, \text{ so}$$

$$\bar{x} = \frac{6}{7}\int_0^1 x(2 - x^2 - x)\, dx = \frac{6}{7}\left[x^2 - \frac{1}{4}x^4 - \frac{1}{3}x^3\right]_0^1 = \frac{5}{14} \text{ and}$$

$$\bar{y} = \frac{6}{7}\int_0^1 \frac{1}{2}[(2 - x^2)^2 - x^2]\, dx = \frac{3}{7}\int_0^1 (4 - 5x^2 + x^4)\, dx = \frac{3}{7}\left[4x - \frac{5}{3}x^3 + \frac{1}{5}x^5\right]_0^1 = \frac{38}{35}.$$

The centroid is $\left(\frac{5}{14}, \frac{38}{35}\right)$.

11. We use Formulas (8) and (9) with $a = 0$, $b = 2$, $f(x) = 1 - \frac{x}{2}$: The area is 1, so

$$\bar{x} = \int_0^2 x\left(1 - \frac{x}{2}\right) dx = \left[\frac{1}{2}x^2 - \frac{1}{6}x^3\right]_0^2 = \frac{2}{3} \text{ and}$$

$$\bar{y} = \int_0^2 \frac{1}{2}\left(1 - \frac{x}{2}\right)^2 dx = \frac{1}{8}\int_0^2 (4 - 4x + x^2)\,dx = \frac{1}{8}\left[4x - 2x^2 + \frac{1}{3}x^3\right]_0^2 = \frac{1}{3}.$$

The centroid is $\left(\dfrac{2}{3}, \dfrac{1}{3}\right)$.

13. The graphs of $y = x^2$ and $y = 6 - x$ meet when $x^2 = 6 - x$, so $x = -3$ or $x = 2$. We use Formulas (10) and (11) with $a = -3$, $b = 2$, $f(x) = 6 - x$, $g(x) = x^2$:

The area is $\displaystyle\int_{-3}^2 (6 - x - x^2)\,dx = \left[6x - \frac{1}{2}x^2 - \frac{1}{3}x^3\right]_{-3}^2 = \frac{125}{6}$, so

$$\bar{x} = \frac{6}{125}\int_{-3}^2 x(6 - x - x^2)\,dx = \frac{6}{125}\left[3x^2 - \frac{1}{3}x^3 - \frac{1}{4}x^4\right]_{-3}^2 = -\frac{1}{2} \text{ and}$$

$$\bar{y} = \frac{6}{125}\int_{-3}^2 \frac{1}{2}[(6 - x)^2 - (x^2)^2]\,dx = \frac{3}{125}\int_{-3}^2 (36 - 12x + x^2 - x^4)\,dx$$

$$= \frac{3}{125}\left[36x - 6x^2 + \frac{1}{3}x^3 - \frac{1}{5}x^5\right]_{-3}^2 = 4. \text{ The centroid is } \left(-\frac{1}{2}, 4\right).$$

15. The curves meet at $(-1, 1)$ and $(2, 4)$. We use Formulas (10) and (11) with $a = -1$, $b = 2$,

$f(x) = x + 2$, $g(x) = x^2$: The area is $\displaystyle\int_{-1}^2 (x + 2 - x^2)\,dx = \left[\frac{1}{2}x^2 + 2x - \frac{1}{3}x^3\right]_{-1}^2 = \frac{9}{2}$, so

$$\bar{x} = \frac{2}{9}\int_{-1}^2 x(x + 2 - x^2)\,dx = \frac{2}{9}\left[\frac{1}{3}x^3 + x^2 - \frac{1}{4}x^4\right]_{-1}^2 = \frac{1}{2} \text{ and}$$

$$\bar{y} = \frac{2}{9}\int_{-1}^2 \frac{1}{2}\left[(x + 2)^2 - (x^2)^2\right]dx = \frac{1}{9}\int_{-1}^2 (x^2 + 4x + 4 - x^4)\,dx$$

$$= \frac{1}{9}\left[\frac{1}{3}x^3 + 2x^2 + 4x - \frac{1}{5}x^5\right]_{-1}^2 = \frac{8}{5}. \text{ The centroid is } \left(\frac{1}{2}, \frac{8}{5}\right).$$

17. By symmetry, $\bar{y} = \bar{x}$. To find $\bar{x}$ we use Formula (10) with $a = 0$, $b = 1$, $f(x) = \sqrt{x}$, $g(x) = x^2$:

The area is $\displaystyle\int_0^1 (\sqrt{x} - x^2)\,dx = \left[\frac{2}{3}x^{3/2} - \frac{1}{3}x^3\right]_0^1 = \frac{1}{3}$, so

$$\bar{x} = 3\int_0^1 x(\sqrt{x} - x^2)\,dx = 3\left[\frac{2}{5}x^{5/2} - \frac{1}{4}x^4\right]_0^1 = \frac{9}{20}. \text{ The centroid is } \left(\frac{9}{20}, \frac{9}{20}\right).$$

19. An isosceles triangle is symmetric across the median to its base. So, if the density is constant, it will balance on a knife-edge under the median. Hence the centroid lies on the median.

21. The region is described by $0 \le x \le 1$, $0 \le y \le \sqrt{x}$. The area is $A = \displaystyle\int_0^1 \sqrt{x}\,dx = \frac{2}{3}$, so the mass

is $M = \delta A = 2 \cdot \dfrac{2}{3} = \dfrac{4}{3}$. By Formulas (8) and (9), $\bar{x} = \dfrac{3}{2}\displaystyle\int_0^1 x\sqrt{x}\,dx = \dfrac{3}{2}\left[\dfrac{2}{5}x^{5/2}\right]_0^1 = \dfrac{3}{5}$ and

$$\bar{y} = \frac{3}{2}\int_0^1 \frac{1}{2}(\sqrt{x})^2\,dx = \frac{3}{4}\int_0^1 x\,dx = \frac{3}{8}. \text{ The center of gravity is } \left(\frac{3}{5}, \frac{3}{8}\right).$$

23. The region is described by $0 \le y \le 1$, $-y \le x \le y$. The area is $A = 1$, so the mass is $M = \delta A = 3 \cdot 1 = 3$. By symmetry, $\bar{x} = 0$. By the analogue of Formula (10) with the roles of x and y reversed,

$$\bar{y} = \int_0^1 y[y - (-y)]\,dy = \int_0^1 2y^2\,dy = \frac{2}{3}y^3\Big]_0^1 = \frac{2}{3}. \text{ The center of gravity is } \left(0, \frac{2}{3}\right).$$

25. The region is described by $0 \le x \le \pi$, $0 \le y \le \sin x$. The area is $A = \displaystyle\int_0^\pi \sin x\,dx = 2$, so the mass is $M = \delta A = 4 \cdot 2 = 8$. By symmetry, $\bar{x} = \dfrac{\pi}{2}$. By Formula (9), $\bar{y} = \dfrac{1}{2} \displaystyle\int_0^\pi \dfrac{1}{2}(\sin x)^2\, dx = \dfrac{\pi}{8}$. The center of gravity is $\left(\dfrac{\pi}{2}, \dfrac{\pi}{8}\right)$.

27. True, by symmetry.

29. True, by symmetry.

31. By symmetry, $\bar{y} = 0$. We use Formula (10) with a replaced by 0, b replaced by a, $f(x) = \dfrac{bx}{a}$, and

$g(x) = -\dfrac{bx}{a}$: The area is ab, so $\bar{x} = \dfrac{1}{ab} \displaystyle\int_0^a x\left(\dfrac{bx}{a} - \left(-\dfrac{bx}{a}\right)\right) dx = \dfrac{2}{a^2} \int_0^a x^2\, dx = \dfrac{2}{a^2} \cdot \dfrac{a^3}{3} = \dfrac{2a}{3}$.

The centroid is $\left(\dfrac{2a}{3}, 0\right)$.

33. We will assume that a, b, and c are positive; the other cases are similar.

The region is described by $0 \le y \le c$, $-a - \dfrac{b-a}{c}y \le x \le a + \dfrac{b-a}{c}y$. By symmetry, $\bar{x} = 0$. To find

$\bar{y}$, we use the analogue of Formula (10) with the roles of x and y reversed. The area is $c(a+b)$, so

$$\bar{y} = \dfrac{1}{c(a+b)} \int_0^c y\left[\left(a + \dfrac{b-a}{c}y\right) - \left(-a - \dfrac{b-a}{c}y\right)\right] dy = \dfrac{1}{c(a+b)} \int_0^c \left(2ay + \dfrac{2(b-a)}{c}y^2\right) dy$$

$$= \dfrac{1}{c(a+b)}\left[ay^2 + \dfrac{2(b-a)}{3c}y^3\right]_0^c = \dfrac{c(a+2b)}{3(a+b)}.$$ The centroid is $\left(0, \dfrac{c(a+2b)}{3(a+b)}\right)$.

35. $\bar{x} = 0$ from the symmetry of the region, $\pi a^2/2$ is the area of the semicircle, $2\pi\bar{y}$ is the distance traveled by the centroid to generate the sphere so $4\pi a^3/3 = (\pi a^2/2)(2\pi\bar{y})$, $\bar{y} = 4a/(3\pi)$

37. $\bar{x} = k$ so $V = (\pi ab)(2\pi k) = 2\pi^2 abk$

39. The region generates a cone of volume $\dfrac{1}{3}\pi ab^2$ when it is revolved about the x-axis, the area of the

region is $\dfrac{1}{2}ab$ so $\dfrac{1}{3}\pi ab^2 = \left(\dfrac{1}{2}ab\right)(2\pi\bar{y})$, $\bar{y} = b/3$. A cone of volume $\dfrac{1}{3}\pi a^2 b$ is generated when the

region is revolved about the y-axis so $\dfrac{1}{3}\pi a^2 b = \left(\dfrac{1}{2}ab\right)(2\pi\bar{x})$, $\bar{x} = a/3$. The centroid is $(a/3, b/3)$.

41. The Theorem of Pappus says that $V = 2\pi Ad$, where A is the area of a region in the plane, d is the distance from the region's centroid to an axis of rotation, and V is the volume of the resulting solid of revolution. In any problem in which 2 of these quantities are given, the Theorem of Pappus can be used to compute the third.

EXERCISE SET 5.8

1. (a) $F = \rho h A = 62.4(5)(100) = 31{,}200$ lb
$P = \rho h = 62.4(5) = 312$ lb/ft^2

   (b) $F = \rho h A = 9810(10)(25) = 2{,}452{,}500$ N
$P = \rho h = 9810(10) = 98.1$ kPa

3. $F = \displaystyle\int_0^2 62.4x(4)\, dx$

$= 249.6 \displaystyle\int_0^2 x\, dx = 499.2$ lb

5. $F = \displaystyle\int_0^5 9810x(2\sqrt{25 - x^2})\, dx$

$= 19{,}620 \displaystyle\int_0^5 x(25 - x^2)^{1/2}\, dx$

$= 8.175 \times 10^5$ N

7. By similar triangles,

$$\frac{w(x)}{6} = \frac{10 - x}{8}, \quad w(x) = \frac{3}{4}(10 - x),$$

$$F = \int_2^{10} 9810x \left[\frac{3}{4}(10 - x)\right] dx$$

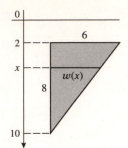

$$= 7357.5 \int_2^{10} (10x - x^2) \, dx = 1,098,720 \text{ N}$$

9. Yes: if $\rho_2 = 2\rho_1$ then $F_2 = \int_a^b \rho_2 h(x)w(x) \, dx = \int_a^b 2\rho_1 h(x)w(x) \, dx = 2 \int_a^b \rho_1 h(x)w(x) \, dx = 2F_1$.

11. Find the forces on the upper and lower halves and add them:

$$\frac{w_1(x)}{\sqrt{2}a} = \frac{x}{\sqrt{2}a/2}, \quad w_1(x) = 2x$$

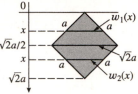

$$F_1 = \int_0^{\sqrt{2}a/2} \rho x(2x) \, dx = 2\rho \int_0^{\sqrt{2}a/2} x^2 \, dx = \sqrt{2}\rho a^3/6,$$

$$\frac{w_2(x)}{\sqrt{2}a} = \frac{\sqrt{2}a - x}{\sqrt{2}a/2}, \quad w_2(x) = 2(\sqrt{2}a - x)$$

$$F_2 = \int_{\sqrt{2}a/2}^{\sqrt{2}a} \rho x[2(\sqrt{2}a - x)] \, dx = 2\rho \int_{\sqrt{2}a/2}^{\sqrt{2}a} (\sqrt{2}ax - x^2) \, dx = \sqrt{2}\rho a^3/3,$$

$$F = F_1 + F_2 = \sqrt{2}\rho a^3/6 + \sqrt{2}\rho a^3/3 = \rho a^3/\sqrt{2} \text{ lb}$$

13. True. By equation (6), the fluid force equals $\rho h A$. For a cylinder, hA is the volume, so ρhA is the weight of the water.

15. False. Let the height of the tank be h, the area of the base be A, and the volume of the tank be V. Then the fluid force on the base is ρhA and the weight of the water is ρV. So if $hA > V$, then the force exceeds the weight. This is true, for example, for a conical tank with its vertex at the top, for which $V = \dfrac{hA}{3}$.

17. Place the x-axis pointing down with its origin at the top of the pool, so that $h(x) = x$ and $w(x) = 10$. Let θ be the angle between the bottom of the pool and the vertical. Then $\tan\theta = 16/(8-4) = 4$, so $\sec\theta = \sqrt{17}$. Hence $F = \int_4^8 62.4h(x)w(x)\sec\theta \, dx = 624\sqrt{17} \int_4^8 x \, dx = 14976\sqrt{17} \approx 61748$ lb.

19. $h(x) = x\sin 60° = \sqrt{3}x/2,$

$\theta = 30°, \sec\theta = 2/\sqrt{3},$

$$F = \int_0^{100} 9810(\sqrt{3}x/2)(200)(2/\sqrt{3}) \, dx$$

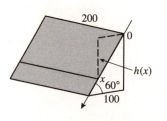

$$= 200 \cdot 9810 \int_0^{100} x \, dx$$

$$= 9810 \cdot 100^3 = 9.81 \times 10^9 \text{ N}$$

21. **(a)** The force on the window is $F = \int_h^{h+2} \rho_0 x(2) \, dx = 4\rho_0(h+1)$ so (assuming that ρ_0 is constant) $dF/dt = 4\rho_0(dh/dt)$ which is a positive constant if dh/dt is a positive constant.

   (b) If $dh/dt = 20$ then $dF/dt = 80\rho_0$ lb/min from part (a).

23. $h = \dfrac{P}{\rho} = \dfrac{14.7\,\text{lb/in}^2}{4.66 \times 10^{-5}\,\text{lb/in}^3} \approx 315{,}000\,\text{in} \approx 5\,\text{mi}$. The answer is not reasonable. In fact the atmosphere is thinner at higher altitudes, and it's difficult to define where the "top" of the atmosphere is.

REVIEW EXERCISES, CHAPTER 5

7. (a) $A = \displaystyle\int_a^b (f(x) - g(x))\,dx + \int_b^c (g(x) - f(x))\,dx + \int_c^d (f(x) - g(x))\,dx$

(b) $A = \displaystyle\int_{-1}^0 (x^3 - x)\,dx + \int_0^1 (x - x^3)\,dx + \int_1^2 (x^3 - x)\,dx = \dfrac{1}{4} + \dfrac{1}{4} + \dfrac{9}{4} = \dfrac{11}{4}$

9. Find where the curves cross: set $x^3 = x^2 + 4$; by observation $x = 2$ is a solution. Then

$$V = \pi \int_0^2 [(x^2 + 4)^2 - (x^3)^2]\,dx = \dfrac{4352}{105}\pi.$$

11. $V = \displaystyle\int_1^4 \left(\sqrt{x} - \dfrac{1}{x}\right)^2 dx = \int_1^4 (x - 2x^{-1/2} + x^{-2})\,dx = \left[\dfrac{1}{2}x^2 - 4x^{1/2} - x^{-1}\right]_1^4 = \dfrac{17}{4}$

13. By implicit differentiation $\dfrac{dy}{dx} = -\left(\dfrac{y}{x}\right)^{1/3}$, so $1 + \left(\dfrac{dy}{dx}\right)^2 = 1 + \left(\dfrac{y}{x}\right)^{2/3} = \dfrac{x^{2/3} + y^{2/3}}{x^{2/3}} = \dfrac{4}{x^{2/3}}$,

$L = \displaystyle\int_{-8}^{-1} \dfrac{2}{(-x)^{1/3}}\,dx = 9.$

15. $A = \displaystyle\int_9^{16} 2\pi\sqrt{25 - x}\,\sqrt{1 + \left(\dfrac{-1}{2\sqrt{25-x}}\right)^2}\,dx = \pi \int_9^{16} \sqrt{101 - 4x}\,dx = \dfrac{\pi}{6}\left(65^{3/2} - 37^{3/2}\right)$

17. (a) $F = kx$, $\dfrac{1}{2} = k\dfrac{1}{4}$, $k = 2$, $W = \displaystyle\int_0^{1/4} kx\,dx = 1/16$ J

(b) $25 = \displaystyle\int_0^L kx\,dx = kL^2/2$, $L = 5$ m

19. The region is described by $-4 \le y \le 4$, $\dfrac{y^2}{4} \le x \le 2 + \dfrac{y^2}{8}$. By symmetry, $\bar{y} = 0$. To find $\bar{x}$, we use the analogue of Formula (11) in Section 5.7. The area is

$A = \displaystyle\int_{-4}^4 \left(2 + \dfrac{y^2}{8} - \dfrac{y^2}{4}\right)dy = \int_{-4}^4 \left(2 - \dfrac{y^2}{8}\right)dy = \left[2y - \dfrac{y^3}{24}\right]_{-4}^4 = \dfrac{32}{3}$. So

$\bar{x} = \dfrac{3}{32}\displaystyle\int_{-4}^4 \dfrac{1}{2}\left[\left(2 + \dfrac{y^2}{8}\right)^2 - \left(\dfrac{y^2}{4}\right)^2\right]dy = \dfrac{3}{64}\int_{-4}^4 \left(4 + \dfrac{y^2}{2} - \dfrac{3y^4}{64}\right)dy$

$= \dfrac{3}{64}\left[4y + \dfrac{y^3}{6} - \dfrac{3y^5}{320}\right]_{-4}^4 = \dfrac{8}{5}$. The centroid is $\left(\dfrac{8}{5}, 0\right)$.

21. **(a)** $F = \int_0^1 \rho x 3 \, dx$ N

(b) By similar triangles, $\dfrac{w(x)}{4} = \dfrac{x}{2}$, $w(x) = 2x$, so

$$F = \int_1^4 \rho(1+x)2x \, dx \text{ lb/ft}^2.$$

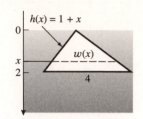

(c) A formula for the parabola is $y = \dfrac{8}{125}x^2 - 10$,

so $F = \int_{-10}^0 9810\,|y|\,2\sqrt{\dfrac{125}{8}(y+10)}\,dy$ N.

MAKING CONNECTIONS, CHAPTER 5

1. **(a)** By equation (2) of Section 5.3, the volume is $V = \int_0^1 2\pi x f(x^2) \, dx$. Making the substitution

$u = x^2$, $du = 2x \, dx$ gives $V = \int_0^1 2\pi f(u) \cdot \dfrac{1}{2} \, du = \pi \int_0^1 f(u) \, du = \pi A_1$.

(b) By the Theorem of Pappus, the volume in (a) equals $2\pi A_2 \bar{x}$, where $\bar{x} = a$ is the x-coordinate of the centroid of R. Hence $a = \dfrac{\pi A_1}{2\pi A_2} = \dfrac{A_1}{2A_2}$.

3. The area of the annulus with inner radius r and outer radius $r + \Delta r$ is $\pi(r + \Delta r)^2 - \pi r^2 \approx 2\pi r \Delta r$,

so its mass is approximately $2\pi r f(r)\Delta r$. Hence the total mass of the lamina is $\int_0^a 2\pi r f(r) \, dr$.

5. **(a)** Consider any solid obtained by sliding a horizontal region, of any shape, some distance vertically. Thus the top and bottom faces, and every horizontal cross-section in between, are all congruent. This includes all of the cases described in part (a) of the problem.

Suppose such a solid, whose base has area A, is floating in a fluid so that its base is a distance h below the surface. The pressure at the base is ρh, so the fluid exerts an upward force on the base of magnitude $\rho h A$. The fluid also exerts forces on the sides of the solid, but those are horizontal, so they don't contribute to the buoyancy. Hence the buoyant force equals $\rho h A$. Since the part of the solid which is below the surface has volume hA, the buoyant force equals the weight of fluid which would fill that volume; i.e. the weight of the fluid displaced by the solid.

(b) Now consider a solid which is the union of finitely many solids of the type described above. The buoyant force on such a solid is the sum of the buoyant forces on its constituents, which equals the sum of the weights of the fluid displaced by them, which equals the weight of the fluid displaced by the whole solid. So Archimedes' Principle applies to the union.

Any solid can be approximated by such unions, so it is plausible that Archimedes' Principle applies to all solids.

CHAPTER 6
Exponential, Logarithmic, and Inverse Trigonometric Functions

EXERCISE SET 6.1

1. (a) -4 (b) 4 (c) $1/4$

3. (a) 2.9690 (b) 0.0341

5. (a) $\log_2 16 = \log_2(2^4) = 4$ (b) $\log_2\left(\dfrac{1}{32}\right) = \log_2(2^{-5}) = -5$

     (c) $\log_4 4 = 1$ (d) $\log_9 3 = \log_9(9^{1/2}) = 1/2$

7. (a) 1.3655 (b) -0.3011

9. (a) $2\ln a + \dfrac{1}{2}\ln b + \dfrac{1}{2}\ln c = 2r + s/2 + t/2$ (b) $\ln b - 3\ln a - \ln c = s - 3r - t$

11. (a) $1 + \log x + \frac{1}{2}\log(x-3)$ (b) $2\ln|x| + 3\ln(\sin x) - \frac{1}{2}\ln(x^2+1)$

13. $\log\dfrac{2^4(16)}{3} = \log(256/3)$ **15.** $\ln\dfrac{\sqrt[3]{x}(x+1)^2}{\cos x}$

17. $\sqrt{x} = 10^{-1} = 0.1,\ x = 0.01$

19. $1/x = e^{-2},\ x = e^2$ **21.** $2x = 8,\ x = 4$

23. $\ln 2x^2 = \ln 3,\ 2x^2 = 3,\ x^2 = 3/2,\ x = \sqrt{3/2}$ (we discard $-\sqrt{3/2}$ because it does not satisfy the original equation)

25. $\ln 5^{-2x} = \ln 3,\ -2x\ln 5 = \ln 3,\ x = -\dfrac{\ln 3}{2\ln 5}$

27. $e^{3x} = 7/2,\ 3x = \ln(7/2),\ x = \dfrac{1}{3}\ln(7/2)$

29. $e^{-x}(x+2) = 0$ so $e^{-x} = 0$ (impossible) or $x + 2 = 0,\ x = -2$

31. (a) Domain: all x; range: $y > -1$ (b) Domain: $x \neq 0$; range: all y

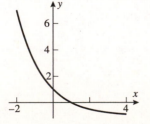

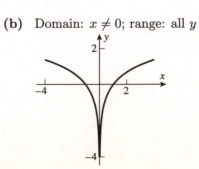

33. **(a)** Domain: $x \neq 0$; range: all y

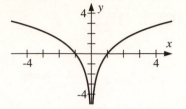

(b) Domain: all x; range: $0 < y \le 1$

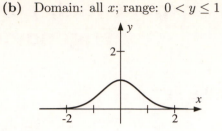

35. False. The graph of an exponential function passes through $(0, 1)$, but the graph of $y = x^3$ does not.

37. True, by definition.

39. $\log_2 7.35 = (\log 7.35)/(\log 2) = (\ln 7.35)/(\ln 2) \approx 2.8777$;
$\log_5 0.6 = (\log 0.6)/(\log 5) = (\ln 0.6)/(\ln 5) \approx -0.3174$

41.

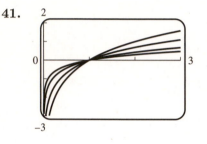

43. **(a)** no, the curve passes through the origin

(b) $y = 2^{x/4}$

(c) $y = 2^{-x}$

(d) $y = (\sqrt{5})^x$

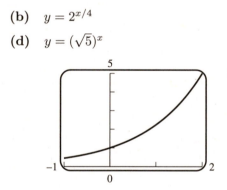

45. $\log(1/2) < 0$ so $3\log(1/2) < 2\log(1/2)$

47. $75e^{-t/125} = 15, t = -125 \ln(1/5) = 125 \ln 5 \approx 201$ days.

49. **(a)** 7.4; basic   **(b)** 4.2; acidic   **(c)** 6.4; acidic   **(d)** 5.9; acidic

51. **(a)** 140 dB; damage   **(b)** 120 dB; damage
(c) 80 dB; no damage   **(d)** 75 dB; no damage

53. Let I_A and I_B be the intensities of the automobile and blender, respectively. Then $\log_{10} I_A/I_0 = 7$ and $\log_{10} I_B/I_0 = 9.3$, $I_A = 10^7 I_0$ and $I_B = 10^{9.3} I_0$, so $I_B/I_A = 10^{2.3} \approx 200$.

55. **(a)** $\log E = 4.4 + 1.5(8.2) = 16.7, E = 10^{16.7} \approx 5 \times 10^{16}$ J

(b) Let M_1 and M_2 be the magnitudes of earthquakes with energies of E and $10E$, respectively. Then $1.5(M_2 - M_1) = \log(10E) - \log E = \log 10 = 1$, $M_2 - M_1 = 1/1.5 = 2/3 \approx 0.67$.

EXERCISE SET 6.2

1. $\dfrac{1}{5x}(5) = \dfrac{1}{x}$

3. $\dfrac{1}{1+x}$

5. $\dfrac{1}{x^2-1}(2x) = \dfrac{2x}{x^2-1}$

7. $\dfrac{d}{dx}\ln x - \dfrac{d}{dx}\ln(1+x^2) = \dfrac{1}{x} - \dfrac{2x}{1+x^2} = \dfrac{1-x^2}{x(1+x^2)}$

9. $\dfrac{d}{dx}(2\ln x) = 2\dfrac{d}{dx}\ln x = \dfrac{2}{x}$

11. $\dfrac{1}{2}(\ln x)^{-1/2}\left(\dfrac{1}{x}\right) = \dfrac{1}{2x\sqrt{\ln x}}$

13. $\ln x + x\dfrac{1}{x} = 1 + \ln x$

15. $2x\log_2(3-2x) + \dfrac{-2x^2}{(\ln 2)(3-2x)}$

17. $\dfrac{2x(1+\log x) - x/(\ln 10)}{(1+\log x)^2}$

19. $\dfrac{1}{\ln x}\left(\dfrac{1}{x}\right) = \dfrac{1}{x\ln x}$

21. $\dfrac{1}{\tan x}(\sec^2 x) = \sec x \csc x$

23. $-\dfrac{1}{x}\sin(\ln x)$

25. $\dfrac{1}{\ln 10 \sin^2 x}(2\sin x \cos x) = 2\dfrac{\cot x}{\ln 10}$

27. $\dfrac{d}{dx}\left[3\ln(x-1) + 4\ln(x^2+1)\right] = \dfrac{3}{x-1} + \dfrac{8x}{x^2+1} = \dfrac{11x^2 - 8x + 3}{(x-1)(x^2+1)}$

29. $\dfrac{d}{dx}\left[\ln \cos x - \dfrac{1}{2}\ln f(4-3x^2)\right] = -\tan x + \dfrac{3x}{4-3x^2}$

31. true

33. true, if $x > 0$ then $\dfrac{d}{dx}\ln|x| = 1/x$; if $x < 0$ then $\dfrac{d}{dx}\ln|x| = 1/x$

35. $\ln|y| = \ln|x| + \dfrac{1}{3}\ln|1+x^2|$, $\dfrac{dy}{dx} = x\sqrt[3]{1+x^2}\left[\dfrac{1}{x} + \dfrac{2x}{3(1+x^2)}\right]$

37. $\ln|y| = \dfrac{1}{3}\ln|x^2-8| + \dfrac{1}{2}\ln|x^3+1| - \ln|x^6 - 7x + 5|$

$\dfrac{dy}{dx} = \dfrac{(x^2-8)^{1/3}\sqrt{x^3+1}}{x^6 - 7x + 5}\left[\dfrac{2x}{3(x^2-8)} + \dfrac{3x^2}{2(x^3+1)} - \dfrac{6x^5 - 7}{x^6 - 7x + 5}\right]$

39. **(a)** $\log_x e = \dfrac{\ln e}{\ln x} = \dfrac{1}{\ln x}$, $\dfrac{d}{dx}[\log_x e] = -\dfrac{1}{x(\ln x)^2}$

(b) $\log_x 2 = \dfrac{\ln 2}{\ln x}$, $\dfrac{d}{dx}[\log_x 2] = -\dfrac{\ln 2}{x(\ln x)^2}$

41. $f'(x_0) = \dfrac{1}{x_0} = e$, $y_0 = -1$, $y - (-1) = e(x - x_0) = ex - 1$, $y = ex - 2$

43. $f(x_0) = f(-e) = 1$, $f'(x)\Big|_{x=-e} = -\dfrac{1}{e}$, $y - 1 = -\dfrac{1}{e}(x + e)$, $y = -\dfrac{1}{e}x$

45. **(a)** Let the equation of the tangent line be $y = mx$ and suppose that it meets the curve at (x_0, y_0). Then $m = \dfrac{1}{x}\Big|_{x=x_0} = \dfrac{1}{x_0}$ and $y_0 = mx_0 + b = \ln x_0$. So $m = \dfrac{1}{x_0} = \dfrac{\ln x_0}{x_0}$ and $\ln x_0 = 1$, $x_0 = e$, $m = \frac{1}{e}$ and the equation of the tangent line is $y = \dfrac{1}{e}x$.

(b) Let $y = mx + b$ be a line tangent to the curve at (x_0, y_0). Then b is the y-intercept and the slope of the tangent line is $m = \dfrac{1}{x_0}$. Moreover, at the point of tangency, $mx_0 + b = \ln x_0$ or $\dfrac{1}{x_0}x_0 + b = \ln x_0$, $b = \ln x_0 - 1$, as required.

47. The area of the triangle PQR, given by $|PQ||QR|/2$ is required. $|PQ| = w$, and, by Exercise 45 Part (b), $|QR| = 1$, so area $= w/2$.

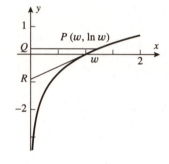

49. If $x = 0$ then $y = \ln e = 1$, and $\dfrac{dy}{dx} = \dfrac{1}{x + e}$. But $e^y = x + e$, so $\dfrac{dy}{dx} = \dfrac{1}{e^y} = e^{-y}$.

51. Let $y = \ln(x + a)$. Following Exercise 49 we get $\dfrac{dy}{dx} = \dfrac{1}{x + a} = e^{-y}$, and when $x = 0$, $y = \ln(a) = 0$ if $a = 1$, so let $a = 1$, then $y = \ln(x + 1)$.

53. **(a)** Set $f(x) = \ln(1 + 3x)$. Then $f'(x) = \dfrac{3}{1 + 3x}$, $f'(0) = 3$. But
$$f'(0) = \lim_{x \to 0} \frac{f(x) - f(0)}{x} = \lim_{x \to 0} \frac{\ln(1 + 3x)}{x}$$

(b) $f(x) = \ln(1 - 5x)$, $f'(x) = \dfrac{-5}{1 - 5x}$, $f'(0) = -5$. But $f'(0) = \lim_{x \to 0} \dfrac{f(x) - f(0)}{x} = \lim_{x \to 0} \dfrac{\ln(1 - 5x)}{x}$

55. **(a)** Let $f(x) = \ln(\cos x)$, then $f(0) = \ln(\cos 0) = \ln 1 = 0$, so $f'(0) = \lim_{x \to 0} \dfrac{f(x) - f(0)}{x} = \lim_{x \to 0} \dfrac{\ln(\cos x)}{x}$, and $f'(0) = -\tan 0 = 0$.

(b) Let $f(x) = x^{\sqrt{2}}$, then $f(1) = 1$, so $f'(1) = \lim_{h \to 0} \dfrac{f(1 + h) - f(1)}{h} = \lim_{h \to 0} \dfrac{(1 + h)^{\sqrt{2}} - 1}{h}$, and $f'(x) = \sqrt{2}x^{\sqrt{2} - 1}$, $f'(1) = \sqrt{2}$.

57. $2\ln|x| - 3\cos x + C$

59. $\displaystyle\int \dfrac{du}{u} = \ln u + C = \ln|\ln x| + C$

61. $u = 1 + x^5, du = 5x^4\, du, \dfrac{1}{5}\displaystyle\int \dfrac{du}{u} = \dfrac{1}{5}\ln|u| + C = \dfrac{1}{5}\ln|1 + x^5| + C$

63. $\displaystyle\int dt + \int \dfrac{dt}{t} = t + \ln|t| + C$

65. $\dfrac{3}{2}\ln(1 + x^2)\Big]_0^2 = \dfrac{3}{2}\ln 5$

67. $du = \dfrac{1}{x}\, dx, \displaystyle\int_e^{e^2} \dfrac{\ln x}{x}\, dx = \int_1^2 u\, du = \dfrac{1}{2}u^2\Big]_1^2 = \dfrac{3}{2}$

69. $\dfrac{1}{2}\ln(2x + e)\Big]_0^e = \dfrac{1}{2}[\ln(3e) - \ln e] = \dfrac{\ln 3}{2}$

71. $y = \ln|t| + C, 5 = y(-1) = C, y = \ln|t| + 5$

EXERCISE SET 6.3

1. **(a)** $f'(x) = 5x^4 + 3x^2 + 1 \geq 1$ so f is one-to-one on $-\infty < x < +\infty$.

    (b) $f(1) = 3$ so $1 = f^{-1}(3)$; $\dfrac{d}{dx}f^{-1}(x) = \dfrac{1}{f'(f^{-1}(x))}, (f^{-1})'(3) = \dfrac{1}{f'(1)} = \dfrac{1}{9}$

3. $f^{-1}(x) = \dfrac{2}{x} - 3$, so directly $\dfrac{d}{dx}f^{-1}(x) = -\dfrac{2}{x^2}$. Using Formula (2),

$f'(x) = \dfrac{-2}{(x + 3)^2}$, so $\dfrac{1}{f'(f^{-1}(x))} = -(1/2)(f^{-1}(x) + 3)^2$,

$\dfrac{d}{dx}f^{-1}(x) = -(1/2)\left(\dfrac{2}{x}\right)^2 = -\dfrac{2}{x^2}$

5. **(a)** $f'(x) = 2x + 8$; $f' < 0$ on $(-\infty, -4)$ and $f' > 0$ on $(-4, +\infty)$; not enough information. By inspection, $f(1) = 10 = f(-9)$, so not one-to-one

    (b) $f'(x) = 10x^4 + 3x^2 + 3 \geq 3 > 0$; $f'(x)$ is positive for all x, so f is one-to-one

    (c) $f'(x) = 2 + \cos x \geq 1 > 0$ for all x, so f is one-to-one

    (d) $f'(x) = -(\ln 2)\left(\dfrac{1}{2}\right)^x < 0$ because $\ln 2 > 0$, so f is one-to-one for all x.

7. $y = f^{-1}(x), x = f(y) = 5y^3 + y - 7, \dfrac{dx}{dy} = 15y^2 + 1, \dfrac{dy}{dx} = \dfrac{1}{15y^2 + 1}$;

    check: $1 = 15y^2\dfrac{dy}{dx} + \dfrac{dy}{dx}, \dfrac{dy}{dx} = \dfrac{1}{15y^2 + 1}$

9. $y = f^{-1}(x), x = f(y) = 2y^5 + y^3 + 1, \dfrac{dx}{dy} = 10y^4 + 3y^2, \dfrac{dy}{dx} = \dfrac{1}{10y^4 + 3y^2}$;

    check: $1 = 10y^4\dfrac{dy}{dx} + 3y^2\dfrac{dy}{dx}, \dfrac{dy}{dx} = \dfrac{1}{10y^4 + 3y^2}$

11. Let $P(a, b)$ be given, not on the line $y = x$. Let Q be its reflection across the line $y = x$, yet to be determined. Let Q_1 have coordinates (b, a).

    (a) Since P does not lie on $y = x$, we have $a \neq b$, i.e. $P \neq Q_1$ since they have different abscissas. The line $\overleftrightarrow{PQ_1}$ has slope $(b - a)/(a - b) = -1$ which is the negative reciprocal of $m = 1$ and so the two lines are perpendicular.

    (b) Let (c, d) be the midpoint of the segment PQ_1. Then $c = (a + b)/2$ and $d = (b + a)/2$ so $c = d$ and the midpoint is on $y = x$.

    (c) Let $Q(c, d)$ be the reflection of P through $y = x$. By definition this means P and Q lie on a line perpendicular to the line $y = x$ and the midpoint of P and Q lies on $y = x$.

(d) Since the line through P and Q_1 is perpendicular to the line $y = x$ it is parallel to the line through P and Q; since both pass through P they are the same line. Finally, since the midpoints of P and Q_1 and of P and Q both lie on $y = x$, they are the same point, and consequently $Q = Q_1$.

13. If $x < y$ then $f(x) \le f(y)$ and $g(x) \le g(y)$; thus $f(x) + g(x) \le f(y) + g(y)$. Moreover, $g(x) \le g(y)$, so $f(g(x)) \le f(g(y))$. Note that $f(x)g(x)$ need not be increasing, e.g. $f(x) = g(x) = x$, both increasing for all x, yet $f(x)g(x) = x^2$, not an increasing function.

15. $7e^{7x}$

17. $x^3 e^x + 3x^2 e^x = x^2 e^x (x + 3)$

19. $\dfrac{dy}{dx} = \dfrac{(e^x + e^{-x})(e^x + e^{-x}) - (e^x - e^{-x})(e^x - e^{-x})}{(e^x + e^{-x})^2}$

$\qquad = \dfrac{(e^{2x} + 2 + e^{-2x}) - (e^{2x} - 2 + e^{-2x})}{(e^x + e^{-x})^2} = 4/(e^x + e^{-x})^2$

21. $(x \sec^2 x + \tan x)e^{x \tan x}$

23. $(1 - 3e^{3x})e^{(x - e^{3x})}$

25. $\dfrac{(x - 1)e^{-x}}{1 - xe^{-x}} = \dfrac{x - 1}{e^x - x}$

27. $f'(x) = 2^{x^2} \ln 2 \cdot 2x = 2(\ln 2) x\, 2^{x^2}$; $y = 2^{x^2}$, $\ln y = x^2 \ln 2$, $\dfrac{1}{y} y' = 2(\ln 2)x$, $y' = 2(\ln 2) x\, 2^{x^2}$

29. $f'(x) = \pi^{\sin x}(\ln \pi) \cos x$;

$\qquad y = \pi^{\sin x}$, $\ln y = (\sin x) \ln \pi$, $\dfrac{1}{y} y' = (\ln \pi) \cos x$, $y' = \pi^{\sin x}(\ln \pi) \cos x$

31. $\ln y = (\ln x) \ln(x^3 - 2x)$, $\dfrac{1}{y} \dfrac{dy}{dx} = \dfrac{3x^2 - 2}{x^3 - 2x} \ln x + \dfrac{1}{x} \ln(x^3 - 2x)$,

$\qquad \dfrac{dy}{dx} = (x^3 - 2x)^{\ln x} \left[\dfrac{3x^2 - 2}{x^3 - 2x} \ln x + \dfrac{1}{x} \ln(x^3 - 2x) \right]$

33. $\ln y = (\tan x) \ln(\ln x)$, $\dfrac{1}{y} \dfrac{dy}{dx} = \dfrac{1}{x \ln x} \tan x + (\sec^2 x) \ln(\ln x)$,

$\qquad \dfrac{dy}{dx} = (\ln x)^{\tan x} \left[\dfrac{\tan x}{x \ln x} + (\sec^2 x) \ln(\ln x) \right]$

35. $\ln y = (\ln x)(\ln(\ln x))$, $\dfrac{dy/dx}{y} = (1/x)(\ln(\ln x)) + (\ln x)\dfrac{1/x}{\ln x} = (1/x)(1 + \ln(\ln x))$

$\qquad dy/dx = \dfrac{1}{x}(\ln x)^{\ln x}(1 + \ln \ln x)$

37. false; $y = Ae^x$ also satisfies $\dfrac{dy}{dx} = y$

39. true; examine the cases $x > 0$ and $x < 0$ separately

41. (a) $f(x) = x^3 - 3x^2 + 2x = x(x - 1)(x - 2)$ so $f(0) = f(1) = f(2) = 0$ thus f is not one-to-one.

(b) $f'(x) = 3x^2 - 6x + 2$, $f'(x) = 0$ when $x = \dfrac{6 \pm \sqrt{36-24}}{6} = 1 \pm \sqrt{3}/3$. $f'(x) > 0$ (f is increasing) if $x < 1 - \sqrt{3}/3$, $f'(x) < 0$ (f is decreasing) if $1 - \sqrt{3}/3 < x < 1 + \sqrt{3}/3$, so $f(x)$ takes on values less than $f(1 - \sqrt{3}/3)$ on both sides of $1 - \sqrt{3}/3$ thus $1 - \sqrt{3}/3$ is the largest value of k.

43. (a) $f'(x) = 4x^3 + 3x^2 = (4x+3)x^2 > 0$ for $0 < x < 2$. So f is increasing on $[0,2]$ and is therefore one-to-one.

(b) $F'(x) = 2f'(2g(x))g'(x)$ so $F'(3) = 2f'(2g(3))g'(3)$. By inspection $f(1) = 3$, so $g(3) = f^{-1}(3) = 1$ and $g'(3) = (f^{-1})'(3) = 1/f'(f^{-1}(3)) = 1/f'(1) = 1/7$ because $f'(x) = 4x^3 + 3x^2$. Thus $F'(3) = 2f'(2)(1/7) = 2(44)(1/7) = 88/7$.
$F(3) = f(2g(3)) = f(2 \cdot 1) = f(2) = 25$, so the line tangent to $F(x)$ at $(3, 25)$ has the equation $y - 25 = (88/7)(x-3)$, $y = (88/7)x - 89/7$.

45. $y = Ae^{kt}$, $dy/dt = kAe^{kt} = k(Ae^{kt}) = ky$

47. (a) $y' = -xe^{-x} + e^{-x} = e^{-x}(1-x)$, $xy' = xe^{-x}(1-x) = y(1-x)$

(b) $y' = -x^2 e^{-x^2/2} + e^{-x^2/2} = e^{-x^2/2}(1-x^2)$, $xy' = xe^{-x^2/2}(1-x^2) = y(1-x^2)$

49. $\ln y = \ln 60 - \ln(5 + 7e^{-t})$, $\dfrac{y'}{y} = \dfrac{7e^{-t}}{5 + 7e^{-t}} = \dfrac{7e^{-t} + 5 - 5}{5 + 7e^{-t}} = 1 - \dfrac{1}{12}y$, so

$\dfrac{dy}{dt} = r\left(1 - \dfrac{y}{K}\right)y$, with $r = 1, K = 12$.

51. $f(x) = e^{3x}$, $f'(0) = \lim\limits_{x \to 0} \dfrac{f(x) - f(0)}{x - 0} = 3e^{3x}\Big]_{x=0} = 3$

53. $\lim\limits_{h \to 0} \dfrac{10^h - 1}{h} = \dfrac{d}{dx}10^x\Big|_{x=0} = \dfrac{d}{dx}e^{x \ln 10}\Big|_{x=0} = \ln 10$

55. $\displaystyle\int \left[\dfrac{2}{x} + 3e^x\right] dx = 2\ln|x| + 3e^x + C$

57. $-\dfrac{1}{5}\displaystyle\int e^u \, du = -\dfrac{1}{5}e^u + C = -\dfrac{1}{5}e^{-5x} + C$

59. $u = 2x$, $du = 2dx$; $\dfrac{1}{2}\displaystyle\int e^u \, du = \dfrac{1}{2}e^u + C = \dfrac{1}{2}e^{2x} + C$

61. $u = \sin x$, $du = \cos x \, dx$; $\displaystyle\int e^u \, du = e^u + C = e^{\sin x} + C$

63. $u = -2x^3$, $du = -6x^2$, $-\dfrac{1}{6}\displaystyle\int e^u du = -\dfrac{1}{6}e^u + C = -\dfrac{1}{6}e^{-2x^3} + C$

65. $\displaystyle\int e^{-x}dx$; $u = -x$, $du = -dx$; $-\displaystyle\int e^u du = -e^u + C = -e^{-x} + C$

67. $\ln(e^x) + \ln(e^{-x}) = \ln(e^x e^{-x}) = \ln 1 = 0$ so $\displaystyle\int [\ln(e^x) + \ln(e^{-x})]dx = C$

69. $5e^x\Big]_{\ln 2}^{3} = 5e^3 - 5(2) = 5e^3 - 10$

71. $\dfrac{1}{2}\displaystyle\int_{-1}^{1} e^u\,du = \dfrac{1}{2}\left(e - e^{-1}\right)$

73. $u = e^x + 4,\ du = e^x dx,\ u = e^{-\ln 3} + 4 = \dfrac{1}{3} + 4 = \dfrac{13}{3}$ when $x = -\ln 3$

$u = e^{\ln 3} + 4 = 3 + 4 = 7$ when $x = \ln 3,\ \displaystyle\int_{13/3}^{7} \dfrac{1}{u}\,du = \ln u\Big]_{13/3}^{7} = \ln(7) - \ln(13/3) = \ln(21/13),$ or

$\ln(e^x + 4)\Big]_{-\ln 3}^{\ln 3} = \ln 7 - \ln(13/3) = \ln(21/13)$

75. $y(t) = (802.137)\displaystyle\int e^{1.528t}\,dt \approx 524.959 e^{1.528t} + C;\ y(0) = 750 \approx 524.959 + C,\ C \approx 225.041,$

$y(t) \approx 524959 e^{1.528t} + 225.041,\ y(12) \approx 48{,}233{,}500{,}000$

77. $\displaystyle\int_0^k e^{2x}\,dx = 3,\ \dfrac{1}{2}e^{2x}\Big]_0^k = 3,\ \dfrac{1}{2}(e^{2k} - 1) = 3,\ e^{2k} = 7,\ k = \dfrac{1}{2}\ln 7$

EXERCISE SET 6.4

1. critical point $x = 0$:
  $x = 0$: relative minimum

f': $\begin{array}{c}---\ \underset{0}{0}\ +++ \\ \hline \\ 0\end{array}$

3. critical points $x = -1, 1$:
  $x = -1$: relative minimum
  $x = 1$: relative maximum

f': $---\ \underset{-1}{0}+\ +\ +\underset{1}{0}---$

5. $f'(x) = x^2(3 - 2x)e^{-2x},\ f'(x) = 0$ for x in $[1, 4]$ when $x = 3/2$; if $x = 1, 3/2, 4$, then $f(x) = e^{-2}, \dfrac{27}{8}e^{-3}, 64e^{-8}$; critical point at $x = 3/2$; absolute maximum of $\dfrac{27}{8}e^{-3}$ at $x = 3/2$, absolute minimum of $64e^{-8}$ at $x = 4$

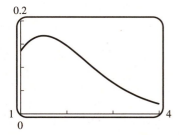

7. $f'(x) = -\dfrac{3x^2 - 10x + 3}{x^2 + 1},\ f'(x) = 0$ when $x = \dfrac{1}{3},\ 3.\ f(0) = 0,$

$f\left(\dfrac{1}{3}\right) = 5\ln\left(\dfrac{10}{9}\right) - 1,\ f(3) = 5\ln 10 - 9,\ f(4) = 5\ln 17 - 12$

and thus f has an absolute minimum of $5(\ln 10 - \ln 9) - 1$ at $x = 1/3$ and an absolute maximum of $5\ln 10 - 9$ at $x = 3$.

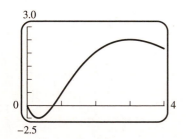

9. (a) $\displaystyle\lim_{x \to +\infty} xe^x = +\infty,\ \lim_{x \to -\infty} xe^x = 0$

  (b) $y = xe^x$;
  $y' = (x + 1)e^x$;
  $y'' = (x + 2)e^x$
  relative minimum at $(-1, -e^{-1}) \approx (-1, -0.37)$
  inflection point at $(-2, -2e^{-2}) \approx (-2, -0.27)$
  horizontal asymptote $y = 0$ as $x \to -\infty$

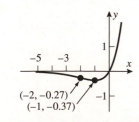

11. **(a)** $\lim\limits_{x\to+\infty}\dfrac{x^2}{e^{2x}}=0,\quad \lim\limits_{x\to-\infty}\dfrac{x^2}{e^{2x}}=+\infty$

(b) $y=x^2/e^{2x}=x^2 e^{-2x}$;

$y'=2x(1-x)e^{-2x}$;

$y''=2(2x^2-4x+1)e^{-2x}$;

$y''=0$ if $2x^2-4x+1=0$, when

$x=\dfrac{4\pm\sqrt{16-8}}{4}=1\pm\sqrt{2}/2\approx 0.29,1.71$

horizontal asymptote $y=0$ as $x\to+\infty$

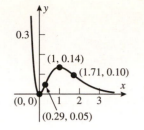

13. **(a)** $\lim\limits_{x\to\pm\infty} x^2 e^{-x^2}=0$

(b) $y=x^2 e^{-x^2}$;

$y'=2x(1-x^2)e^{-x^2}$;

$y'=0$ if $x=0,\pm1$;

$y''=2(1-5x^2+2x^4)e^{-x^2}$

$y''=0$ if $2x^4-5x^2+1=0$,

$x^2=\dfrac{5\pm\sqrt{17}}{4}$,

$x=\pm\tfrac{1}{2}\sqrt{5+\sqrt{17}}\approx\pm1.51$,

$x=\pm\tfrac{1}{2}\sqrt{5-\sqrt{17}}\approx\pm0.47$

horizontal asymptote $y=0$ as $x\to\pm\infty$

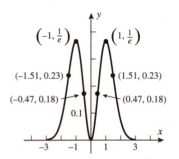

15. **(a)** $\lim\limits_{x\to-\infty} f(x)=0,\quad \lim\limits_{x\to+\infty} f(x)=-\infty$

(b) $f'(x)=-\dfrac{e^x(x-2)}{(x-1)^2}$ so

$f'(x)=0$ when $x=2$

$f''(x)=-\dfrac{e^x(x^2-4x+5)}{(x-1)^3}$ so

$f''(x)\ne 0$ always

relative maximum when $x=2$, no point of inflection
vertical asymptote $x=1$
horizontal asymptote $y=0$ as $x\to-\infty$

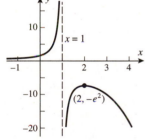

17. **(a)** $\lim\limits_{x\to+\infty} f(x)=0,\quad \lim\limits_{x\to-\infty} f(x)=+\infty$

(b) $f'(x)=x(2-x)e^{1-x},\ f''(x)=(x^2-4x+2)e^{1-x}$
critical points at $x=0,2$;
relative minimum at $x=0$,
relative maximum at $x=2$
points of inflection at $x=2\pm\sqrt{2}$
horizontal asymptote $y=0$ as $x\to+\infty$

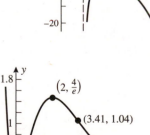

19. **(a)** $\lim\limits_{x\to0^+} y=\lim\limits_{x\to0^+} x\ln x=\lim\limits_{x\to0^+}\dfrac{\ln x}{1/x}=\lim\limits_{x\to0^+}\dfrac{1/x}{-1/x^2}=0$;

$\lim\limits_{x\to+\infty} y=+\infty$

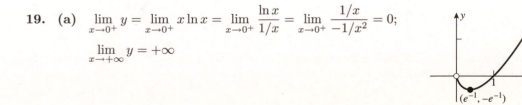

(b) $y = x \ln x$,
$y' = 1 + \ln x$,
$y'' = 1/x$,
$y' = 0$ when $x = e^{-1}$

21. (a) $\lim\limits_{x \to 0^+} x^2 \ln(2x) = \lim\limits_{x \to 0^+} (x^2 \ln 2) + \lim\limits_{x \to 0^+} (x^2 \ln x) = 0$

by the rule given, $\lim\limits_{x \to +\infty} x^2 \ln x = +\infty$ by inspection

(b) $y = x^2 \ln(2x)$,
$y' = 2x \ln(2x) + x$,
$y'' = 2 \ln(2x) + 3$,
$y' = 0$ if $x = 1/(2\sqrt{e})$,
$y'' = 0$ if $x = 1/(2e^{3/2})$

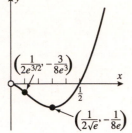

23. (a) $\lim\limits_{x \to +\infty} f(x) = +\infty$, $\lim\limits_{x \to 0^+} f(x) = 0$

(b) $y = x^{2/3} \ln x$

$y' = \dfrac{2 \ln x + 3}{3 x^{1/3}}$

$y' = 0$ when $\ln x = -\dfrac{3}{2}$, $x = e^{-3/2}$

$y'' = \dfrac{-3 + 2 \ln x}{9 x^{4/3}}$,

$y'' = 0$ when $\ln x = \dfrac{3}{2}$, $x = e^{3/2}$

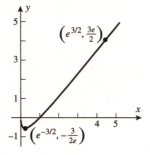

25. (a)

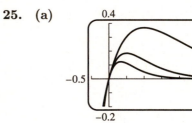

(b) $y' = (1 - bx)e^{-bx}$, $y'' = b^2(x - 2/b)e^{-bx}$; relative maximum at $x = 1/b$, $y = 1/(be)$; point of inflection at $x = 2/b$, $y = 2/(be^2)$. Increasing b moves the relative maximum and the point of inflection to the left and down, i.e. towards the origin.

27. (a) The oscillations of $e^x \cos x$ about zero increase as $x \to +\infty$ so the limit does not exist, and $\lim\limits_{x \to -\infty} e^x \cos x = 0$.

(b) $y = e^x$ and $y = e^x \cos x$ intersect for $x = 2\pi n$ for any integer n. $y = -e^x$ and $y = e^x \cos x$ intersect for $x = 2\pi n + \pi$ for any integer n.

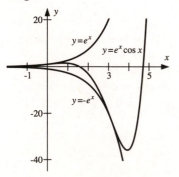

(c) The curve $y = e^{ax} \cos bx$ oscillates between $y = e^{ax}$ and $y = -e^{ax}$. The frequency of oscillation increases when b increases.

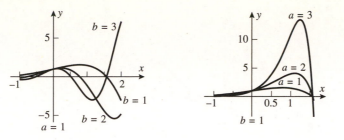

29. false; it is the reflection of the graph over the line $y = x$

31. true, it is $(e^2 + 1)/(e^2 - 1)$

33. **(a)** $y'(t) = \dfrac{LAke^{-kt}}{(1 + Ae^{-kt})^2} S$, so $y'(0) = \dfrac{LAk}{(1 + A)^2}$

  (b) The rate of growth increases to its maximum, which occurs when y is halfway between 0 and L, or when $t = \dfrac{1}{k} \ln A$; it then decreases back towards zero.

  (c) From (2) one sees that $\dfrac{dy}{dt}$ is maximized when y lies half way between 0 and L, i.e. $y = L/2$. This follows since the right side of (2) is a parabola (with y as independent variable) with y-intercepts $y = 0, L$. The value $y = L/2$ corresponds to $t = \dfrac{1}{k} \ln A$, from (4).

35. $t \approx 7.67$

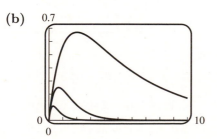

37. Since $0 < y < L$ the right-hand side of (5) of Example 3 can change sign only if the factor $L - 2y$ changes sign, which it does when $y = L/2$, at which point we have $\dfrac{L}{2} = \dfrac{L}{1 + Ae^{-kt}}$, $1 = Ae^{-kt}$, $t = \dfrac{1}{k} \ln A$.

39. **(a)** $\dfrac{dC}{dt} = \dfrac{K}{a - b} \left(ae^{-at} - be^{-bt} \right)$ so $\dfrac{dC}{dt} = 0$ at $t = \dfrac{\ln(a/b)}{a - b}$. This is the only stationary point and $C(0) = 0$, $\lim_{t \to +\infty} C(t) = 0$, $C(t) > 0$ for $0 < t < +\infty$, so it is an absolute maximum.

  (b) 0.7

41. 3/2

43. $A = \displaystyle\int_0^{\ln 2} \left(e^{2x} - e^x\right) dx$

$= \left(\dfrac{1}{2}e^{2x} - e^x\right)\Bigg]_0^{\ln 2} = 1/2$

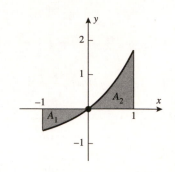

45. Area $= -\displaystyle\int_{-1}^0 \left(e^x - 1\right) dx + \int_0^1 \left(e^x - 1\right) dx = 1/e + e - 2$

47. $f_{\text{ave}} = \dfrac{1}{e-1}\displaystyle\int_1^e \dfrac{1}{x}\, dx = \dfrac{1}{e-1}(\ln e - \ln 1) = \dfrac{1}{e-1}$

49. $f_{\text{ave}} = \dfrac{1}{4}\displaystyle\int_0^4 e^{-2x}\, dx = -\dfrac{1}{8}e^{-2x}\Bigg]_0^4 = \dfrac{1}{8}(1 - e^{-8})$

51. **(a)** the displacement is positive on $(0,5)$

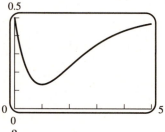

(b) The displacement is $\displaystyle\int_0^5 \left(\dfrac{1}{2} - te^{-t}\right) dt = \left[\dfrac{t}{2} + (t+1)e^{-t}\right]_0^5 = \dfrac{3}{2} + 6e^{-5}$.

53. The graphs of $y = 1$ and $y = e^x \sin x$ intersect twice, near $x = 1$ and $x = 3$. Let $f(x) = 1 - e^x \sin x$, $f'(x) = -e^x(\cos x + \sin x)$,

and $x_{n+1} = x_n + \dfrac{1 - e_n^x \sin x_n}{e_n^x(\cos x_n + \sin x_n)}$.

If $x_1 = 1$ then $x_2 \approx 0.65725814$, $x_3 \approx 0.59118311, \ldots$,
$x_5 \approx x_6 \approx 0.58853274$,
and if $x_1 = 3$ then $x_2 \approx 3.10759324$, $x_3 \approx 3.09649396$,
$\ldots, x_5 \approx x_6 \approx 3.09636393$.

55. A graphing utility shows that there are two inflection points at $x \approx 0.25, -1.25$. These points

are the zeros of $f''(x) = (x^4 + 4x^3 + 8x^2 + 4x - 1)\dfrac{e^{-x}}{(x^2+1)^3}$. It is equivalent to find the zeros of

$g(x) = x^4 + 4x^3 + 8x^2 + 4x - 1$. One root is $x = -1$ by inspection. Since $g'(x) = 4x^3 + 12x^2 + 16x + 4$,

Newton's Method becomes

$$x_{n+1} = x_n - \dfrac{x_n^4 + 4x_n^3 + 8x_n^2 + 4x_n - 1}{4x_n^3 + 12x_n^2 + 16x_n + 4}$$

With $x_0 = 0.25$, $x_1 \approx 0.18572695$, $x_2 \approx 0.179563312$, $x_3 \approx 0.179509029$, $x_4 \approx x_5 \approx 0.179509025$. So the points of inflection are at $x \approx 0.17951$, $x = -1$.

57. $V = \pi \displaystyle\int_0^{\ln 3} e^{2x}\, dx = \dfrac{\pi}{2} e^{2x}\Big]_0^{\ln 3} = 4\pi$

59. $V = 2\pi \displaystyle\int_0^1 \dfrac{x}{x^2+1}\, dx$

$$= \pi \ln(x^2+1)\Big]_0^1 = \pi \ln 2$$

61. $(dx/dt)^2 + (dy/dt)^2 = [e^t(\cos t - \sin t)]^2 + [e^t(\cos t + \sin t)]^2 = 2e^{2t}$,

$$L = \int_0^{\pi/2} \sqrt{2}\, e^t dt = \sqrt{2}(e^{\pi/2} - 1)$$

63. $dy/dx = \dfrac{\sec x \tan x}{\sec x} = \tan x$, $\sqrt{1 + (y')^2} = \sqrt{1 + \tan^2 x} = \sec x$ when $0 < x < \pi/4$, so

$$L = \int_0^{\pi/4} \sec x\, dx = \ln(1 + \sqrt{2})$$

65. $f'(x) = e^x$, $1 + [f'(x)]^2 = 1 + e^{2x}$, $S = \displaystyle\int_0^1 2\pi e^x \sqrt{1 + e^{2x}}\, dx \approx 22.94$

EXERCISE SET 6.5

1. (a) $\displaystyle\lim_{x\to 2} \dfrac{x^2 - 4}{x^2 + 2x - 8} = \lim_{x\to 2} \dfrac{(x-2)(x+2)}{(x+4)(x-2)} = \lim_{x\to 2} \dfrac{x+2}{x+4} = \dfrac{2}{3}$, or, using L'Hôpital's Rule,

$$\lim_{x\to 2} \dfrac{x^2 - 4}{x^2 + 2x - 8} = \lim_{x\to 2} \dfrac{2x}{2x + 2} = \dfrac{2}{3}$$

(b) $\displaystyle\lim_{x\to +\infty} \dfrac{2x - 5}{3x + 7} = \dfrac{2 - \lim\limits_{x\to +\infty} \dfrac{5}{x}}{3 + \lim\limits_{x\to +\infty} \dfrac{7}{x}} = \dfrac{2}{3}$ or, using L'Hôpital's Rule, $\displaystyle\lim_{x\to +\infty} \dfrac{2x - 5}{3x + 7} = \lim_{x\to +\infty} \dfrac{2}{3} = \dfrac{2}{3}$

3. true; $\ln x$ is not defined for negative x

5. false

7. $\displaystyle\lim_{x\to 0} \dfrac{e^x}{\cos x} = 1$

9. $\displaystyle\lim_{\theta\to 0} \dfrac{\sec^2 \theta}{1} = 1$

11. $\displaystyle\lim_{x\to \pi^+} \dfrac{\cos x}{1} = -1$

13. $\displaystyle\lim_{x\to +\infty} \dfrac{1/x}{1} = 0$

15. $\displaystyle\lim_{x\to 0^+} \dfrac{-\csc^2 x}{1/x} = \lim_{x\to 0^+} \dfrac{-x}{\sin^2 x} = \lim_{x\to 0^+} \dfrac{-1}{2\sin x \cos x} = -\infty$

17. $\displaystyle\lim_{x\to +\infty} \dfrac{100x^{99}}{e^x} = \lim_{x\to +\infty} \dfrac{(100)(99)x^{98}}{e^x} = \cdots = \lim_{x\to +\infty} \dfrac{(100)(99)(98)\cdots(1)}{e^x} = 0$

19. $\displaystyle\lim_{x\to+\infty} xe^{-x} = \lim_{x\to+\infty} \frac{x}{e^x} = \lim_{x\to+\infty} \frac{1}{e^x} = 0$

21. $\displaystyle\lim_{x\to+\infty} x\sin(\pi/x) = \lim_{x\to+\infty} \frac{\sin(\pi/x)}{1/x} = \lim_{x\to+\infty} \frac{(-\pi/x^2)\cos(\pi/x)}{-1/x^2} = \lim_{x\to+\infty} \pi\cos(\pi/x) = \pi$

23. $\displaystyle\lim_{x\to(\pi/2)^-} \sec 3x\cos 5x = \lim_{x\to(\pi/2)^-} \frac{\cos 5x}{\cos 3x} = \lim_{x\to(\pi/2)^-} \frac{-5\sin 5x}{-3\sin 3x} = \frac{-5(+1)}{(-3)(-1)} = -\frac{5}{3}$

25. $y = (1-3/x)^x$, $\displaystyle\lim_{x\to+\infty} \ln y = \lim_{x\to+\infty} \frac{\ln(1-3/x)}{1/x} = \lim_{x\to+\infty} \frac{-3}{1-3/x} = -3$, $\displaystyle\lim_{x\to+\infty} y = e^{-3}$

27. $y = (e^x + x)^{1/x}$, $\displaystyle\lim_{x\to 0} \ln y = \lim_{x\to 0} \frac{\ln(e^x + x)}{x} = \lim_{x\to 0} \frac{e^x + 1}{e^x + x} = 2$, $\displaystyle\lim_{x\to 0} y = e^2$

29. $y = (2-x)^{\tan(\pi x/2)}$, $\displaystyle\lim_{x\to 1} \ln y = \lim_{x\to 1} \frac{\ln(2-x)}{\cot(\pi x/2)} = \lim_{x\to 1} \frac{2\sin^2(\pi x/2)}{\pi(2-x)} = 2/\pi$, $\displaystyle\lim_{x\to 1} y = e^{2/\pi}$

31. $\displaystyle\lim_{x\to 0} \left(\frac{1}{\sin x} - \frac{1}{x}\right) = \lim_{x\to 0} \frac{x - \sin x}{x\sin x} = \lim_{x\to 0} \frac{1-\cos x}{x\cos x + \sin x} = \lim_{x\to 0} \frac{\sin x}{2\cos x - x\sin x} = 0$

33. $\displaystyle\lim_{x\to+\infty} \frac{(x^2 + x) - x^2}{\sqrt{x^2 + x} + x} = \lim_{x\to+\infty} \frac{x}{\sqrt{x^2 + x} + x} = \lim_{x\to+\infty} \frac{1}{\sqrt{1 + 1/x} + 1} = 1/2$

35. $\displaystyle\lim_{x\to+\infty} [x - \ln(x^2 + 1)] = \lim_{x\to+\infty} [\ln e^x - \ln(x^2 + 1)] = \lim_{x\to+\infty} \ln\frac{e^x}{x^2 + 1}$,

$\displaystyle\lim_{x\to+\infty} \frac{e^x}{x^2 + 1} = \lim_{x\to+\infty} \frac{e^x}{2x} = \lim_{x\to+\infty} \frac{e^x}{2} = +\infty$ so $\displaystyle\lim_{x\to+\infty} [x - \ln(x^2 + 1)] = +\infty$

37. $y = x^{\sin x}$, $\ln y = \sin x\ln x$, $\displaystyle\lim_{x\to 0^+} \ln y = \lim_{x\to 0^+} \frac{\ln x}{\csc x} = \lim_{x\to 0^+} \frac{1/x}{-\csc x\cot x} = \lim_{x\to 0^+} \left(\frac{\sin x}{x}\right)(-\tan x) =$

$1(-0) = 0$, so $\displaystyle\lim_{x\to 0^+} x^{\sin x} = \lim_{x\to 0^+} y = e^0 = 1$.

39. $y = \left[-\frac{1}{\ln x}\right]^x$, $\ln y = x\ln\left[-\frac{1}{\ln x}\right]$,

$\displaystyle\lim_{x\to 0^+} \ln y = \lim_{x\to 0^+} \frac{\ln\left[-\frac{1}{\ln x}\right]}{1/x} = \lim_{x\to 0^+} \left(-\frac{1}{x\ln x}\right)(-x^2) = -\lim_{x\to 0^+} \frac{x}{\ln x} = 0$, $\displaystyle\lim_{x\to 0^+} y = e^0 = 1$

41. $y = (\ln x)^{1/x}$, $\ln y = (1/x)\ln\ln x$,

$\displaystyle\lim_{x\to+\infty} \ln y = \lim_{x\to+\infty} \frac{\ln\ln x}{x} = \lim_{x\to+\infty} \frac{1/(x\ln x)}{1} = 0$, $\displaystyle\lim_{x\to+\infty} y = 1$

43. $y = (\tan x)^{\pi/2 - x}$, $\ln y = (\pi/2 - x)\ln\tan x$, $\displaystyle\lim_{x\to(\pi/2)^-} \ln y = \lim_{x\to(\pi/2)^-} \frac{\ln\tan x}{1/(\pi/2 - x)}$

$= \displaystyle\lim_{x\to(\pi/2)^-} \frac{(\sec^2 x/\tan x)}{1/(\pi/2 - x)^2} = \lim_{x\to(\pi/2)^-} \frac{(\pi/2 - x)(\pi/2 - x)}{\cos x\sin x}$

$= \displaystyle\lim_{x\to(\pi/2)^-} \frac{(\pi/2 - x)}{\cos x}\lim_{x\to(\pi/2)^-} \frac{(\pi/2 - x)}{\sin x} = 1\cdot 0 = 0$, $\displaystyle\lim_{x\to(\pi/2)^-} y = 1$

45. (a) L'Hôpital's Rule does not apply to the problem $\displaystyle\lim_{x\to 1} \frac{3x^2 - 2x + 1}{3x^2 - 2x}$ because it is not a $\dfrac{0}{0}$ form.

(b) $\lim\limits_{x \to 1} \dfrac{3x^2 - 2x + 1}{3x^2 - 2x} = 2$

47. $\lim\limits_{x \to +\infty} \dfrac{1/(x \ln x)}{1/(2\sqrt{x})} = \lim\limits_{x \to +\infty} \dfrac{2}{\sqrt{x}\,\ln x} = 0$

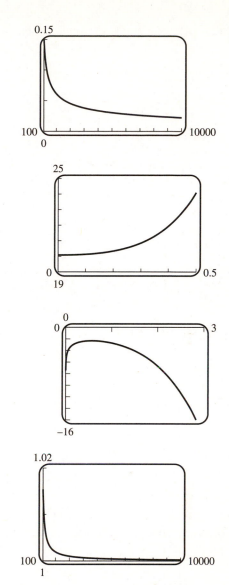

49. $y = (\sin x)^{3/\ln x}$,

$\lim\limits_{x \to 0^+} \ln y = \lim\limits_{x \to 0^+} \dfrac{3 \ln \sin x}{\ln x} = \lim\limits_{x \to 0^+} (3 \cos x) \dfrac{x}{\sin x} = 3$,

$\lim\limits_{x \to 0^+} y = e^3$

51. $\ln x - e^x = \ln x - \dfrac{1}{e^{-x}} = \dfrac{e^{-x} \ln x - 1}{e^{-x}}$;

$\lim\limits_{x \to +\infty} e^{-x} \ln x = \lim\limits_{x \to +\infty} \dfrac{\ln x}{e^x} = \lim\limits_{x \to +\infty} \dfrac{1/x}{e^x} = 0$

by L'Hôpital's Rule, so

$\lim\limits_{x \to +\infty} [\ln x - e^x] = \lim\limits_{x \to +\infty} \dfrac{e^{-x} \ln x - 1}{e^{-x}} = -\infty$

53. $y = (\ln x)^{1/x}$,

$\lim\limits_{x \to +\infty} \ln y = \lim\limits_{x \to +\infty} \dfrac{\ln(\ln x)}{x} = \lim\limits_{x \to +\infty} \dfrac{1}{x \ln x} = 0$;

$\lim\limits_{x \to +\infty} y = 1$; $y = 1$ is the horizontal asymptote

55. (a) 0   **(b)** $+\infty$ **(c)** 0   **(d)** $-\infty$ **(e)** $+\infty$ **(f)** $-\infty$

57. $\lim\limits_{x \to +\infty} \dfrac{1 + 2\cos 2x}{1}$ does not exist, nor is it $\pm\infty$; $\lim\limits_{x \to +\infty} \dfrac{x + \sin 2x}{x} = \lim\limits_{x \to +\infty} \left(1 + \dfrac{\sin 2x}{x}\right) = 1$

59. $\lim\limits_{x \to +\infty} (2 + 2x \cos 2x + \sin 2x)$ does not exist, nor is it $\pm\infty$; $\lim\limits_{x \to +\infty} \dfrac{x(2 + \sin 2x)}{x + 1} = \lim\limits_{x \to +\infty} \dfrac{2 + \sin 2x}{1 + 1/x}$,

which does not exist because $\sin 2x$ oscillates between -1 and 1 as $x \to +\infty$

61. $\lim\limits_{R \to 0^+} \dfrac{\dfrac{Vt}{L} e^{-Rt/L}}{1} = \dfrac{Vt}{L}$

63. (b) $\lim\limits_{x \to +\infty} x(k^{1/x} - 1) = \lim\limits_{t \to 0^+} \dfrac{k^t - 1}{t} = \lim\limits_{t \to 0^+} \dfrac{(\ln k)k^t}{1} = \ln k$

(c) $\ln 0.3 = -1.20397$, $1024 \left(\sqrt[1024]{0.3} - 1 \right) = -1.20327$;

$\ln 2 = 0.69315$, $1024 \left(\sqrt[1024]{2} - 1 \right) = 0.69338$

65. (a) No; $\sin(1/x)$ oscillates as $x \to 0$.

(b)

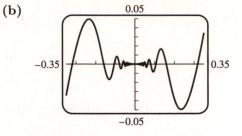

(c) For the limit as $x \to 0^+$ use the Squeezing Theorem together with the inequalities
$-x^2 \le x^2 \sin(1/x) \le x^2$. For $x \to 0^-$ do the same; thus $\lim\limits_{x \to 0} f(x) = 0$.

67. $\lim\limits_{x \to 0^+} \dfrac{\sin(1/x)}{(\sin x)/x}$, $\lim\limits_{x \to 0^+} \dfrac{\sin x}{x} = 1$ but $\lim\limits_{x \to 0^+} \sin(1/x)$ does not exist because $\sin(1/x)$ oscillates between

-1 and 1 as $x \to +\infty$, so $\lim\limits_{x \to 0^+} \dfrac{x \sin(1/x)}{\sin x}$ does not exist.

EXERCISE SET 6.6

1. (a)   **(b)**   **(c)**

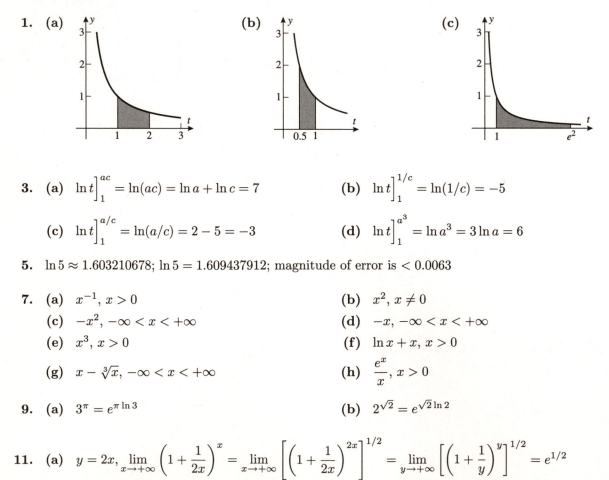

3. (a) $\ln t \Big]_1^{ac} = \ln(ac) = \ln a + \ln c = 7$

(b) $\ln t \Big]_1^{1/c} = \ln(1/c) = -5$

(c) $\ln t \Big]_1^{a/c} = \ln(a/c) = 2 - 5 = -3$

(d) $\ln t \Big]_1^{a^3} = \ln a^3 = 3 \ln a = 6$

5. $\ln 5 \approx 1.603210678$; $\ln 5 = 1.609437912$; magnitude of error is < 0.0063

7. (a) x^{-1}, $x > 0$ **(b)** x^2, $x \ne 0$

(c) $-x^2$, $-\infty < x < +\infty$ **(d)** $-x$, $-\infty < x < +\infty$

(e) x^3, $x > 0$ **(f)** $\ln x + x$, $x > 0$

(g) $x - \sqrt[3]{x}$, $-\infty < x < +\infty$ **(h)** $\dfrac{e^x}{x}$, $x > 0$

9. (a) $3^\pi = e^{\pi \ln 3}$ **(b)** $2^{\sqrt 2} = e^{\sqrt 2 \ln 2}$

11. (a) $y = 2x$, $\lim\limits_{x \to +\infty} \left(1 + \dfrac{1}{2x} \right)^x = \lim\limits_{x \to +\infty} \left[\left(1 + \dfrac{1}{2x} \right)^{2x} \right]^{1/2} = \lim\limits_{y \to +\infty} \left[\left(1 + \dfrac{1}{y} \right)^y \right]^{1/2} = e^{1/2}$

(b) $y = 2x$, $\lim_{y \to 0} (1+y)^{2/y} = \lim_{y \to 0} \left[(1+y)^{1/y}\right]^2 = e^2$

13. $g'(x) = (x^6 - x^3) \cdot 3x^2 = 3x^8 - 3x^5$

$g(x) = \left[\dfrac{1}{3}t^3 - \dfrac{1}{2}t^2\right]_1^{x^3} = \dfrac{1}{3}x^9 - \dfrac{1}{2}x^6 + \dfrac{1}{6}$, so $g'(x) = 3x^8 - 3x^5$

15. **(a)** $\dfrac{1}{x^3}(3x^2) = \dfrac{3}{x}$ **(b)** $e^{\ln x} \dfrac{1}{x} = 1$

17. $F'(x) = \dfrac{\sin x}{x^2 + 1}$, $F''(x) = \dfrac{(x^2 + 1)\cos x - 2x \sin x}{(x^2 + 1)^2}$

    (a) 0   **(b)** 0   **(c)** 1

19. true

21. false; integral does not exist

23. **(a)** $\dfrac{d}{dx} \displaystyle\int_1^{x^2} t\sqrt{1+t}\, dt = x^2\sqrt{1+x^2}(2x) = 2x^3\sqrt{1+x^2}$

    (b) $\displaystyle\int_1^{x^2} t\sqrt{1+t}\, dt = -\dfrac{2}{3}(x^2+1)^{3/2} + \dfrac{2}{5}(x^2+1)^{5/2} - \dfrac{4\sqrt{2}}{15}$

25. **(a)** $-\cos x^3$ **(b)** $-\dfrac{\tan^2 x}{1 + \tan^2 x}\sec^2 x = -\tan^2 x$

27. $-3\dfrac{3x - 1}{9x^2 + 1} + 2x\dfrac{x^2 - 1}{x^4 + 1}$

29. **(a)** $\sin^2(x^3)(3x^2) - \sin^2(x^2)(2x) = 3x^2\sin^2(x^3) - 2x\sin^2(x^2)$

    (b) $\dfrac{1}{1+x}(1) - \dfrac{1}{1-x}(-1) = \dfrac{2}{1-x^2}$

31. From geometry, $\displaystyle\int_0^3 f(t)\, dt = 0$, $\displaystyle\int_3^5 f(t)\, dt = 6$, and $\displaystyle\int_5^7 f(t)\, dt = 0$.

    Also, $\displaystyle\int_7^{10} f(t)\, dt = \displaystyle\int_7^{10} (4t - 37)/3\, dt = -3$

    (a) $F(0) = 0$, $F(3) = 0$, $F(5) = 6$, $F(7) = 6$, $F(10) = 3$

    (b) F is increasing where $F' = f$ is positive, so on $[3/2, 6]$ and $[37/4, 10]$, decreasing on $[0, 3/2]$ and $[6, 37/4]$

    (c) critical points when $F'(x) = f(x) = 0$, so $x = 3/2, 6, 37/4$; maximum $15/2$ at $x = 6$, minimum $-9/4$ at $x = 3/2$

    (d)

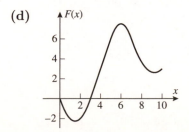

33. $x < 0 : F(x) = \int_{-1}^{x} (-t)dt = -\frac{1}{2}t^2 \Big]_{-1}^{x} = \frac{1}{2}(1 - x^2),$

$x \geq 0 : F(x) = \int_{-1}^{0} (-t)dt + \int_{0}^{x} t\,dt = \frac{1}{2} + \frac{1}{2}x^2;\ F(x) = \begin{cases} (1 - x^2)/2, & x < 0 \\ (1 + x^2)/2, & x \geq 0 \end{cases}$

35. $y(x) = 2 + \int_{1}^{x} \frac{2t^2 + 1}{t}\,dt = 2 + (t^2 + \ln t)\Big]_{1}^{x} = x^2 + \ln x + 1$

37. $y(x) = 1 + \int_{\pi/4}^{x} (\sec^2 t - \sin t)dt = \tan x + \cos x - \sqrt{2}/2$

39. $P(x) = P_0 + \int_{0}^{x} r(t)dt$ individuals

41. II has a minimum at $x = 12$, and I has a zero there, so I could be the derivative of II; on the other hand I has a minimum near $x = 1/3$, but II is not zero there, so II could not be the derivative of I, so I is the graph of $f(x)$ and II is the graph of $\int_{0}^{x} f(t)\,dt$.

43. (a)   where $f(t) = 0$; by the First Derivative Test, at $t = 3$

   (b)   where $f(t) = 0$; by the First Derivative Test, at $t = 1, 5$

   (c)   at $t = 0, 1$ or 5; from the graph it is evident that it is at $t = 5$

   (d)   at $t = 0, 3$ or 5; from the graph it is evident that it is at $t = 3$

   (e) F is concave up when $F'' = f'$ is positive, i.e. where f is increasing, so on $(0, 1/2)$ and $(2, 4)$; it is concave down on $(1/2, 2)$ and $(4, 5)$

   (f)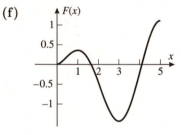

45. $C'(x) = \cos(\pi x^2/2)$, $C''(x) = -\pi x \sin(\pi x^2/2)$

   (a) $\cos t$ goes from negative to positive at $2k\pi - \pi/2$, and from positive to negative at $t = 2k\pi + \pi/2$, so $C(x)$ has relative minima when $\pi x^2/2 = 2k\pi - \pi/2$, $x = \pm\sqrt{4k - 1}$, $k = 1, 2, \ldots$, and $C(x)$ has relative maxima when $\pi x^2/2 = (4k + 1)\pi/2$, $x = \pm\sqrt{4k + 1}$, $k = 0, 1, \ldots$.

   (b) $\sin t$ changes sign at $t = k\pi$, so $C(x)$ has inflection points at $\pi x^2/2 = k\pi$, $x = \pm\sqrt{2k}$, $k = 1, 2, \ldots$; the case $k = 0$ is distinct due to the factor of x in $C''(x)$, but x changes sign at $x = 0$ and $\sin(\pi x^2/2)$ does not, so there is also a point of inflection at $x = 0$

47. Differentiate: $f(x) = 2e^{2x}$, so $4 + \int_{a}^{x} f(t)dt = 4 + \int_{a}^{x} 2e^{2t}\,dt = 4 + e^{2t}\Big]_{a}^{x} = 4 + e^{2x} - e^{2a} = e^{2x}$ provided $e^{2a} = 4$, $a = (\ln 4)/2 = \ln 2$.

49. From Exercise 48(d) $\left| e - \left(1 + \frac{1}{50}\right)^{50} \right| < y(50)$, and from the graph $y(50) < 0.06$

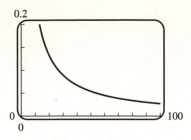

51. $F'(x) = f(x)$, thus $F'(x)$ has a value at each x in I because f is continuous on I so F is continuous on I because a function that is differentiable at a point is also continuous at that point

EXERCISE SET 6.7

1. $\tan\theta = 4/3$, $0 < \theta < \pi/2$; use the triangle shown to get $\sin\theta = 4/5$, $\cos\theta = 3/5$, $\cot\theta = 3/4$, $\sec\theta = 5/3$, $\csc\theta = 5/4$

3. (a) $0 \le x \le \pi$ (b) $-1 \le x \le 1$
   (c) $-\pi/2 < x < \pi/2$ (d) $-\infty < x < +\infty$

5. Let $\theta = \cos^{-1}(3/5)$, $\sin 2\theta = 2\sin\theta\cos\theta = 2(4/5)(3/5) = 24/25$

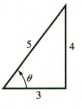

7. (a) $\cos(\tan^{-1} x) = \dfrac{1}{\sqrt{1+x^2}}$ (b) $\tan(\cos^{-1} x) = \dfrac{\sqrt{1-x^2}}{x}$

   (c) $\sin(\sec^{-1} x) = \dfrac{\sqrt{x^2-1}}{x}$ (d) $\cot(\sec^{-1} x) = \dfrac{1}{\sqrt{x^2-1}}$

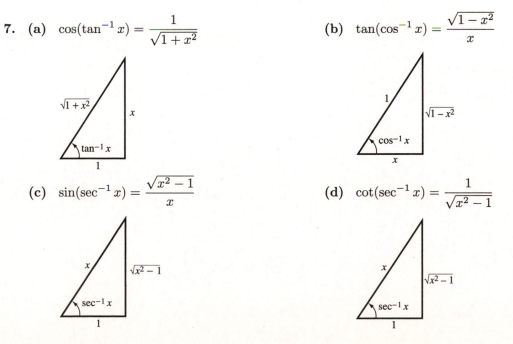

9. (a)

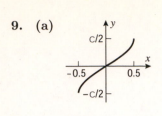

(b)

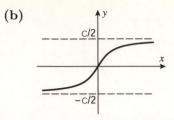

11. (a) $\sin^{-1}(\sin^{-1} 0.25) \approx 0.25545$. $\sin^{-1} 0.9 > 1$, so it is not in the domain of $\sin^{-1} x$.
   (b) $-1 \le \sin^{-1} x \le 1$ is necessary, or $|x| \le \sin 1 \approx 0.84147$.

13. $\dfrac{3}{\sqrt{1-(3x)^2}} = \dfrac{3}{\sqrt{1-9x^2}}$

15. $\dfrac{1}{\sqrt{1-1/x^2}}(-1/x^2) = -\dfrac{1}{|x|\sqrt{x^2-1}}$

17. $\dfrac{3x^2}{1+(x^3)^2} = \dfrac{3x^2}{1+x^6}$

19. $y = 1/\tan x = \cot x$, $dy/dx = -\csc^2 x$

21. $\dfrac{e^x}{|x|\sqrt{x^2-1}} + e^x \sec^{-1} x$

23. 0

25. $x^3 + x \tan^{-1} y = e^y$, $3x^2 + \dfrac{x}{1+y^2}y' + \tan^{-1} y = e^y y'$, $y' = \dfrac{(3x^2 + \tan^{-1} y)(1+y^2)}{(1+y^2)e^y - x}$

27. $\displaystyle\int \left[\dfrac{1}{2\sqrt{1-x^2}} - \dfrac{3}{1+x^2} \right] dx = \dfrac{1}{2}\sin^{-1} x - 3\tan^{-1} x + C$

29. $u = 2x$, $\dfrac{1}{2}\displaystyle\int \dfrac{1}{\sqrt{1-u^2}} du = \dfrac{1}{2}\sin^{-1}(2x) + C$

31. $u = e^x$, $\displaystyle\int \dfrac{1}{1+u^2} du = \tan^{-1}(e^x) + C$

33. $u = \tan x$, $\displaystyle\int \dfrac{1}{\sqrt{1-u^2}} du = \sin^{-1}(\tan x) + C$

35. $\sin^{-1} x \Big]_0^{1/\sqrt{2}} = \sin^{-1}(1/\sqrt{2}) - \sin^{-1} 0 = \pi/4$

37. $\sec^{-1} x \Big]_{\sqrt{2}}^{2} = \sec^{-1} 2 - \sec^{-1}\sqrt{2} = \pi/3 - \pi/4 = \pi/12$

39. $u = \tan^{-1} x$; $\displaystyle\int_{\pi/4}^{\pi/3} \sqrt{u}\, du = \dfrac{2}{3}u^{3/2} \Big]_{\pi/4}^{\pi/3} = \pi^{3/2}\left(\dfrac{2}{9\sqrt{3}} - \dfrac{1}{12} \right)$

41. $u = \sqrt{x}$, $2\displaystyle\int_1^{\sqrt{3}} \dfrac{1}{u^2+1} du = 2\tan^{-1} u \Big]_1^{\sqrt{3}} = 2(\tan^{-1}\sqrt{3} - \tan^{-1} 1) = 2(\pi/3 - \pi/4) = \pi/6$

43. $u = \sqrt{3}x^2$, $\dfrac{1}{2\sqrt{3}}\displaystyle\int_0^{\sqrt{3}} \dfrac{1}{\sqrt{4-u^2}} du = \dfrac{1}{2\sqrt{3}}\sin^{-1}\dfrac{u}{2} \Big]_0^{\sqrt{3}} = \dfrac{1}{2\sqrt{3}}\left(\dfrac{\pi}{3} \right) = \dfrac{\pi}{6\sqrt{3}}$

45. $u = 3x$, $\dfrac{1}{3}\displaystyle\int_0^{\sqrt{3}} \dfrac{1}{1+u^2} du = \dfrac{1}{3}\tan^{-1} u \Big]_0^{\sqrt{3}} = \dfrac{1}{3}\dfrac{\pi}{3} = \dfrac{\pi}{9}$

47. (a) $\sin^{-1}(x/3) + C$ (b) $(1/\sqrt{5})\tan^{-1}(x/\sqrt{5}) + C$
(c) $(1/\sqrt{\pi})\sec^{-1}(x/\sqrt{\pi}) + C$

49. false; the range of $\sin^{-1}$ is $[-\pi/2, \pi/2]$, so the equation is only true for x in this range.

51. true; the line $y = \pi/2$ is a horizontal asymptote as $x \to \infty$ and as $x \to -\infty$.

53. (a)

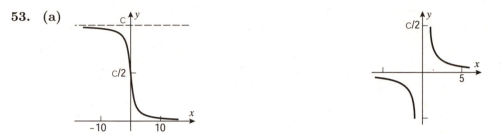

(b) The domain of $\cot^{-1} x$ is $(-\infty, +\infty)$, the range is $(0, \pi)$; the domain of $\csc^{-1} x$ is $(-\infty, -1] \cup [1, +\infty)$, the range is $[-\pi/2, 0) \cup (0, \pi/2]$.

55. (a) $55.0°$ (b) $33.6°$ (c) $25.8°$

57. $\dfrac{d}{dx}\left[\cos^{-1} x\right] = \dfrac{d}{dx}\left[\dfrac{\pi}{2} - \sin^{-1} x\right] = -\dfrac{d}{dx}\left[\sin^{-1} x\right] = -\dfrac{1}{\sqrt{1 - x^2}}$

59. (a) By the chain rule, $\dfrac{d}{dx}[\csc^{-1} x] = \dfrac{d}{dx}\sin^{-1}\dfrac{1}{x} = -\dfrac{1}{x^2}\dfrac{1}{\sqrt{1 - (1/x)^2}} = \dfrac{-1}{|x|\sqrt{x^2 - 1}}$

(b) By the chain rule, $\dfrac{d}{dx}[\csc^{-1} u] = \dfrac{du}{dx}\dfrac{d}{du}[\csc^{-1} u] = \dfrac{-1}{|u|\sqrt{u^2 - 1}}\dfrac{du}{dx}$

(c) $\sec^{-1} x + \csc^{-1} x = \cos^{-1}(1/x) + \sin^{-1}(1/x) = \pi/2$, so
$\dfrac{d}{dx}\sec^{-1} x = -\dfrac{d}{dx}\csc^{-1} x = \dfrac{1}{|x|\sqrt{x^2 - 1}}$ by part (a).

(d) By the chain rule, $\dfrac{d}{dx}[\sec^{-1} u] = \dfrac{du}{dx}\dfrac{d}{du}[\sec^{-1} u] = \dfrac{1}{|u|\sqrt{u^2 - 1}}\dfrac{du}{dx}$

61. 0

63. $-\dfrac{1}{1 + x}\left(\dfrac{1}{2}x^{-1/2}\right) = -\dfrac{1}{2(1 + x)\sqrt{x}}$

65. (a) If $\gamma = 90°$, then $\sin\gamma = 1$, $\sqrt{1 - \sin^2\phi\sin^2\gamma} = \sqrt{1 - \sin^2\phi} = \cos\phi$,
$D = \tan\phi\tan\lambda = (\tan 23.45°)(\tan 65°) \approx 0.93023374$ so $h \approx 21.1$ hours.

(b) If $\gamma = 270°$, then $\sin\gamma = -1$, $D = -\tan\phi\tan\lambda \approx -0.93023374$ so $h \approx 2.9$ hours.

67. $y = 0$ when $x^2 = 6000v^2/g$, $x = 10v\sqrt{60/g} = 1000\sqrt{30}$ for $v = 400$ and $g = 32$;
$\tan\theta = 3000/x = 3/\sqrt{30}$, $\theta = \tan^{-1}(3/\sqrt{30}) \approx 29°$.

69. The area is given by $\displaystyle\int_0^k (1/\sqrt{1 - x^2} - x)\,dx = \sin^{-1} k - k^2/2 = 1$; solve for k to get
$k \approx 0.997301$.

71. $V = \displaystyle\int_{-2}^{2} \pi\dfrac{1}{4 + x^2}\,dx = \dfrac{\pi}{2}\tan^{-1}(x/2)\bigg]_{-2}^{2} = \pi^2/4$

73. **(a)** $V = \pi \int_0^1 (\sin^{-1} x)^2 \, dx$. Substitute $x = \sin u$ to get

$$V = \pi \int_0^{\pi/2} u^2 \cos u \, du = \pi \left[2u \cos u + (u^2 - 2) \sin u \right]_0^{\pi/2} = \frac{\pi^3}{4} - 2\pi,$$

by Endpaper Integral Table Formula 47.

(b) $V = 2\pi \int_0^{\pi/2} y(1 - \sin y) \, dy = 2\pi \left[\frac{y^2}{2} - \sin y + y \cos y \right]_0^{\pi/2} = 2\pi \left(\frac{\pi^2}{8} - 1 \right) = \frac{\pi^3}{4} - 2\pi$, by Endpaper Integral Table Formula 44.

75. $f_{\text{ave}} = \frac{1}{\sqrt{3} - 1} \int_1^{\sqrt{3}} \frac{dx}{1 + x^2} = \frac{1}{\sqrt{3} - 1} \tan^{-1} x \Big]_1^{\sqrt{3}} = \frac{1}{\sqrt{3} - 1} \left(\frac{\pi}{3} - \frac{\pi}{4} \right) = \frac{1}{\sqrt{3} - 1} \frac{\pi}{12}$

77. By the Mean-Value Theorem on the interval $[0, x]$,

$$\frac{\tan^{-1} x - \tan^{-1} 0}{x - 0} = \frac{\tan^{-1} x}{x} = \frac{1}{1 + c^2} \text{ for } c \text{ in } (0, x), \text{ but}$$

$$\frac{1}{1 + x^2} < \frac{1}{1 + c^2} < 1 \text{ for } c \text{ in } (0, x) \text{ so } \frac{1}{1 + x^2} < \frac{\tan^{-1} x}{x} < 1, \frac{x}{1 + x^2} < \tan^{-1} x < x.$$

79. $y = \int \frac{3}{\sqrt{1 - t^2}} \, dt = 3 \sin^{-1} t + C, y\left(\frac{\sqrt{3}}{2} \right) = 0 = \pi + C, C = -\pi, y = 3 \sin^{-1} t - \pi$

81. $y = \int \frac{1}{25 + 9t^2} \, dt = \frac{1}{15} \tan^{-1} \left(\frac{3}{5} t \right) + C, \frac{\pi}{30} = y\left(-\frac{5}{3} \right) = -\frac{1}{15} \frac{\pi}{4} + C,$

$C = \frac{\pi}{20}, y = \frac{1}{15} \tan^{-1} \left(\frac{3}{5} t \right) + \frac{\pi}{20}$

83. **(a)** Let $\theta = \sin^{-1}(-x)$ then $\sin \theta = -x$, $-\pi/2 \leq \theta \leq \pi/2$. But $\sin(-\theta) = -\sin \theta$ and $-\pi/2 \leq -\theta \leq \pi/2$ so $\sin(-\theta) = -(-x) = x$, $-\theta = \sin^{-1} x$, $\theta = -\sin^{-1} x$.

(b) proof is similar to that in Part (a)

85. **(a)** $\sin^{-1} x = \tan^{-1} \frac{x}{\sqrt{1 - x^2}}$ (see figure)

(b) $\sin^{-1} x + \cos^{-1} x = \pi/2$; $\cos^{-1} x = \pi/2 - \sin^{-1} x = \pi/2 - \tan^{-1} \frac{x}{\sqrt{1 - x^2}}$

87. **(a)** $\tan^{-1} \frac{1}{2} + \tan^{-1} \frac{1}{3} = \tan^{-1} \frac{1/2 + 1/3}{1 - (1/2)(1/3)} = \tan^{-1} 1 = \pi/4$

(b) $2 \tan^{-1} \frac{1}{3} = \tan^{-1} \frac{1}{3} + \tan^{-1} \frac{1}{3} = \tan^{-1} \frac{1/3 + 1/3}{1 - (1/3)(1/3)} = \tan^{-1} \frac{3}{4},$

$2 \tan^{-1} \frac{1}{3} + \tan^{-1} \frac{1}{7} = \tan^{-1} \frac{3}{4} + \tan^{-1} \frac{1}{7} = \tan^{-1} \frac{3/4 + 1/7}{1 - (3/4)(1/7)} = \tan^{-1} 1 = \pi/4$

EXERCISE SET 6.8

1. **(a)** $\sinh 3 \approx 10.0179$

   (b) $\cosh(-2) \approx 3.7622$

   (c) $\tanh(\ln 4) = 15/17 \approx 0.8824$

   (d) $\sinh^{-1}(-2) \approx -1.4436$

   (e) $\cosh^{-1} 3 \approx 1.7627$

   (f) $\tanh^{-1} \dfrac{3}{4} \approx 0.9730$

3. **(a)** $\sinh(\ln 3) = \dfrac{1}{2}(e^{\ln 3} - e^{-\ln 3}) = \dfrac{1}{2}\left(3 - \dfrac{1}{3}\right) = \dfrac{4}{3}$

   (b) $\cosh(-\ln 2) = \dfrac{1}{2}(e^{-\ln 2} + e^{\ln 2}) = \dfrac{1}{2}\left(\dfrac{1}{2} + 2\right) = \dfrac{5}{4}$

   (c) $\tanh(2\ln 5) = \dfrac{e^{2\ln 5} - e^{-2\ln 5}}{e^{2\ln 5} + e^{-2\ln 5}} = \dfrac{25 - 1/25}{25 + 1/25} = \dfrac{312}{313}$

   (d) $\sinh(-3\ln 2) = \dfrac{1}{2}(e^{-3\ln 2} - e^{3\ln 2}) = \dfrac{1}{2}\left(\dfrac{1}{8} - 8\right) = -\dfrac{63}{16}$

5.

	$\sinh x_0$	$\cosh x_0$	$\tanh x_0$	$\coth x_0$	$\operatorname{sech} x_0$	$\operatorname{csch} x_0$
(a)	2	$\sqrt{5}$	$2/\sqrt{5}$	$\sqrt{5}/2$	$1/\sqrt{5}$	$1/2$
(b)	$3/4$	$5/4$	$3/5$	$5/3$	$4/5$	$4/3$
(c)	$4/3$	$5/3$	$4/5$	$5/4$	$3/5$	$3/4$

   (a) $\cosh^2 x_0 = 1 + \sinh^2 x_0 = 1 + (2)^2 = 5,\ \cosh x_0 = \sqrt{5}$

   (b) $\sinh^2 x_0 = \cosh^2 x_0 - 1 = \dfrac{25}{16} - 1 = \dfrac{9}{16},\ \sinh x_0 = \dfrac{3}{4}$ (because $x_0 > 0$)

   (c) $\operatorname{sech}^2 x_0 = 1 - \tanh^2 x_0 = 1 - \left(\dfrac{4}{5}\right)^2 = 1 - \dfrac{16}{25} = \dfrac{9}{25},\ \operatorname{sech} x_0 = \dfrac{3}{5},$

   $\cosh x_0 = \dfrac{1}{\operatorname{sech} x_0} = \dfrac{5}{3},$ from $\dfrac{\sinh x_0}{\cosh x_0} = \tanh x_0$ we get $\sinh x_0 = \left(\dfrac{5}{3}\right)\left(\dfrac{4}{5}\right) = \dfrac{4}{3}$

7. $\dfrac{d}{dx}\cosh^{-1}x = \dfrac{d}{dx}\ln(x + \sqrt{x^2 - 1}) = \dfrac{1}{x + \sqrt{x^2 - 1}}\left(1 + \dfrac{2x}{2\sqrt{x^2 - 1}}\right) = \dfrac{1}{x + \sqrt{x^2 - 1}}\dfrac{\sqrt{x^2 - 1} + x}{\sqrt{x^2 - 1}}$

   $\qquad = \dfrac{1}{\sqrt{x^2 - 1}}$

   $\dfrac{d}{dx}\tanh^{-1}x = \dfrac{d}{dx}\left[\dfrac{1}{2}\ln\left(\dfrac{1+x}{1-x}\right)\right] = \dfrac{1}{2}\cdot\dfrac{1}{\frac{1+x}{1-x}}\cdot\dfrac{(1-x)\cdot 1 - (1+x)(-1)}{(1-x)^2}$

   $\qquad = \dfrac{2}{2(1+x)(1-x)} = \dfrac{1}{1-x^2}$

9. $4\cosh(4x - 8)$

11. $-\dfrac{1}{x}\operatorname{csch}^2(\ln x)$

13. $\dfrac{1}{x^2}\operatorname{csch}(1/x)\coth(1/x)$

15. $\dfrac{2 + 5\cosh(5x)\sinh(5x)}{\sqrt{4x + \cosh^2(5x)}}$

17. $x^{5/2}\tanh(\sqrt{x})\operatorname{sech}^2(\sqrt{x}) + 3x^2\tanh^2(\sqrt{x})$

19. $\dfrac{1}{\sqrt{1+x^2/9}}\left(\dfrac{1}{3}\right) = 1/\sqrt{9+x^2}$

21. $1/\left[(\cosh^{-1} x)\sqrt{x^2-1}\right]$

23. $-(\tanh^{-1} x)^{-2}/(1-x^2)$

25. $\dfrac{\sinh x}{\sqrt{\cosh^2 x - 1}} = \dfrac{\sinh x}{|\sinh x|} = \begin{cases} 1, & x > 0 \\ -1, & x < 0 \end{cases}$

27. $-\dfrac{e^x}{2x\sqrt{1-x}} + e^x \operatorname{sech}^{-1}\sqrt{x}$

29. $\dfrac{1}{7}\sinh^7 x + C$

31. $\dfrac{2}{3}(\tanh x)^{3/2} + C$

33. $\dfrac{1}{2}\ln(\cosh(2x)) + C$

35. $-\dfrac{1}{3}\operatorname{sech}^3 x\,\bigg]_{\ln 2}^{\ln 3} = 37/375$

37. $u = 3x,\ \dfrac{1}{3}\displaystyle\int \dfrac{1}{\sqrt{1+u^2}}\,du = \dfrac{1}{3}\sinh^{-1} 3x + C$

39. $u = e^x,\ \displaystyle\int \dfrac{1}{u\sqrt{1-u^2}}\,du = -\operatorname{sech}^{-1}(e^x) + C$

41. $u = 2x,\ \displaystyle\int \dfrac{du}{u\sqrt{1+u^2}} = -\operatorname{csch}^{-1}|u| + C = -\operatorname{csch}^{-1}|2x| + C$

43. $\tanh^{-1} x\,\bigg]_0^{1/2} = \tanh^{-1}(1/2) - \tanh^{-1}(0) = \dfrac{1}{2}\ln\dfrac{1+1/2}{1-1/2} = \dfrac{1}{2}\ln 3$

45. True. $\cosh x - \sinh x = \dfrac{e^x + e^{-x}}{2} - \dfrac{e^x - e^{-x}}{2} = e^{-x}$ is positive for all x.

47. True. Only $\sinh x$ has this property.

49. $A = \displaystyle\int_0^{\ln 3} \sinh 2x\,dx = \dfrac{1}{2}\cosh 2x\,\bigg]_0^{\ln 3} = \dfrac{1}{2}[\cosh(2\ln 3) - 1]$,

but $\cosh(2\ln 3) = \cosh(\ln 9) = \dfrac{1}{2}(e^{\ln 9} + e^{-\ln 9}) = \dfrac{1}{2}(9 + 1/9) = 41/9$ so $A = \dfrac{1}{2}[41/9 - 1] = 16/9$.

51. $V = \pi \displaystyle\int_0^5 (\cosh^2 2x - \sinh^2 2x)\,dx = \pi \int_0^5 dx = 5\pi$

53. $y' = \sinh x,\ 1 + (y')^2 = 1 + \sinh^2 x = \cosh^2 x$

$L = \displaystyle\int_0^{\ln 2} \cosh x\,dx = \sinh x\,\bigg]_0^{\ln 2} = \sinh(\ln 2) = \dfrac{1}{2}(e^{\ln 2} - e^{-\ln 2}) = \dfrac{1}{2}\left(2 - \dfrac{1}{2}\right) = \dfrac{3}{4}$

55. **(a)** $\displaystyle\lim_{x \to +\infty} \sinh x = \lim_{x \to +\infty} \dfrac{1}{2}(e^x - e^{-x}) = +\infty - 0 = +\infty$

(b) $\displaystyle\lim_{x \to -\infty} \sinh x = \lim_{x \to -\infty} \dfrac{1}{2}(e^x - e^{-x}) = 0 - \infty = -\infty$

(c) $\displaystyle\lim_{x \to +\infty} \tanh x = \lim_{x \to +\infty} \dfrac{e^x - e^{-x}}{e^x + e^{-x}} = 1$

(d) $\displaystyle\lim_{x \to -\infty} \tanh x = \lim_{x \to -\infty} \dfrac{e^x - e^{-x}}{e^x + e^{-x}} = -1$

(e) $\displaystyle\lim_{x \to +\infty} \sinh^{-1} x = \lim_{x \to +\infty} \ln(x + \sqrt{x^2 + 1}) = +\infty$

(f) $\lim\limits_{x \to 1^-} \tanh^{-1} x = \lim\limits_{x \to 1^-} \frac{1}{2}[\ln(1+x) - \ln(1-x)] = +\infty$

57. $\sinh(-x) = \frac{1}{2}(e^{-x} - e^x) = -\frac{1}{2}(e^x - e^{-x}) = -\sinh x$

$\cosh(-x) = \frac{1}{2}(e^{-x} + e^x) = \frac{1}{2}(e^x + e^{-x}) = \cosh x$

59. (a) Divide $\cosh^2 x - \sinh^2 x = 1$ by $\cosh^2 x$.

(b) $\tanh(x+y) = \dfrac{\sinh x \cosh y + \cosh x \sinh y}{\cosh x \cosh y + \sinh x \sinh y} = \dfrac{\dfrac{\sinh x}{\cosh x} + \dfrac{\sinh y}{\cosh y}}{1 + \dfrac{\sinh x \sinh y}{\cosh x \cosh y}} = \dfrac{\tanh x + \tanh y}{1 + \tanh x \tanh y}$

(c) Let $y = x$ in part (b).

61. (a) $\dfrac{d}{dx}(\cosh^{-1} x) = \dfrac{1 + x/\sqrt{x^2-1}}{x + \sqrt{x^2-1}} = 1/\sqrt{x^2-1}$

(b) $\dfrac{d}{dx}(\tanh^{-1} x) = \dfrac{d}{dx}\left[\frac{1}{2}(\ln(1+x) - \ln(1-x))\right] = \frac{1}{2}\left(\dfrac{1}{1+x} + \dfrac{1}{1-x}\right) = 1/(1-x^2)$

63. If $|u| < 1$ then, by Theorem 6.8.6, $\displaystyle\int \dfrac{du}{1-u^2} = \tanh^{-1} u + C$.

For $|u| > 1$, $\displaystyle\int \dfrac{du}{1-u^2} = \coth^{-1} u + C = \tanh^{-1}(1/u) + C$.

65. (a) $\lim\limits_{x \to +\infty}(\cosh^{-1} x - \ln x) = \lim\limits_{x \to +\infty}[\ln(x + \sqrt{x^2-1}) - \ln x]$

$= \lim\limits_{x \to +\infty} \ln \dfrac{x + \sqrt{x^2-1}}{x} = \lim\limits_{x \to +\infty} \ln(1 + \sqrt{1 - 1/x^2}) = \ln 2$

(b) $\lim\limits_{x \to +\infty} \dfrac{\cosh x}{e^x} = \lim\limits_{x \to +\infty} \dfrac{e^x + e^{-x}}{2e^x} = \lim\limits_{x \to +\infty} \frac{1}{2}(1 + e^{-2x}) = 1/2$

67. Let $x = -u/a$, $\displaystyle\int \dfrac{1}{\sqrt{u^2 - a^2}}\, du = -\int \dfrac{a}{a\sqrt{x^2-1}}\, dx = -\cosh^{-1} x + C = -\cosh^{-1}(-u/a) + C$.

$-\cosh^{-1}(-u/a) = -\ln(-u/a + \sqrt{u^2/a^2 - 1}) = \ln\left[\dfrac{a}{-u + \sqrt{u^2 - a^2}} \dfrac{u + \sqrt{u^2 - a^2}}{u + \sqrt{u^2 - a^2}}\right]$

$= \ln\left|u + \sqrt{u^2 - a^2}\right| - \ln a = \ln|u + \sqrt{u^2 - a^2}| + C_1$

so $\displaystyle\int \dfrac{1}{\sqrt{u^2 - a^2}}\, du = \ln\left|u + \sqrt{u^2 - a^2}\right| + C_2$.

69. $\displaystyle\int_{-a}^{a} e^{tx}\, dx = \frac{1}{t} e^{tx}\Big]_{-a}^{a} = \frac{1}{t}(e^{at} - e^{-at}) = \dfrac{2\sinh at}{t}$ for $t \neq 0$.

71. From part (b) of Exercise 70, $S = a\cosh(b/a) - a$ so $30 = a\cosh(200/a) - a$. Let $u = 200/a$, then $a = 200/u$ so $30 = (200/u)[\cosh u - 1], \cosh u - 1 = 0.15u$. If $f(u) = \cosh u - 0.15u - 1$, then $u_{n+1} = u_n - \dfrac{\cosh u_n - 0.15u_n - 1}{\sinh u_n - 0.15}; u_1 = 0.3, \ldots, u_4 \approx u_5 \approx 0.297792782 \approx 200/a$ so $a \approx 671.6079505$. From part (a), $L = 2a\sinh(b/a) \approx 2(671.6079505)\sinh(0.297792782) \approx 405.9\,\text{ft}$.

73. Set $a = 68.7672$, $b = 0.0100333$, $c = 693.8597$, $d = 299.2239$.

(a)

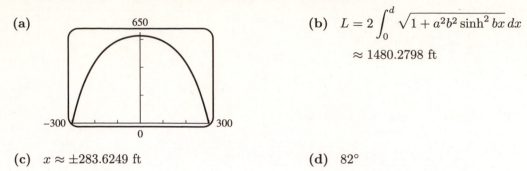

650

−300　　　300

0

(b) $L = 2 \int_0^d \sqrt{1 + a^2 b^2 \sinh^2 bx}\, dx$

≈ 1480.2798 ft

(c) $x \approx \pm 283.6249$ ft

(d) $82°$

75. **(a)** When the bow of the boat is at the point (x, y) and the person has walked a distance D, then the person is located at the point $(0, D)$, the line segment connecting $(0, D)$ and (x, y) has length a; thus $a^2 = x^2 + (D - y)^2$, $D = y + \sqrt{a^2 - x^2} = a\,\mathrm{sech}^{-1}(x/a)$.

(b) Find D when $a = 15$, $x = 10$: $D = 15\,\mathrm{sech}^{-1}(10/15) = 15\ln\left(\dfrac{1 + \sqrt{5/9}}{2/3}\right) \approx 14.44$ m.

(c) $dy/dx = -\dfrac{a^2}{x\sqrt{a^2 - x^2}} + \dfrac{x}{\sqrt{a^2 - x^2}} = \dfrac{1}{\sqrt{a^2 - x^2}}\left[-\dfrac{a^2}{x} + x\right] = -\dfrac{1}{x}\sqrt{a^2 - x^2}$,

$1 + [y']^2 = 1 + \dfrac{a^2 - x^2}{x^2} = \dfrac{a^2}{x^2}$; with $a = 15$, $L = \int_5^{15}\sqrt{\dfrac{225}{x^2}}\, dx = \int_5^{15}\dfrac{15}{x}\, dx = 15\ln x\Big]_5^{15} =$

$15\ln 3 \approx 16.48$ m.

REVIEW EXERCISES, CHAPTER 6

1. **(a)** $x = f(y) = (e^y)^2 + 1$; $y = f^{-1}(x) = \ln\sqrt{x - 1} = \tfrac{1}{2}\ln(x - 1)$

(b) $x = f(y) = \sin\left(\dfrac{1 - 2y}{y}\right)$; $y = f^{-1}(x) = \dfrac{1}{2 + \sin^{-1}x}$

(c) $x = f(y) = \dfrac{1}{1 + 3\tan^{-1}y}$; $y = f^{-1}(x) = \tan\left(\dfrac{1 - x}{3x}\right)$. Since $-\pi/2 \le \tan^{-1}y \le \pi/2$, we

must restrict the domain of f^{-1} to the intervals $x \le -\dfrac{2}{3\pi - 2}$ and $x \ge \dfrac{2}{3\pi + 2}$.

3. Draw triangles of sides 5, 12, 13, and 3, 4, 5. Then $\sin[\cos^{-1}(4/5)] = 3/5$, $\sin[\cos^{-1}(5/13)] = 12/13$, $\cos[\sin^{-1}(4/5)] = 3/5$, $\cos[\sin^{-1}(5/13)] = 12/13$

(a) $\cos[\cos^{-1}(4/5) + \sin^{-1}(5/13)] = \cos(\cos^{-1}(4/5))\cos(\sin^{-1}(5/13))$
$$- \sin(\cos^{-1}(4/5))\sin(\sin^{-1}(5/13))$$
$$= \frac{4}{5}\frac{12}{13} - \frac{3}{5}\frac{5}{13} = \frac{33}{65}.$$

(b) $\sin[\sin^{-1}(4/5) + \cos^{-1}(5/13)] = \sin(\sin^{-1}(4/5))\cos(\cos^{-1}(5/13))$
$$+ \cos(\sin^{-1}(4/5))\sin(\cos^{-1}(5/13))$$
$$= \frac{4}{5}\frac{5}{13} + \frac{3}{5}\frac{12}{13} = \frac{56}{65}.$$

5. $y = 5$ ft $= 60$ in, so $60 = \log x$, $x = 10^{60}$ in $\approx 1.58 \times 10^{55}$ mi.

7. $3\ln\left(e^{2x}(e^x)^3\right) + 2\exp(\ln 1) = 3\ln e^{2x} + 3\ln(e^x)^3 + 2\cdot 1 = 3(2x) + (3\cdot 3)x + 2 = 15x + 2$

9. (a)

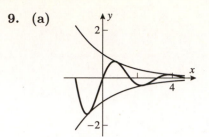

(b) The curve $y = e^{-x/2} \sin 2x$ has x-intercepts at $x = -\pi/2, 0, \pi/2, \pi, 3\pi/2$. It intersects the curve $y = e^{-x/2}$ at $x = \pi/4, 5\pi/4$ and it intersects the curve $y = -e^{-x/2}$ at $x = -\pi/4, 3\pi/4$.

11. (a)

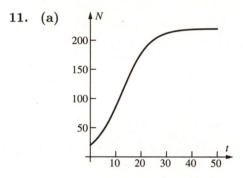

(b) $N = 80$ when $t = 9.35$ yrs
(c) 220 sheep

13. (a) The function $\ln x - x^{0.2}$ is negative at $x = 1$ and positive at $x = 4$, so by the Intermediate Value Theorem it must be zero somewhere in between.
(b) There are two roots: $x \approx 3.654$ and $x \approx 332105.108$.

15. 0

17. $\left(1 + \dfrac{3}{x}\right)^{-x} = \left[\left(1 + \dfrac{3}{x}\right)^{x/3}\right]^{(-3)}$ so the limit is e^{-3}

19. $y = \ln(x+1) + 2\ln(x+2) - 3\ln(x+3) - 4\ln(x+4)$, $dy/dx = \dfrac{1}{x+1} + \dfrac{2}{x+2} - \dfrac{3}{x+3} - \dfrac{4}{x+4}$

21. $\dfrac{1}{2x}(2) = 1/x$

23. $\dfrac{1}{3x(\ln x + 1)^{2/3}}$

25. $\log_{10} \ln x = \dfrac{\ln \ln x}{\ln 10}, y' = \dfrac{1}{(\ln 10)(x \ln x)}$

27. $y = \dfrac{3}{2}\ln x + \dfrac{1}{2}\ln(1 + x^4)$, $y' = \dfrac{3}{2x} + \dfrac{2x^3}{1 + x^4}$

29. $y = x^2 + 1$ so $y' = 2x$.

31. $y' = 2e^{\sqrt{x}} + 2xe^{\sqrt{x}}\dfrac{d}{dx}\sqrt{x} = 2e^{\sqrt{x}} + \sqrt{x}e^{\sqrt{x}}$

33. $y' = \dfrac{2}{\pi(1 + 4x^2)}$

35. $\ln y = e^x \ln x$, $\dfrac{y'}{y} = e^x\left(\dfrac{1}{x} + \ln x\right)$, $\dfrac{dy}{dx} = x^{e^x}e^x\left(\dfrac{1}{x} + \ln x\right) = e^x\left[x^{e^x-1} + x^{e^x}\ln x\right]$

37. $y' = \dfrac{2}{|2x+1|\sqrt{(2x+1)^2 - 1}} = \dfrac{1}{|2x+1|\sqrt{x^2 + x}}$

39. $\ln y = 3 \ln x - \dfrac{1}{2} \ln(x^2 + 1)$, $y'/y = \dfrac{3}{x} - \dfrac{x}{x^2 + 1}$, $y = \dfrac{3x^2}{\sqrt{x^2 + 1}} - \dfrac{x^4}{(x^2 + 1)^{3/2}}$

41. **(b)**

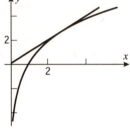

(c) $\dfrac{dy}{dx} = \dfrac{1}{2} - \dfrac{1}{x}$ so $\dfrac{dy}{dx} < 0$ at $x = 1$ and $\dfrac{dy}{dx} > 0$ at $x = e$

(d) The slope is a continuous function which goes from a negative value to a positive value; therefore it must take the value zero between, by the Intermediate Value Theorem.

(e) $\dfrac{dy}{dx} = 0$ when $x = 2$

43. Solve $\dfrac{dy}{dt} = 3\dfrac{dx}{dt}$ given $y = x \ln x$. Then $\dfrac{dy}{dt} = \dfrac{dy}{dx}\dfrac{dx}{dt} = (1 + \ln x)\dfrac{dx}{dt}$, so $1 + \ln x = 3$, $\ln x = 2$, $x = e^2$.

45. Set $y = \log_b x$ and solve $y' = 1$: $y' = \dfrac{1}{x \ln b} = 1$ so $x = \dfrac{1}{\ln b}$. The curves intersect when (x, x) lies on the graph of $y = \log_b x$, so $x = \log_b x$. From Formula (8), Section 1.6, $\log_b x = \dfrac{\ln x}{\ln b}$ from which $\ln x = 1$, $x = e$, $\ln b = 1/e$, $b = e^{1/e} \approx 1.4447$.

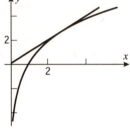

47. Yes, g must be differentiable (where $f' \neq 0$); this can be inferred from the graphs. Note that if $f' = 0$ at a point then g' cannot exist (infinite slope).

49. Let $P(x_0, y_0)$ be a point on $y = e^{3x}$ then $y_0 = e^{3x_0}$. $dy/dx = 3e^{3x}$ so $m_{\text{tan}} = 3e^{3x_0}$ at P and an equation of the tangent line at P is $y - y_0 = 3e^{3x_0}(x - x_0)$, $y - e^{3x_0} = 3e^{3x_0}(x - x_0)$. If the line passes through the origin then $(0, 0)$ must satisfy the equation so $-e^{3x_0} = -3x_0 e^{3x_0}$ which gives $x_0 = 1/3$ and thus $y_0 = e$. The point is $(1/3, e)$.

51. $y' = ae^{ax}\sin bx + be^{ax}\cos bx$ and $y'' = (a^2 - b^2)e^{ax}\sin bx + 2abe^{ax}\cos bx$, so $y'' - 2ay' + (a^2 + b^2)y$
$= (a^2 - b^2)e^{ax}\sin bx + 2abe^{ax}\cos bx - 2a(ae^{ax}\sin bx + be^{ax}\cos bx) + (a^2 + b^2)e^{ax}\sin bx = 0$.

53. **(a)**

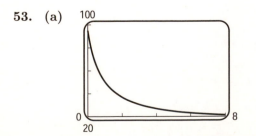

(b) as t tends to $+\infty$, the population tends to 19

$$\lim_{t\to+\infty} P(t) = \lim_{t\to+\infty} \frac{95}{5 - 4e^{-t/4}} = \frac{95}{5 - 4 \lim_{t\to+\infty} e^{-t/4}} = \frac{95}{5} = 19$$

(c) the rate of population growth tends to zero

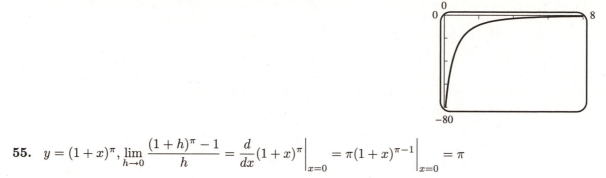

55. $y = (1+x)^\pi$, $\displaystyle\lim_{h\to0} \frac{(1+h)^\pi - 1}{h} = \frac{d}{dx}(1+x)^\pi\Big|_{x=0} = \pi(1+x)^{\pi-1}\Big|_{x=0} = \pi$

57. In the case $+\infty - (-\infty)$ the limit is $+\infty$; in the case $-\infty - (+\infty)$ the limit is $-\infty$, because large positive (negative) quantities are added to large positive (negative) quantities. The cases $+\infty - (+\infty)$ and $-\infty - (-\infty)$ are indeterminate; large numbers of opposite sign are subtracted, and more information about the sizes is needed.

59. $\displaystyle\lim_{x\to+\infty} (e^x - x^2) = \lim_{x\to+\infty} x^2(e^x/x^2 - 1)$, but $\displaystyle\lim_{x\to+\infty} \frac{e^x}{x^2} = \lim_{x\to+\infty} \frac{e^x}{2x} = \lim_{x\to+\infty} \frac{e^x}{2} = +\infty$

so $\displaystyle\lim_{x\to+\infty} (e^x/x^2 - 1) = +\infty$ and thus $\displaystyle\lim_{x\to+\infty} x^2(e^x/x^2 - 1) = +\infty$

61. $= \displaystyle\lim_{x\to0} \frac{(x^2 + 2x)e^x}{6\sin 3x\cos 3x} = \lim_{x\to0} \frac{(x^2 + 2x)e^x}{3\sin 6x} = \lim_{x\to0} \frac{(x^2 + 4x + 2)e^x}{18\cos 6x} = \frac{1}{9}$

63. $f'(x) = -\dfrac{2x}{e^{x^2}}$

$f''(x) = \dfrac{2(2x^2 - 1)}{e^{x^2}}$

(a) $(-\infty, 0]$ **(b)** $[0, +\infty)$

(c) $(-\infty, -\sqrt{2}/2), (\sqrt{2}/2, +\infty)$ **(d)** $(-\sqrt{2}/2, \sqrt{2}/2)$

(e) $-\sqrt{2}/2, \sqrt{2}/2$

65. $f'(x) = 2x/(1+x^2)$; critical point at $x = 0$; relative minimum of 0 at $x = 0$ (first derivative test)

67. $\displaystyle\lim_{x\to0^+} f(x) = \lim_{x\to+\infty} f(x) = +\infty$ and $f'(x) = \dfrac{e^x(x-2)}{x^3}$, stationary point at $x = 2$; by Theorem 3.4.4 $f(x)$ has an absolute minimum at $x = 2$, and $m = e^2/4$.

69. $f'(x) = 1/2 + 2x/(x^2 + 1)$,

$f'(x) = 0$ on $[-4, 0]$ for $x = -2 \pm \sqrt{3}$

if $x = -2 - \sqrt{3}, -2 + \sqrt{3}$ then

$f(x) = -1 - \sqrt{3}/2 + \ln 4 + \ln(2 + \sqrt{3}) \approx 0.84$,

$-1 + \sqrt{3}/2 + \ln 4 + \ln(2 - \sqrt{3}) \approx -0.06$,

absolute maximum at $x = -2 - \sqrt{3}$,

absolute minimum at $x = -2 + \sqrt{3}$

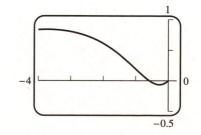

71. $3x^{1/3} - 5e^x + C$ **73.** $\tan^{-1} x + 2\sin^{-1} x + C$

75. $0.351220577, 0.420535296, 0.386502483$

77. $f(x) = e^x, [a, b] = [0, 1], \Delta x = \dfrac{1}{n}; \displaystyle\lim_{n \to +\infty} \sum_{k=1}^{n} f(x_k^*)\dfrac{1}{n} = \int_0^1 e^x\,dx = e - 1$

79. $\displaystyle\int_1^3 e^x\,dx = e^x \Big]_1^3 = e^3 - e$

81. (a) $y(x) = \sin x - 5e^x + C,\ y(0) = 0 = -5 + C,\ C = 5,\ y(x) = \sin x - 5e^x + 5$

    (b) $y(x) = \displaystyle\int xe^{x^2}\,dx = \dfrac{1}{2}e^{x^2} + C,\ y(0) = 0,\ y(x) = \dfrac{1}{2}e^{x^2} - \dfrac{1}{2}$

83. $\displaystyle\int_0^1 e^{-x/2}\,dx = 2(1 - 1/\sqrt{e})$

85. $V = \displaystyle\int_1^4 \left(\sqrt{x} - \dfrac{1}{\sqrt{x}}\right)^2 dx = 2\ln 2 + \dfrac{3}{2}$

87. (a) $\cosh 3x = \cosh(2x + x) = \cosh 2x \cosh x + \sinh 2x \sinh x$

$$= (2\cosh^2 x - 1)\cosh x + (2\sinh x \cosh x)\sinh x$$
$$= 2\cosh^3 x - \cosh x + 2\sinh^2 x \cosh x$$
$$= 2\cosh^3 x - \cosh x + 2(\cosh^2 x - 1)\cosh x = 4\cosh^3 x - 3\cosh x$$

    (b) from Theorem 6.8.2 with x replaced by $\dfrac{x}{2}$: $\cosh x = 2\cosh^2\dfrac{x}{2} - 1,$

$$2\cosh^2\dfrac{x}{2} = \cosh x + 1,\ \cosh^2\dfrac{x}{2} = \dfrac{1}{2}(\cosh x + 1),$$

$$\cosh\dfrac{x}{2} = \sqrt{\dfrac{1}{2}(\cosh x + 1)}\ \text{(because } \cosh\dfrac{x}{2} > 0\text{)}$$

    (c) from Theorem 6.8.2 with x replaced by $\dfrac{x}{2}$: $\cosh x = 2\sinh^2\dfrac{x}{2} + 1,$

$$2\sinh^2\dfrac{x}{2} = \cosh x - 1,\ \sinh^2\dfrac{x}{2} = \dfrac{1}{2}(\cosh x - 1),\ \sinh\dfrac{x}{2} = \pm\sqrt{\dfrac{1}{2}(\cosh x - 1)}$$

MAKING CONNECTIONS, CHAPTER 6

1. **(a)** If $t > 0$ then $A(-t)$ is the amount K there was t time-units ago in order that there be 1 unit now, i.e. $K \cdot A(t) = 1$, so $K = \dfrac{1}{A(t)}$. But, as said above, $K = A(-t)$. So $A(-t) = \dfrac{1}{A(t)}$.

   (b) **(i)** If s and t are positive, then the amount 1 becomes $A(s)$ after s seconds, and that in turn is $A(s)A(t)$ after another t seconds, i.e. 1 becomes $A(s)A(t)$ after $s + t$ seconds. But this amount is also $A(s+t)$, so $A(s)A(t) = A(s+t)$.

   (ii) If $0 \leq -s \leq t$ then $A(-s)A(s+t) = A(t)$. From Part (i) we get $A(s+t) = A(s)A(t)$.

   (iii) If $0 \leq t \leq -s$ then $A(s+t) = \dfrac{1}{A(-s-t)} = \dfrac{1}{A(-s)A(-t)} = A(s)A(t)$

   by Parts (i) and (ii).

   (iv) If s and t are both negative then by Part (i),
   $$A(s+t) = \frac{1}{A(-s-t)} = \frac{1}{A(-s)A(-t)} = A(s)A(t).$$

   (c) If $n > 0$ then $A\left(\dfrac{1}{n}\right)A\left(\dfrac{1}{n}\right)\ldots A\left(\dfrac{1}{n}\right) = A\left(n\dfrac{1}{n}\right) = A(1)$, so $A\left(\dfrac{1}{n}\right) = A(1)^{1/n} = b^{1/n}$

   from Part (b). If $n < 0$ then by Part (a), $A\left(\dfrac{1}{n}\right) = \dfrac{1}{A\left(-\frac{1}{n}\right)} = \dfrac{1}{A(1)^{-1/n}} = A(1)^{1/n} = b^{1/n}$

   (d) Let m, n be integers. Assume $n \neq 0$ and $m > 0$. Then $A\left(\dfrac{m}{n}\right) = A\left(\dfrac{1}{n}\right)^m = A(1)^{m/n} = b^{m/n}$

   (e) If f, g are continuous functions of t and f and g are equal on the rational numbers $\left\{\dfrac{m}{n} : n \neq 0\right\}$, then $f(t) = g(t)$ for all t. Because if x is irrational, then let t_n be a sequence of rational numbers which converges to x. Then for all $n > 0$, $f(t_n) = g(t_n)$ and thus $f(x) = \lim\limits_{n \to +\infty} f(t_n) = \lim\limits_{n \to +\infty} g(t_n) = g(x)$

3. Since $y = e^x$ and $y = \ln x$ are inverse functions, their graphs are symmetric with respect to the line $y = x$; consequently the areas A_1 and A_3 are equal (see figure). But $A_1 + A_2 = e$, so
   $$\int_1^e \ln x\, dx + \int_0^1 e^x\, dx = A_2 + A_3 = A_2 + A_1 = e$$

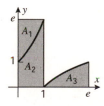

CHAPTER 7
Principles of Integral Evaluation

EXERCISE SET 7.1

1. $u = 4 - 2x, du = -2dx, -\frac{1}{2}\int u^3\,du = -\frac{1}{8}u^4 + C = -\frac{1}{8}(4 - 2x)^4 + C$

3. $u = x^2, du = 2x\,dx, \quad \frac{1}{2}\int \sec^2 u\,du = \frac{1}{2}\tan u + C = \frac{1}{2}\tan(x^2) + C$

5. $u = 2 + \cos 3x, du = -3\sin 3x\,dx, \quad -\frac{1}{3}\int \frac{du}{u} = -\frac{1}{3}\ln|u| + C = -\frac{1}{3}\ln(2 + \cos 3x) + C$

7. $u = e^x, du = e^x\,dx, \quad \int \sinh u\,du = \cosh u + C = \cosh e^x + C$

9. $u = \tan x, du = \sec^2 x\,dx, \quad \int e^u\,du = e^u + C = e^{\tan x} + C$

11. $u = \cos 5x, du = -5\sin 5x\,dx, \quad -\frac{1}{5}\int u^5\,du = -\frac{1}{30}u^6 + C = -\frac{1}{30}\cos^6 5x + C$

13. $u = e^x, du = e^x\,dx, \quad \int \frac{du}{\sqrt{4 + u^2}} = \ln\left(u + \sqrt{u^2 + 4}\right) + C = \ln\left(e^x + \sqrt{e^{2x} + 4}\right) + C$

15. $u = \sqrt{x - 1}, du = \frac{1}{2\sqrt{x - 1}}\,dx, \quad 2\int e^u\,du = 2e^u + C = 2e^{\sqrt{x-1}} + C$

17. $u = \sqrt{x}, du = \frac{1}{2\sqrt{x}}\,dx, \quad \int 2\cosh u\,du = 2\sinh u + C = 2\sinh\sqrt{x} + C$

19. $u = \sqrt{x}, du = \frac{1}{2\sqrt{x}}\,dx, \quad \int \frac{2\,du}{3^u} = 2\int e^{-u\ln 3}\,du = -\frac{2}{\ln 3}e^{-u\ln 3} + C = -\frac{2}{\ln 3}3^{-\sqrt{x}} + C$

21. $u = \frac{2}{x}, du = -\frac{2}{x^2}\,dx, \quad -\frac{1}{2}\int \operatorname{csch}^2 u\,du = \frac{1}{2}\coth u + C = \frac{1}{2}\coth\frac{2}{x} + C$

23. $u = e^{-x}, du = -e^{-x}\,dx, \quad -\int \frac{du}{4 - u^2} = -\frac{1}{4}\ln\left|\frac{2 + u}{2 - u}\right| + C = -\frac{1}{4}\ln\left|\frac{2 + e^{-x}}{2 - e^{-x}}\right| + C$

25. $u = e^x, du = e^x\,dx, \quad \int \frac{e^x\,dx}{\sqrt{1 - e^{2x}}} = \int \frac{du}{\sqrt{1 - u^2}} = \sin^{-1} u + C = \sin^{-1} e^x + C$

27. $u = x^2, du = 2x\,dx, \quad \frac{1}{2}\int \frac{du}{\csc u} = \frac{1}{2}\int \sin u\,du = -\frac{1}{2}\cos u + C = -\frac{1}{2}\cos(x^2) + C$

29. $4^{-x^2} = e^{-x^2\ln 4}, u = -x^2\ln 4, du = -2x\ln 4\,dx = -x\ln 16\,dx,$

$$-\frac{1}{\ln 16}\int e^u\,du = -\frac{1}{\ln 16}e^u + C = -\frac{1}{\ln 16}e^{-x^2\ln 4} + C = -\frac{1}{\ln 16}4^{-x^2} + C$$

31. **(a)** $u = \sin x, du = \cos x\,dx, \quad \int u\,du = \frac{1}{2}u^2 + C = \frac{1}{2}\sin^2 x + C$

(b) $\int \sin x\cos x\,dx = \frac{1}{2}\int \sin 2x\,dx = -\frac{1}{4}\cos 2x + C = -\frac{1}{4}(\cos^2 x - \sin^2 x) + C$

(c) $-\dfrac{1}{4}(\cos^2 x - \sin^2 x) + C = -\dfrac{1}{4}(1 - \sin^2 x - \sin^2 x) + C = -\dfrac{1}{4} + \dfrac{1}{2}\sin^2 x + C,$
and this is the same as the answer in part (a) except for the constants.

33. (a) $\dfrac{\sec^2 x}{\tan x} = \dfrac{1}{\cos^2 x \tan x} = \dfrac{1}{\cos x \sin x}$

(b) $\csc 2x = \dfrac{1}{\sin 2x} = \dfrac{1}{2\sin x \cos x} = \dfrac{1}{2}\dfrac{\sec^2 x}{\tan x}$, so $\displaystyle\int \csc 2x\,dx = \dfrac{1}{2}\ln\tan x + C$

(c) $\sec x = \dfrac{1}{\cos x} = \dfrac{1}{\sin(\pi/2 - x)} = \csc(\pi/2 - x)$, so

$$\int \sec x\,dx = \int \csc(\pi/2 - x)\,dx = -\dfrac{1}{2}\ln\tan(\pi/2 - x) + C$$

EXERCISE SET 7.2

1. $u = x$, $dv = e^{-2x}dx$, $du = dx$, $v = -\tfrac{1}{2}e^{-2x}$;

$$\int xe^{-2x}dx = -\dfrac{1}{2}xe^{-2x} + \int \dfrac{1}{2}e^{-2x}dx = -\dfrac{1}{2}xe^{-2x} - \dfrac{1}{4}e^{-2x} + C$$

3. $u = x^2$, $dv = e^x dx$, $du = 2x\,dx$, $v = e^x$; $\displaystyle\int x^2 e^x dx = x^2 e^x - 2\int xe^x dx.$

For $\displaystyle\int xe^x dx$ use $u = x$, $dv = e^x dx$, $du = dx$, $v = e^x$ to get

$$\int xe^x dx = xe^x - e^x + C_1 \text{ so } \int x^2 e^x dx = x^2 e^x - 2xe^x + 2e^x + C$$

5. $u = x$, $dv = \sin 3x\,dx$, $du = dx$, $v = -\dfrac{1}{3}\cos 3x$;

$$\int x\sin 3x\,dx = -\dfrac{1}{3}x\cos 3x + \dfrac{1}{3}\int \cos 3x\,dx = -\dfrac{1}{3}x\cos 3x + \dfrac{1}{9}\sin 3x + C$$

7. $u = x^2$, $dv = \cos x\,dx$, $du = 2x\,dx$, $v = \sin x$; $\displaystyle\int x^2 \cos x\,dx = x^2 \sin x - 2\int x\sin x\,dx$

For $\displaystyle\int x\sin x\,dx$ use $u = x$, $dv = \sin x\,dx$ to get

$$\int x\sin x\,dx = -x\cos x + \sin x + C_1 \text{ so } \int x^2 \cos x\,dx = x^2 \sin x + 2x\cos x - 2\sin x + C$$

9. $u = \ln x$, $dv = x\,dx$, $du = \dfrac{1}{x}dx$, $v = \dfrac{1}{2}x^2$; $\displaystyle\int x\ln x\,dx = \dfrac{1}{2}x^2 \ln x - \dfrac{1}{2}\int x\,dx = \dfrac{1}{2}x^2 \ln x - \dfrac{1}{4}x^2 + C$

11. $u = (\ln x)^2$, $dv = dx$, $du = 2\dfrac{\ln x}{x}dx$, $v = x$; $\displaystyle\int (\ln x)^2 dx = x(\ln x)^2 - 2\int \ln x\,dx.$

Use $u = \ln x$, $dv = dx$ to get $\displaystyle\int \ln x\,dx = x\ln x - \int dx = x\ln x - x + C_1$ so

$$\int (\ln x)^2 dx = x(\ln x)^2 - 2x\ln x + 2x + C$$

13. $u = \ln(3x - 2)$, $dv = dx$, $du = \dfrac{3}{3x - 2}dx$, $v = x$; $\displaystyle\int \ln(3x - 2)dx = x\ln(3x - 2) - \int \dfrac{3x}{3x - 2}dx$

but $\displaystyle\int \frac{3x}{3x-2}dx = \int\left(1+\frac{2}{3x-2}\right)dx = x + \frac{2}{3}\ln(3x-2)+C_1$ so

$$\int \ln(3x-2)dx = x\ln(3x-2) - x - \frac{2}{3}\ln(3x-2)+C$$

15. $u = \sin^{-1}x,\ dv = dx,\ du = 1/\sqrt{1-x^2}dx,\ v = x;$

$$\int \sin^{-1}x\,dx = x\sin^{-1}x - \int x/\sqrt{1-x^2}dx = x\sin^{-1}x + \sqrt{1-x^2}+C$$

17. $u = \tan^{-1}(3x),\ dv = dx,\ du = \dfrac{3}{1+9x^2}dx,\ v = x;$

$$\int \tan^{-1}(3x)dx = x\tan^{-1}(3x) - \int \frac{3x}{1+9x^2}dx = x\tan^{-1}(3x) - \frac{1}{6}\ln(1+9x^2)+C$$

19. $u = e^x,\ dv = \sin x\,dx,\ du = e^x dx,\ v = -\cos x;\ \displaystyle\int e^x \sin x\,dx = -e^x \cos x + \int e^x \cos x\,dx.$

For $\displaystyle\int e^x \cos x\,dx$ use $u = e^x,\ dv = \cos x\,dx$ to get $\displaystyle\int e^x \cos x = e^x \sin x - \int e^x \sin x\,dx$ so

$$\int e^x \sin x\,dx = -e^x \cos x + e^x \sin x - \int e^x \sin x\,dx,$$

$$2\int e^x \sin x\,dx = e^x(\sin x - \cos x) + C_1, \int e^x \sin x\,dx = \frac{1}{2}e^x(\sin x - \cos x)+C$$

21. $u = \sin(\ln x),\ dv = dx,\ du = \dfrac{\cos(\ln x)}{x}dx,\ v = x;$

$$\int \sin(\ln x)dx = x\sin(\ln x) - \int \cos(\ln x)dx.\ \text{Use } u = \cos(\ln x),\ dv = dx \text{ to get}$$

$$\int \cos(\ln x)dx = x\cos(\ln x) + \int \sin(\ln x)dx \text{ so}$$

$$\int \sin(\ln x)dx = x\sin(\ln x) - x\cos(\ln x) - \int \sin(\ln x)dx,$$

$$\int \sin(\ln x)dx = \frac{1}{2}x[\sin(\ln x) - \cos(\ln x)]+C$$

23. $u = x,\ dv = \sec^2 x\,dx,\ du = dx,\ v = \tan x;$

$$\int x\sec^2 x\,dx = x\tan x - \int \tan x\,dx = x\tan x - \int \frac{\sin x}{\cos x}dx = x\tan x + \ln|\cos x|+C$$

25. $u = x^2,\ dv = xe^{x^2}dx,\ du = 2x\,dx,\ v = \dfrac{1}{2}e^{x^2};$

$$\int x^3 e^{x^2}dx = \frac{1}{2}x^2 e^{x^2} - \int xe^{x^2}dx = \frac{1}{2}x^2 e^{x^2} - \frac{1}{2}e^{x^2}+C$$

27. $u = x,\ dv = e^{2x}dx,\ du = dx,\ v = \dfrac{1}{2}e^{2x};$

$$\int_0^2 xe^{2x}dx = \frac{1}{2}xe^{2x}\Big]_0^2 - \frac{1}{2}\int_0^2 e^{2x}dx = e^4 - \frac{1}{4}e^{2x}\Big]_0^2 = e^4 - \frac{1}{4}(e^4 - 1) = (3e^4 + 1)/4$$

29. $u = \ln x,\ dv = x^2 dx,\ du = \dfrac{1}{x}dx,\ v = \dfrac{1}{3}x^3;$

$$\int_1^e x^2 \ln x\, dx = \frac{1}{3}x^3 \ln x \Big]_1^e - \frac{1}{3}\int_1^e x^2 dx = \frac{1}{3}e^3 - \frac{1}{9}x^3 \Big]_1^e = \frac{1}{3}e^3 - \frac{1}{9}(e^3 - 1) = (2e^3 + 1)/9$$

31. $u = \ln(x+2),\ dv = dx,\ du = \dfrac{1}{x+2}dx,\ v = x;$

$$\int_{-1}^1 \ln(x+2)dx = x\ln(x+2)\Big]_{-1}^1 - \int_{-1}^1 \frac{x}{x+2}dx = \ln 3 + \ln 1 - \int_{-1}^1 \left[1 - \frac{2}{x+2}\right]dx$$

$$= \ln 3 - [x - 2\ln(x+2)]\Big]_{-1}^1 = \ln 3 - (1 - 2\ln 3) + (-1 - 2\ln 1) = 3\ln 3 - 2$$

33. $u = \sec^{-1}\sqrt{\theta},\ dv = d\theta,\ du = \dfrac{1}{2\theta\sqrt{\theta-1}}d\theta,\ v = \theta;$

$$\int_2^4 \sec^{-1}\sqrt{\theta}\,d\theta = \theta\sec^{-1}\sqrt{\theta}\Big]_2^4 - \frac{1}{2}\int_2^4 \frac{1}{\sqrt{\theta-1}}d\theta = 4\sec^{-1}2 - 2\sec^{-1}\sqrt{2} - \sqrt{\theta-1}\Big]_2^4$$

$$= 4\left(\frac{\pi}{3}\right) - 2\left(\frac{\pi}{4}\right) - \sqrt{3} + 1 = \frac{5\pi}{6} - \sqrt{3} + 1$$

35. $u = x,\ dv = \sin 2x\, dx,\ du = dx,\ v = -\dfrac{1}{2}\cos 2x;$

$$\int_0^\pi x\sin 2x\, dx = -\frac{1}{2}x\cos 2x \Big]_0^\pi + \frac{1}{2}\int_0^\pi \cos 2x\, dx = -\pi/2 + \frac{1}{4}\sin 2x\Big]_0^\pi = -\pi/2$$

37. $u = \tan^{-1}\sqrt{x},\ dv = \sqrt{x}dx,\ du = \dfrac{1}{2\sqrt{x}(1+x)}dx,\ v = \dfrac{2}{3}x^{3/2};$

$$\int_1^3 \sqrt{x}\tan^{-1}\sqrt{x}dx = \frac{2}{3}x^{3/2}\tan^{-1}\sqrt{x}\Big]_1^3 - \frac{1}{3}\int_1^3 \frac{x}{1+x}dx$$

$$= \frac{2}{3}x^{3/2}\tan^{-1}\sqrt{x}\Big]_1^3 - \frac{1}{3}\int_1^3 \left[1 - \frac{1}{1+x}\right]dx$$

$$= \left[\frac{2}{3}x^{3/2}\tan^{-1}\sqrt{x} - \frac{1}{3}x + \frac{1}{3}\ln|1+x|\right]_1^3 = (2\sqrt{3}\pi - \pi/2 - 2 + \ln 2)/3$$

39. true

41. false; e^x is not a factor of the integrand

43. $t = \sqrt{x},\ t^2 = x,\ dx = 2t\, dt$

$$\int e^{\sqrt{x}}dx = 2\int te^t\, dt;\ u = t, dv = e^t dt, du = dt, v = e^t,$$

$$\int e^{\sqrt{x}}dx = 2te^t - 2\int e^t\, dt = 2(t-1)e^t + C = 2(\sqrt{x}-1)e^{\sqrt{x}} + C$$

45. Let $f_1(x), f_2(x), f_3(x)$ denote successive antiderivatives of $f(x)$,
so that $f_3'(x) = f_2(x), f_2'(x) = f_1(x), f_1'(x) = f(x)$. Let $p(x) = ax^2 + bx + c$.

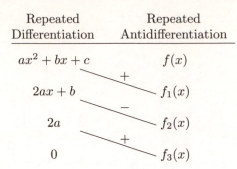

Repeated Differentiation	Repeated Antidifferentiation
$ax^2 + bx + c$	$f(x)$
$2ax + b$	$f_1(x)$
$2a$	$f_2(x)$
0	$f_3(x)$

Then $\displaystyle\int p(x)f(x)\,dx = (ax^2 + bx + c)f_1(x) - (2ax + b)f_2(x) + 2af_3(x) + C$. Check:

$$\frac{d}{dx}[(ax^2 + bx + c)f_1(x) - (2ax + b)f_2(x) + 2af_3(x)]$$

$$= (2ax + b)f_1(x) + (ax^2 + bx + c)f(x) - 2af_2(x) - (2ax + b)f_1(x) + 2af_2(x) = p(x)f(x)$$

47.

Repeated Differentiation	Repeated Antidifferentiation
$3x^2 - x + 2$	e^{-x}
$6x - 1$	$-e^{-x}$
6	e^{-x}
0	$-e^{-x}$

$$\int (3x^2 - x + 2)e^{-x} = -(3x^2 - x + 2)e^{-x} - (6x - 1)e^{-x} - 6e^{-x} + C = -e^{-x}[3x^2 + 5x + 7] + C$$

49.

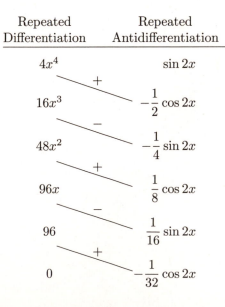

Repeated Differentiation	Repeated Antidifferentiation
$4x^4$	$\sin 2x$
$16x^3$	$-\dfrac{1}{2}\cos 2x$
$48x^2$	$-\dfrac{1}{4}\sin 2x$
$96x$	$\dfrac{1}{8}\cos 2x$
96	$\dfrac{1}{16}\sin 2x$
0	$-\dfrac{1}{32}\cos 2x$

$$\int 4x^4 \sin 2x\,dx = (-2x^4 + 6x^2 - 3)\cos 2x + (4x^3 - 6x)\sin 2x + C$$

51.

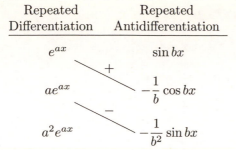

$$I = \int e^{ax} \sin bx \, dx = -\frac{1}{b} e^{ax} \cos bx + \frac{a}{b^2} e^{ax} \sin bx - \frac{a^2}{b^2} I, \text{ so}$$

$$I = \frac{e^{ax}}{a^2 + b^2} (a \sin bx - b \cos bx) + C$$

53. **(a)** We perform a single integration by parts:

$$u = \cos x, dv = \sin x \, dx, du = -\sin x \, dx, v = -\cos x, \int \sin x \cos x \, dx = -\cos^2 x - \int \sin x \cos x \, dx.$$

Thus

$$2 \int \sin x \cos x \, dx = -\cos^2 x + C, \int \sin x \cos x \, dx = -\frac{1}{2} \cos^2 x + C$$

Alternatively, $u = \sin x, du = \cos x \, dx, \int \sin x \cos x \, dx = \int u \, du = \frac{1}{2} u^2 + C = \frac{1}{2} \sin^2 x + C$

(b) Since $\sin^2 x + \cos^2 x = 1$, they are equal (although the symbol 'C' refers to different constants in the two equations).

55. **(a)** $A = \int_1^e \ln x \, dx = (x \ln x - x) \Big]_1^e = 1$

(b) $V = \pi \int_1^e (\ln x)^2 dx = \pi \left[(x(\ln x)^2 - 2x \ln x + 2x) \right]_1^e = \pi(e - 2)$

57. $V = 2\pi \int_0^\pi x \sin x \, dx = 2\pi (-x \cos x + \sin x) \Big]_0^\pi = 2\pi^2$

59. distance $= \int_0^\pi t^3 \sin t \, dt$

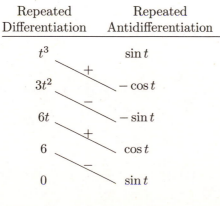

$$\int_0^\pi t^3 \sin t \, dx = [(-t^3 \cos t + 3t^2 \sin t + 6t \cos t - 6 \sin t)] \Big]_0^\pi = \pi^3 - 6\pi$$

61. (a) $\displaystyle\int \sin^4 x\,dx = -\frac{1}{4}\sin^3 x \cos x + \frac{3}{4}\int \sin^2 x\,dx$

$$= -\frac{1}{4}\sin^3 x \cos x + \frac{3}{4}\left[-\frac{1}{2}\sin x \cos x + \frac{1}{2}x\right] + C$$

$$= -\frac{1}{4}\sin^3 x \cos x - \frac{3}{8}\sin x \cos x + \frac{3}{8}x + C$$

(b) $\displaystyle\int_0^{\pi/2} \sin^5 x\,dx = -\frac{1}{5}\sin^4 x \cos x\Big]_0^{\pi/2} + \frac{4}{5}\int_0^{\pi/2} \sin^3 x\,dx$

$$= \frac{4}{5}\left[-\frac{1}{3}\sin^2 x \cos x\Big]_0^{\pi/2} + \frac{2}{3}\int_0^{\pi/2} \sin x\,dx\right]$$

$$= -\frac{8}{15}\cos x\Big]_0^{\pi/2} = \frac{8}{15}$$

63. $u = \sin^{n-1} x,\ dv = \sin x\,dx,\ du = (n-1)\sin^{n-2} x \cos x\,dx,\ v = -\cos x;$

$$\int \sin^n x\,dx = -\sin^{n-1} x \cos x + (n-1)\int \sin^{n-2} x \cos^2 x\,dx$$

$$= -\sin^{n-1} x \cos x + (n-1)\int \sin^{n-2} x\,(1-\sin^2 x)dx$$

$$= -\sin^{n-1} x \cos x + (n-1)\int \sin^{n-2} x\,dx - (n-1)\int \sin^n x\,dx,$$

$$n\int \sin^n x\,dx = -\sin^{n-1} x \cos x + (n-1)\int \sin^{n-2} x\,dx,$$

$$\int \sin^n x\,dx = -\frac{1}{n}\sin^{n-1} x \cos x + \frac{n-1}{n}\int \sin^{n-2} x\,dx$$

65. (a) $\displaystyle\int \tan^4 x\,dx = \frac{1}{3}\tan^3 x - \int \tan^2 x\,dx = \frac{1}{3}\tan^3 x - \tan x + \int dx = \frac{1}{3}\tan^3 x - \tan x + x + C$

(b) $\displaystyle\int \sec^4 x\,dx = \frac{1}{3}\sec^2 x \tan x + \frac{2}{3}\int \sec^2 x\,dx = \frac{1}{3}\sec^2 x \tan x + \frac{2}{3}\tan x + C$

(c) $\displaystyle\int x^3 e^x\,dx = x^3 e^x - 3\int x^2 e^x\,dx = x^3 e^x - 3\left[x^2 e^x - 2\int x e^x\,dx\right]$

$$= x^3 e^x - 3x^2 e^x + 6\left[x e^x - \int e^x\,dx\right] = x^3 e^x - 3x^2 e^x + 6x e^x - 6e^x + C$$

67. $u = x,\ dv = f''(x)dx,\ du = dx,\ v = f'(x);$

$$\int_{-1}^1 x f''(x)dx = x f'(x)\Big]_{-1}^1 - \int_{-1}^1 f'(x)dx$$

$$= f'(1) + f'(-1) - f(x)\Big]_{-1}^1 = f'(1) + f'(-1) - f(1) + f(-1)$$

69. $u = \ln(x+1),\ dv = dx,\ du = \dfrac{dx}{x+1},\ v = x+1;$

$$\int \ln(x+1)\,dx = \int u\,dv = uv - \int v\,du = (x+1)\ln(x+1) - \int dx = (x+1)\ln(x+1) - x + C$$

71. $u = \tan^{-1} x,\ dv = x\,dx,\ du = \dfrac{1}{1+x^2}\,dx,\ v = \dfrac{1}{2}(x^2+1)$

$$\int x\tan^{-1}x\,dx = \int u\,dv = uv - \int v\,du = \frac{1}{2}(x^2+1)\tan^{-1}x - \frac{1}{2}\int dx$$

$$= \frac{1}{2}(x^2+1)\tan^{-1}x - \frac{1}{2}x + C$$

EXERCISE SET 7.3

1. $u = \cos x, \; -\int u^3 du = -\frac{1}{4}\cos^4 x + C$

3. $\int \sin^2 5\theta = \frac{1}{2}\int(1-\cos 10\theta)\,d\theta = \frac{1}{2}\theta - \frac{1}{20}\sin 10\theta + C$

5. $\int \sin^3 a\theta\,d\theta = \int \sin a\theta(1-\cos^2 a\theta)\,d\theta = -\frac{1}{a}\cos a\theta - \frac{1}{3a}\cos^3 a\theta + C \quad (a \neq 0)$

7. $u = \sin ax, \quad \frac{1}{a}\int u\,du = \frac{1}{2a}\sin^2 ax + C, a \neq 0$

9. $\int \sin^2 t\cos^3 t\,dt = \int \sin^2 t(1-\sin^2 t)\cos t\,dt = \int(\sin^2 t - \sin^4 t)\cos t\,dt$

$$= \frac{1}{3}\sin^3 t - \frac{1}{5}\sin^5 t + C$$

11. $\int \sin^2 x\cos^2 x\,dx = \frac{1}{4}\int \sin^2 2x\,dx = \frac{1}{8}\int(1-\cos 4x)dx = \frac{1}{8}x - \frac{1}{32}\sin 4x + C$

13. $\int \sin 2x\cos 3x\,dx = \frac{1}{2}\int(\sin 5x - \sin x)dx = -\frac{1}{10}\cos 5x + \frac{1}{2}\cos x + C$

15. $\int \sin x\cos(x/2)dx = \frac{1}{2}\int[\sin(3x/2) + \sin(x/2)]dx = -\frac{1}{3}\cos(3x/2) - \cos(x/2) + C$

17. $\int_0^{\pi/2} \cos^3 x\,dx = \int_0^{\pi/2}(1-\sin^2 x)\cos x\,dx$

$$= \left[\sin x - \frac{1}{3}\sin^3 x\right]_0^{\pi/2} = \frac{2}{3}$$

19. $\int_0^{\pi/3} \sin^4 3x\cos^3 3x\,dx = \int_0^{\pi/3}\sin^4 3x(1-\sin^2 3x)\cos 3x\,dx = \left[\frac{1}{15}\sin^5 3x - \frac{1}{21}\sin^7 3x\right]_0^{\pi/3} = 0$

21. $\int_0^{\pi/6} \sin 4x\cos 2x\,dx = \frac{1}{2}\int_0^{\pi/6}(\sin 2x + \sin 6x)dx = \left[-\frac{1}{4}\cos 2x - \frac{1}{12}\cos 6x\right]_0^{\pi/6}$

$$= [(-1/4)(1/2) - (1/12)(-1)] - [-1/4 - 1/12] = 7/24$$

23. $\frac{1}{2}\tan(2x-1) + C$

25. $u = e^{-x}, du = -e^{-x}\,dx; \; -\int \tan u\,du = \ln|\cos u| + C = \ln|\cos(e^{-x})| + C$

27. $\dfrac{1}{4}\ln|\sec 4x + \tan 4x| + C$

29. $u = \tan x, \quad \displaystyle\int u^2\,du = \dfrac{1}{3}\tan^3 x + C$

31. $\displaystyle\int \tan 4x(1+\tan^2 4x)\sec^2 4x\,dx = \int (\tan 4x + \tan^3 4x)\sec^2 4x\,dx = \dfrac{1}{8}\tan^2 4x + \dfrac{1}{16}\tan^4 4x + C$

33. $\displaystyle\int \sec^4 x(\sec^2 x - 1)\sec x \tan x\,dx = \int (\sec^6 x - \sec^4 x)\sec x \tan x\,dx = \dfrac{1}{7}\sec^7 x - \dfrac{1}{5}\sec^5 x + C$

35. $\displaystyle\int (\sec^2 x - 1)^2 \sec x\,dx = \int (\sec^5 x - 2\sec^3 x + \sec x)dx = \int \sec^5 x\,dx - 2\int \sec^3 x\,dx + \int \sec x\,dx$

$\qquad = \dfrac{1}{4}\sec^3 x \tan x + \dfrac{3}{4}\displaystyle\int \sec^3 x\,dx - 2\int \sec^3 x\,dx + \ln|\sec x + \tan x|$

$\qquad = \dfrac{1}{4}\sec^3 x \tan x - \dfrac{5}{4}\left[\dfrac{1}{2}\sec x \tan x + \dfrac{1}{2}\ln|\sec x + \tan x|\right] + \ln|\sec x + \tan x| + C$

$\qquad = \dfrac{1}{4}\sec^3 x \tan x - \dfrac{5}{8}\sec x \tan x + \dfrac{3}{8}\ln|\sec x + \tan x| + C$

37. $\displaystyle\int \sec^2 t(\sec t \tan t)dt = \dfrac{1}{3}\sec^3 t + C$

39. $\displaystyle\int \sec^4 x\,dx = \int (1+\tan^2 x)\sec^2 x\,dx = \int (\sec^2 x + \tan^2 x \sec^2 x)dx = \tan x + \dfrac{1}{3}\tan^3 x + C$

41. $u = 4x$, use equation (19) to get

$\qquad \dfrac{1}{4}\displaystyle\int \tan^3 u\,du = \dfrac{1}{4}\left[\dfrac{1}{2}\tan^2 u + \ln|\cos u|\right] + C = \dfrac{1}{8}\tan^2 4x + \dfrac{1}{4}\ln|\cos 4x| + C$

43. $\displaystyle\int \sqrt{\tan x}(1+\tan^2 x)\sec^2 x\,dx = \dfrac{2}{3}\tan^{3/2} x + \dfrac{2}{7}\tan^{7/2} x + C$

45. $\displaystyle\int_0^{\pi/8} (\sec^2 2x - 1)dx = \left[\dfrac{1}{2}\tan 2x - x\right]_0^{\pi/8} = 1/2 - \pi/8$

47. $u = x/2$,

$\qquad 2\displaystyle\int_0^{\pi/4} \tan^5 u\,du = \left[\dfrac{1}{2}\tan^4 u - \tan^2 u - 2\ln|\cos u|\right]_0^{\pi/4} = 1/2 - 1 - 2\ln(1/\sqrt{2}) = -1/2 + \ln 2$

49. $\displaystyle\int (\csc^2 x - 1)\csc^2 x(\csc x \cot x)dx = \int (\csc^4 x - \csc^2 x)(\csc x \cot x)dx = -\dfrac{1}{5}\csc^5 x + \dfrac{1}{3}\csc^3 x + C$

51. $\displaystyle\int (\csc^2 x - 1)\cot x\,dx = \int \csc x(\csc x \cot x)dx - \int \dfrac{\cos x}{\sin x}dx = -\dfrac{1}{2}\csc^2 x - \ln|\sin x| + C$

53. true

55. false

57. (a) $\displaystyle \int_0^{2\pi} \sin mx \cos nx \, dx = \frac{1}{2} \int_0^{2\pi} [\sin(m+n)x + \sin(m-n)x] dx = \left[-\frac{\cos(m+n)x}{2(m+n)} - \frac{\cos(m-n)x}{2(m-n)} \right]_0^{2\pi}$

but $\cos(m+n)x \Big]_0^{2\pi} = 0, \cos(m-n)x \Big]_0^{2\pi} = 0.$

(b) $\displaystyle \int_0^{2\pi} \cos mx \cos nx \, dx = \frac{1}{2} \int_0^{2\pi} [\cos(m+n)x + \cos(m-n)x] dx;$

since $m \neq n$, evaluate sin at integer multiples of 2π to get 0.

(c) $\displaystyle \int_0^{2\pi} \sin mx \sin nx \, dx = \frac{1}{2} \int_0^{2\pi} [\cos(m-n)x - \cos(m+n)x] \, dx;$

since $m \neq n$, evaluate sin at integer multiples of 2π to get 0.

59. $y' = \tan x, \; 1 + (y')^2 = 1 + \tan^2 x = \sec^2 x,$

$$L = \int_0^{\pi/4} \sqrt{\sec^2 x} \, dx = \int_0^{\pi/4} \sec x \, dx = \ln |\sec x + \tan x| \Big]_0^{\pi/4} = \ln(\sqrt{2} + 1)$$

61. $\displaystyle V = \pi \int_0^{\pi/4} (\cos^2 x - \sin^2 x) dx = \pi \int_0^{\pi/4} \cos 2x \, dx = \frac{1}{2} \pi \sin 2x \Big]_0^{\pi/4} = \pi/2$

63. With $0 < \alpha < \beta, D = D_\beta - D_\alpha = \dfrac{L}{2\pi} \displaystyle\int_\alpha^\beta \sec x \, dx = \dfrac{L}{2\pi} \ln |\sec x + \tan x| \Big]_\alpha^\beta = \dfrac{L}{2\pi} \ln \left| \dfrac{\sec \beta + \tan \beta}{\sec \alpha + \tan \alpha} \right|$

65. (a) $\displaystyle \int \csc x \, dx = \int \sec(\pi/2 - x) dx = -\ln |\sec(\pi/2 - x) + \tan(\pi/2 - x)| + C$

$$= -\ln |\csc x + \cot x| + C$$

(b) $-\ln |\csc x + \cot x| = \ln \dfrac{1}{|\csc x + \cot x|} = \ln \dfrac{|\csc x - \cot x|}{|\csc^2 x - \cot^2 x|} = \ln |\csc x - \cot x|,$

$$-\ln |\csc x + \cot x| = -\ln \left| \dfrac{1}{\sin x} + \dfrac{\cos x}{\sin x} \right| = \ln \left| \dfrac{\sin x}{1 + \cos x} \right|$$

$$= \ln \left| \dfrac{2 \sin(x/2) \cos(x/2)}{2 \cos^2(x/2)} \right| = \ln |\tan(x/2)|$$

67. $a \sin x + b \cos x = \sqrt{a^2 + b^2} \left[\dfrac{a}{\sqrt{a^2 + b^2}} \sin x + \dfrac{b}{\sqrt{a^2 + b^2}} \cos x \right] = \sqrt{a^2 + b^2} (\sin x \cos \theta + \cos x \sin \theta)$

where $\cos \theta = a/\sqrt{a^2 + b^2}$ and $\sin \theta = b/\sqrt{a^2 + b^2}$ so $a \sin x + b \cos x = \sqrt{a^2 + b^2} \sin(x + \theta)$

and $\displaystyle \int \dfrac{dx}{a \sin x + b \cos x} = \dfrac{1}{\sqrt{a^2 + b^2}} \int \csc(x + \theta) dx = -\dfrac{1}{\sqrt{a^2 + b^2}} \ln |\csc(x + \theta) + \cot(x + \theta)| + C$

$$= -\dfrac{1}{\sqrt{a^2 + b^2}} \ln \left| \dfrac{\sqrt{a^2 + b^2} + a \cos x - b \sin x}{a \sin x + b \cos x} \right| + C$$

69. (a) $\displaystyle \int_0^{\pi/2} \sin^3 x \, dx = \frac{2}{3}$ **(b)** $\displaystyle \int_0^{\pi/2} \sin^4 x \, dx = \frac{1 \cdot 3}{2 \cdot 4} \cdot \frac{\pi}{2} = 3\pi/16$

(c) $\displaystyle \int_0^{\pi/2} \sin^5 x \, dx = \frac{2 \cdot 4}{3 \cdot 5} = 8/15$ **(d)** $\displaystyle \int_0^{\pi/2} \sin^6 x \, dx = \frac{1 \cdot 3 \cdot 5}{2 \cdot 4 \cdot 6} \cdot \frac{\pi}{2} = 5\pi/32$

EXERCISE SET 7.4

1. $x = 2\sin\theta$, $dx = 2\cos\theta\,d\theta$,

$$4\int\cos^2\theta\,d\theta = 2\int(1+\cos 2\theta)d\theta = 2\theta + \sin 2\theta + C$$

$$= 2\theta + 2\sin\theta\cos\theta + C = 2\sin^{-1}(x/2) + \frac{1}{2}x\sqrt{4-x^2} + C$$

3. $x = 4\sin\theta$, $dx = 4\cos\theta\,d\theta$,

$$16\int\sin^2\theta\,d\theta = 8\int(1-\cos 2\theta)d\theta = 8\theta - 4\sin 2\theta + C = 8\theta - 8\sin\theta\cos\theta + C$$

$$= 8\sin^{-1}(x/4) - \frac{1}{2}x\sqrt{16-x^2} + C$$

5. $x = 2\tan\theta$, $dx = 2\sec^2\theta\,d\theta$,

$$\frac{1}{8}\int\frac{1}{\sec^2\theta}d\theta = \frac{1}{8}\int\cos^2\theta\,d\theta = \frac{1}{16}\int(1+\cos 2\theta)d\theta = \frac{1}{16}\theta + \frac{1}{32}\sin 2\theta + C$$

$$= \frac{1}{16}\theta + \frac{1}{16}\sin\theta\cos\theta + C = \frac{1}{16}\tan^{-1}\frac{x}{2} + \frac{x}{8(4+x^2)} + C$$

7. $x = 3\sec\theta$, $dx = 3\sec\theta\tan\theta\,d\theta$,

$$3\int\tan^2\theta\,d\theta = 3\int(\sec^2\theta - 1)d\theta = 3\tan\theta - 3\theta + C = \sqrt{x^2-9} - 3\sec^{-1}\frac{x}{3} + C$$

9. $x = \sin\theta$, $dx = \cos\theta\,d\theta$,

$$3\int\sin^3\theta\,d\theta = 3\int\left[1-\cos^2\theta\right]\sin\theta\,d\theta$$

$$= 3\left(-\cos\theta + \cos^3\theta\right) + C = -3\sqrt{1-x^2} + (1-x^2)^{3/2} + C$$

11. $x = \frac{2}{3}\sec\theta$, $dx = \frac{2}{3}\sec\theta\tan\theta\,d\theta$, $\frac{3}{4}\int\frac{1}{\sec\theta}d\theta = \frac{3}{4}\int\cos\theta\,d\theta = \frac{3}{4}\sin\theta + C = \frac{1}{4x}\sqrt{9x^2-4} + C$

13. $x = \sin\theta$, $dx = \cos\theta\,d\theta$, $\int\frac{1}{\cos^2\theta}d\theta = \int\sec^2\theta\,d\theta = \tan\theta + C = x/\sqrt{1-x^2} + C$

15. $x = 3\sec\theta$, $dx = 3\sec\theta\tan\theta\,d\theta$, $\int\sec\theta\,d\theta = \ln|\sec\theta + \tan\theta| + C = \ln\left|\frac{1}{3}x + \frac{1}{3}\sqrt{x^2-9}\right| + C$

17. $x = \frac{3}{2}\sec\theta$, $dx = \frac{3}{2}\sec\theta\tan\theta\,d\theta$,

$$\frac{3}{2}\int\frac{\sec\theta\tan\theta\,d\theta}{27\tan^3\theta} = \frac{1}{18}\int\frac{\cos\theta}{\sin^2\theta}d\theta = -\frac{1}{18}\frac{1}{\sin\theta} + C = -\frac{1}{18}\csc\theta + C = -\frac{x}{9\sqrt{4x^2-9}} + C$$

19. $e^x = \sin\theta$, $e^x dx = \cos\theta\,d\theta$,

$$\int\cos^2\theta\,d\theta = \frac{1}{2}\int(1+\cos 2\theta)d\theta = \frac{1}{2}\theta + \frac{1}{4}\sin 2\theta + C = \frac{1}{2}\sin^{-1}(e^x) + \frac{1}{2}e^x\sqrt{1-e^{2x}} + C$$

21. $x = \sin\theta$, $dx = \cos\theta\,d\theta$,

$$5\int_0^1\sin^3\theta\cos^2\theta\,d\theta = 5\left[-\frac{1}{3}\cos^3\theta + \frac{1}{5}\cos^5\theta\right]_0^{\pi/2} = 5(1/3 - 1/5) = 2/3$$

23. $x = \sec\theta$, $dx = \sec\theta\tan\theta\,d\theta$, $\displaystyle\int_{\pi/4}^{\pi/3}\frac{1}{\sec\theta}\,d\theta = \int_{\pi/4}^{\pi/3}\cos\theta\,d\theta = \sin\theta\Big]_{\pi/4}^{\pi/3} = (\sqrt{3} - \sqrt{2})/2$

25. $x = \sqrt{3}\tan\theta$, $dx = \sqrt{3}\sec^2\theta\,d\theta$,

$$\frac{1}{9}\int_{\pi/6}^{\pi/3}\frac{\sec\theta}{\tan^4\theta}\,d\theta = \frac{1}{9}\int_{\pi/6}^{\pi/3}\frac{\cos^3\theta}{\sin^4\theta}\,d\theta = \frac{1}{9}\int_{\pi/6}^{\pi/3}\frac{1-\sin^2\theta}{\sin^4\theta}\cos\theta\,d\theta = \frac{1}{9}\int_{1/2}^{\sqrt{3}/2}\frac{1-u^2}{u^4}\,du\ (u = \sin\theta)$$

$$= \frac{1}{9}\int_{1/2}^{\sqrt{3}/2}(u^{-4} - u^{-2})\,du = \frac{1}{9}\left[-\frac{1}{3u^3} + \frac{1}{u}\right]_{1/2}^{\sqrt{3}/2} = \frac{10\sqrt{3}+18}{243}$$

27. true

29. false; $x = a\sec\theta$

31. $u = x^2 + 4$, $du = 2x\,dx$,

$$\frac{1}{2}\int\frac{1}{u}\,du = \frac{1}{2}\ln|u| + C = \frac{1}{2}\ln(x^2 + 4) + C;\ \text{or } x = 2\tan\theta,\ dx = 2\sec^2\theta\,d\theta,$$

$$\int\tan\theta\,d\theta = \ln|\sec\theta| + C_1 = \ln\frac{\sqrt{x^2+4}}{2} + C_1 = \ln(x^2+4)^{1/2} - \ln 2 + C_1$$

$$= \frac{1}{2}\ln(x^2 + 4) + C \text{ with } C = C_1 - \ln 2$$

33. $y' = \dfrac{1}{x}$, $1 + (y')^2 = 1 + \dfrac{1}{x^2} = \dfrac{x^2 + 1}{x^2}$,

$$L = \int_1^2\sqrt{\frac{x^2+1}{x^2}}\,dx;\qquad x = \tan\theta,\ dx = \sec^2\theta\,d\theta,$$

$$L = \int_{\pi/4}^{\tan^{-1}(2)}\frac{\sec^3\theta}{\tan\theta}\,d\theta = \int_{\pi/4}^{\tan^{-1}(2)}\frac{\tan^2\theta + 1}{\tan\theta}\sec\theta\,d\theta = \int_{\pi/4}^{\tan^{-1}(2)}(\sec\theta\tan\theta + \csc\theta)\,d\theta$$

$$= \Big[\sec\theta + \ln|\csc\theta - \cot\theta|\Big]_{\pi/4}^{\tan^{-1}(2)} = \sqrt{5} + \ln\left(\frac{\sqrt{5}}{2} - \frac{1}{2}\right) - \left[\sqrt{2} + \ln|\sqrt{2} - 1|\right]$$

$$= \sqrt{5} - \sqrt{2} + \ln\frac{2 + 2\sqrt{2}}{1 + \sqrt{5}}$$

35. $y' = 2x$, $1 + (y')^2 = 1 + 4x^2$,

$$S = 2\pi\int_0^1 x^2\sqrt{1 + 4x^2}\,dx;\ x = \frac{1}{2}\tan\theta,\ dx = \frac{1}{2}\sec^2\theta\,d\theta,$$

$$S = \frac{\pi}{4}\int_0^{\tan^{-1}2}\tan^2\theta\sec^3\theta\,d\theta = \frac{\pi}{4}\int_0^{\tan^{-1}2}(\sec^2\theta - 1)\sec^3\theta\,d\theta = \frac{\pi}{4}\int_0^{\tan^{-1}2}(\sec^5\theta - \sec^3\theta)\,d\theta$$

$$= \frac{\pi}{4}\left[\frac{1}{4}\sec^3\theta\tan\theta - \frac{1}{8}\sec\theta\tan\theta - \frac{1}{8}\ln|\sec\theta + \tan\theta|\right]_0^{\tan^{-1}2} = \frac{\pi}{32}[18\sqrt{5} - \ln(2 + \sqrt{5})]$$

37. $\displaystyle\int\frac{1}{(x-2)^2 + 1}\,dx = \tan^{-1}(x - 2) + C$

39. $\displaystyle\int \frac{1}{\sqrt{4-(x-1)^2}}\,dx = \sin^{-1}\left(\frac{x-1}{2}\right) + C$

41. $\displaystyle\int \frac{1}{\sqrt{(x-3)^2+1}}\,dx = \ln\left(x-3+\sqrt{(x-3)^2+1}\right) + C$

43. $\displaystyle\int \sqrt{4-(x+1)^2}\,dx,\quad$ let $x+1 = 2\sin\theta,$

$\displaystyle= \int 4\cos^2\theta\,d\theta = \int 2(1+\cos 2\theta)\,d\theta$

$\displaystyle= 2\theta + \sin 2\theta + C = 2\sin^{-1}\left(\frac{x+1}{2}\right) + \frac{1}{2}(x+1)\sqrt{3-2x-x^2} + C$

45. $\displaystyle\int \frac{1}{2(x+1)^2+5}\,dx = \frac{1}{2}\int \frac{1}{(x+1)^2+5/2}\,dx = \frac{1}{\sqrt{10}}\tan^{-1}\sqrt{2/5}(x+1) + C$

47. $\displaystyle\int_1^2 \frac{1}{\sqrt{4x-x^2}}\,dx = \int_1^2 \frac{1}{\sqrt{4-(x-2)^2}}\,dx = \sin^{-1}\frac{x-2}{2}\bigg]_1^2 = \pi/6$

49. $u = \sin^2 x, du = 2\sin x\cos x\,dx;$

$\displaystyle\frac{1}{2}\int \sqrt{1-u^2}\,du = \frac{1}{4}\left[u\sqrt{1-u^2} + \sin^{-1}u\right] + C = \frac{1}{4}\left[\sin^2 x\sqrt{1-\sin^4 x} + \sin^{-1}(\sin^2 x)\right] + C$

51. **(a)** $\displaystyle x = 3\sinh u, dx = 3\cosh u\,du, \int du = u + C = \sinh^{-1}(x/3) + C$

(b) $x = 3\tan\theta, dx = 3\sec^2\theta\,d\theta,$

$\displaystyle\int \sec\theta\,d\theta = \ln|\sec\theta + \tan\theta| + C = \ln\left(\sqrt{x^2+9}/3 + x/3\right) + C$

but $\sinh^{-1}(x/3) = \ln\left(x/3 + \sqrt{x^2/9+1}\right) = \ln\left(x/3 + \sqrt{x^2+9}/3\right)$ so the results agree.

EXERCISE SET 7.5

1. $\displaystyle\frac{A}{(x-3)} + \frac{B}{(x+4)}$

3. $\displaystyle\frac{2x-3}{x^2(x-1)} = \frac{A}{x} + \frac{B}{x^2} + \frac{C}{x-1}$

5. $\displaystyle\frac{A}{x} + \frac{B}{x^2} + \frac{C}{x^3} + \frac{Dx+E}{x^2+2}$

7. $\displaystyle\frac{Ax+B}{x^2+5} + \frac{Cx+D}{(x^2+5)^2}$

9. $\displaystyle\frac{1}{(x-4)(x+1)} = \frac{A}{x-4} + \frac{B}{x+1}; A = \frac{1}{5}, B = -\frac{1}{5}$ so

$\displaystyle\frac{1}{5}\int \frac{1}{x-4}\,dx - \frac{1}{5}\int \frac{1}{x+1}\,dx = \frac{1}{5}\ln|x-4| - \frac{1}{5}\ln|x+1| + C = \frac{1}{5}\ln\left|\frac{x-4}{x+1}\right| + C$

11. $\displaystyle\frac{11x+17}{(2x-1)(x+4)} = \frac{A}{2x-1} + \frac{B}{x+4}; A = 5, B = 3$ so

$\displaystyle 5\int \frac{1}{2x-1}\,dx + 3\int \frac{1}{x+4}\,dx = \frac{5}{2}\ln|2x-1| + 3\ln|x+4| + C$

13. $\dfrac{2x^2 - 9x - 9}{x(x+3)(x-3)} = \dfrac{A}{x} + \dfrac{B}{x+3} + \dfrac{C}{x-3}$; $A = 1$, $B = 2$, $C = -1$ so

$$\int \dfrac{1}{x}\,dx + 2\int \dfrac{1}{x+3}\,dx - \int \dfrac{1}{x-3}\,dx = \ln|x| + 2\ln|x+3| - \ln|x-3| + C = \ln\left|\dfrac{x(x+3)^2}{x-3}\right| + C$$

Note that the symbol C has been recycled; to save space this recycling is usually not mentioned.

15. $\dfrac{x^2 - 8}{x+3} = x - 3 + \dfrac{1}{x+3}$, $\displaystyle\int\left(x - 3 + \dfrac{1}{x+3}\right)dx = \dfrac{1}{2}x^2 - 3x + \ln|x+3| + C$

17. $\dfrac{3x^2 - 10}{x^2 - 4x + 4} = 3 + \dfrac{12x - 22}{x^2 - 4x + 4}$, $\dfrac{12x - 22}{(x-2)^2} = \dfrac{A}{x-2} + \dfrac{B}{(x-2)^2}$; $A = 12$, $B = 2$ so

$$\int 3\,dx + 12\int \dfrac{1}{x-2}\,dx + 2\int \dfrac{1}{(x-2)^2}\,dx = 3x + 12\ln|x-2| - 2/(x-2) + C$$

19. $u = x^2 - 3x - 10$, $du = (2x - 3)\,dx$,

$$\int \dfrac{du}{u} = \ln|u| + C = \ln|x^2 - 3x - 10| + C$$

21. $\dfrac{x^5 + x^2 + 2}{x^3 - x} = x^2 + 1 + \dfrac{x^2 + x + 2}{x^3 - x}$,

$\dfrac{x^2 + x + 2}{x(x+1)(x-1)} = \dfrac{A}{x} + \dfrac{B}{x+1} + \dfrac{C}{x-1}$; $A = -2$, $B = 1$, $C = 2$ so

$$\int (x^2 + 1)\,dx - \int \dfrac{2}{x}\,dx + \int \dfrac{1}{x+1}\,dx + \int \dfrac{2}{x-1}\,dx$$

$$= \dfrac{1}{3}x^3 + x - 2\ln|x| + \ln|x+1| + 2\ln|x-1| + C = \dfrac{1}{3}x^3 + x + \ln\left|\dfrac{(x+1)(x-1)^2}{x^2}\right| + C$$

23. $\dfrac{2x^2 + 3}{x(x-1)^2} = \dfrac{A}{x} + \dfrac{B}{x-1} + \dfrac{C}{(x-1)^2}$; $A = 3$, $B = -1$, $C = 5$ so

$$3\int \dfrac{1}{x}\,dx - \int \dfrac{1}{x-1}\,dx + 5\int \dfrac{1}{(x-1)^2}\,dx = 3\ln|x| - \ln|x-1| - 5/(x-1) + C$$

25. $\dfrac{2x^2 - 10x + 4}{(x+1)(x-3)^2} = \dfrac{A}{x+1} + \dfrac{B}{x-3} + \dfrac{C}{(x-3)^2}$; $A = 1$, $B = 1$, $C = -2$ so

$$\int \dfrac{1}{x+1}\,dx + \int \dfrac{1}{x-3}\,dx - \int \dfrac{2}{(x-3)^2}\,dx = \ln|x+1| + \ln|x-3| + \dfrac{2}{x-3} + C_1$$

27. $\dfrac{x^2}{(x+1)^3} = \dfrac{A}{x+1} + \dfrac{B}{(x+1)^2} + \dfrac{C}{(x+1)^3}$; $A = 1$, $B = -2$, $C = 1$ so

$$\int \dfrac{1}{x+1}\,dx - \int \dfrac{2}{(x+1)^2}\,dx + \int \dfrac{1}{(x+1)^3}\,dx = \ln|x+1| + \dfrac{2}{x+1} - \dfrac{1}{2(x+1)^2} + C$$

29. $\dfrac{2x^2 - 1}{(4x-1)(x^2+1)} = \dfrac{A}{4x-1} + \dfrac{Bx + C}{x^2+1}$; $A = -14/17$, $B = 12/17$, $C = 3/17$ so

$$\int \dfrac{2x^2 - 1}{(4x-1)(x^2+1)}\,dx = -\dfrac{7}{34}\ln|4x-1| + \dfrac{6}{17}\ln(x^2+1) + \dfrac{3}{17}\tan^{-1}x + C$$

31. $\dfrac{x^3 + 3x^2 + x + 9}{(x^2 + 1)(x^2 + 3)} = \dfrac{Ax + B}{x^2 + 1} + \dfrac{Cx + D}{x^2 + 3}$; $A = 0$, $B = 3$, $C = 1$, $D = 0$ so

$$\int \frac{x^3 + 3x^2 + x + 9}{(x^2 + 1)(x^2 + 3)}\, dx = 3\tan^{-1} x + \frac{1}{2}\ln(x^2 + 3) + C$$

33. $\dfrac{x^3 - 2x^2 + 2x - 2}{x^2 + 1} = x - 2 + \dfrac{x}{x^2 + 1}$,

$$\int \frac{x^3 - 3x^2 + 2x - 3}{x^2 + 1}\, dx = \frac{1}{2}x^2 - 2x + \frac{1}{2}\ln(x^2 + 1) + C$$

35. true

37. true

39. Let $x = \sin\theta$ to get $\displaystyle\int \frac{1}{x^2 + 4x - 5}\, dx$, and $\dfrac{1}{(x + 5)(x - 1)} = \dfrac{A}{x + 5} + \dfrac{B}{x - 1}$; $A = -1/6$,

$B = 1/6$ so we get $-\dfrac{1}{6}\displaystyle\int \frac{1}{x + 5}\, dx + \frac{1}{6}\int \frac{1}{x - 1}\, dx = \frac{1}{6}\ln\left|\frac{x - 1}{x + 5}\right| + C = \frac{1}{6}\ln\left(\frac{1 - \sin\theta}{5 + \sin\theta}\right) + C.$

41. $u = e^x$, $du = e^x\, dx$, $\displaystyle\int \frac{e^{3x}}{e^{2x} + 4}\, dx = \int \frac{u^2}{u^2 + 4}\, du = u - 2\tan^{-1}\frac{u}{2} + C = e^x - \tan^{-1}(e^x/2) + C$

43. $V = \pi \displaystyle\int_0^2 \frac{x^4}{(9 - x^2)^2}\, dx$, $\dfrac{x^4}{x^4 - 18x^2 + 81} = 1 + \dfrac{18x^2 - 81}{x^4 - 18x^2 + 81}$,

$$\frac{18x^2 - 81}{(9 - x^2)^2} = \frac{18x^2 - 81}{(x + 3)^2(x - 3)^2} = \frac{A}{x + 3} + \frac{B}{(x + 3)^2} + \frac{C}{x - 3} + \frac{D}{(x - 3)^2};$$

$A = -\dfrac{9}{4}$, $B = \dfrac{9}{4}$, $C = \dfrac{9}{4}$, $D = \dfrac{9}{4}$ so

$$V = \pi \left[x - \frac{9}{4}\ln|x + 3| - \frac{9/4}{x + 3} + \frac{9}{4}\ln|x - 3| - \frac{9/4}{x - 3} \right]_0^2 = \pi\left(\frac{19}{5} - \frac{9}{4}\ln 5\right)$$

45. $\dfrac{x^2 + 1}{(x^2 + 2x + 3)^2} = \dfrac{Ax + B}{x^2 + 2x + 3} + \dfrac{Cx + D}{(x^2 + 2x + 3)^2}$; $A = 0$, $B = 1$, $C = D = -2$ so

$$\int \frac{x^2 + 1}{(x^2 + 2x + 3)^2}\, dx = \int \frac{1}{(x + 1)^2 + 2}\, dx - \int \frac{2x + 2}{(x^2 + 2x + 3)^2}\, dx$$

$$= \frac{1}{\sqrt{2}}\tan^{-1}\frac{x + 1}{\sqrt{2}} + 1/(x^2 + 2x + 3) + C$$

47. $x^4 - 3x^3 - 7x^2 + 27x - 18 = (x - 1)(x - 2)(x - 3)(x + 3)$,

$$\frac{1}{(x - 1)(x - 2)(x - 3)(x + 3)} = \frac{A}{x - 1} + \frac{B}{x - 2} + \frac{C}{x - 3} + \frac{D}{x + 3};$$

$A = 1/8$, $B = -1/5$, $C = 1/12$, $D = -1/120$ so

$$\int \frac{dx}{x^4 - 3x^3 - 7x^2 + 27x - 18} = \frac{1}{8}\ln|x - 1| - \frac{1}{5}\ln|x - 2| + \frac{1}{12}\ln|x - 3| - \frac{1}{120}\ln|x + 3| + C$$

49. Let $u = x^2$, $du = 2x\, dx$, $\displaystyle\int_0^1 \frac{x}{x^4 + 1}\, dx = \frac{1}{2}\int_0^1 \frac{1}{1 + u^2}\, du = \frac{1}{2}\tan^{-1} u \Big]_0^1 = \frac{1}{2}\frac{\pi}{4} = \frac{\pi}{8}.$

51. If the polynomial has distinct roots $r_1, r_2, r_1 \neq r_2$, then the partial fraction decomposition will contain terms of the form $\dfrac{A}{x - r_1}, \dfrac{B}{x - r_2}$, and they will give logarithms and no inverse tangents. If there are two roots not distinct, say $x = r$, then the terms $\dfrac{A}{x - r}, \dfrac{B}{(x - r)^2}$ will appear, and neither will give an inverse tangent term. The only other possibility is no real roots, and the integrand can be written in the form $\dfrac{1}{a\left(x + \frac{b}{2a}\right)^2 + c - \frac{b^2}{4a}}$, which will yield an inverse tangent, specifically of the form $\tan^{-1}\left[A\left(x + \dfrac{b}{2a}\right)\right]$ for some constant A.

53. Yes, for instance the integrand $\dfrac{1}{x^2 + 1}$, whose integral is precisely $\tan^{-1} x + C$.

EXERCISE SET 7.6

1. Formula (60): $\dfrac{4}{9}\left[3x + \ln|-1 + 3x|\right] + C$ **3.** Formula (65): $\dfrac{1}{5}\ln\left|\dfrac{x}{5 + 2x}\right| + C$

5. Formula (102): $\dfrac{1}{5}(x - 1)(2x + 3)^{3/2} + C$ **7.** Formula (108): $\dfrac{1}{2}\ln\left|\dfrac{\sqrt{4 - 3x} - 2}{\sqrt{4 - 3x} + 2}\right| + C$

9. Formula (69): $\dfrac{1}{8}\ln\left|\dfrac{x + 4}{x - 4}\right| + C$

11. Formula (73): $\dfrac{x}{2}\sqrt{x^2 - 3} - \dfrac{3}{2}\ln\left|x + \sqrt{x^2 - 3}\right| + C$

13. Formula (95): $\dfrac{x}{2}\sqrt{x^2 + 4} - 2\ln(x + \sqrt{x^2 + 4}) + C$

15. Formula (74): $\dfrac{x}{2}\sqrt{9 - x^2} + \dfrac{9}{2}\sin^{-1}\dfrac{x}{3} + C$

17. Formula (79): $\sqrt{4 - x^2} - 2\ln\left|\dfrac{2 + \sqrt{4 - x^2}}{x}\right| + C$

19. Formula (38): $-\dfrac{1}{14}\sin(7x) + \dfrac{1}{2}\sin x + C$ **21.** Formula (50): $\dfrac{x^4}{16}[4\ln x - 1] + C$

23. Formula (42): $\dfrac{e^{-2x}}{13}(-2\sin(3x) - 3\cos(3x)) + C$

25. $u = e^{2x}, du = 2e^{2x}dx$, Formula (62): $\dfrac{1}{2}\displaystyle\int \dfrac{u\,du}{(4 - 3u)^2} = \dfrac{1}{18}\left[\dfrac{4}{4 - 3e^{2x}} + \ln|4 - 3e^{2x}|\right] + C$

27. $u = 3\sqrt{x}, du = \dfrac{3}{2\sqrt{x}}dx$, Formula (68): $\dfrac{2}{3}\displaystyle\int \dfrac{du}{u^2 + 4} = \dfrac{1}{3}\tan^{-1}\dfrac{3\sqrt{x}}{2} + C$

29. $u = 2x, du = 2dx$, Formula (76): $\dfrac{1}{2}\displaystyle\int \dfrac{du}{\sqrt{u^2 - 9}} = \dfrac{1}{2}\ln\left|2x + \sqrt{4x^2 - 9}\right| + C$

31. $u = 2x^2, du = 4xdx, u^2 du = 16x^5\,dx$, Formula (81):

$\dfrac{1}{4}\displaystyle\int \dfrac{u^2\,du}{\sqrt{2 - u^2}} = -\dfrac{x^2}{4}\sqrt{2 - 4x^4} + \dfrac{1}{4}\sin^{-1}(\sqrt{2}x^2) + C$

33. $u = \ln x, du = dx/x$, Formula (26): $\displaystyle\int \sin^2 u \, du = \frac{1}{2}\ln x + \frac{1}{4}\sin(2\ln x) + C$

35. $u = -2x, du = -2dx$, Formula (51): $\displaystyle\frac{1}{4}\int ue^u \, du = \frac{1}{4}(-2x - 1)e^{-2x} + C$

37. $u = \sin 3x, du = 3\cos 3x \, dx$, Formula (67): $\displaystyle\frac{1}{3}\int \frac{du}{u(u+1)^2} = \frac{1}{3}\left[\frac{1}{1+\sin 3x} + \ln\left|\frac{\sin 3x}{1+\sin 3x}\right|\right] + C$

39. $u = 4x^2, du = 8x\,dx$, Formula (70): $\displaystyle\frac{1}{8}\int \frac{du}{u^2 - 1} = \frac{1}{16}\ln\left|\frac{4x^2 - 1}{4x^2 + 1}\right| + C$

41. $u = 2e^x, du = 2e^x dx$, Formula (74):

$$\frac{1}{2}\int \sqrt{3 - u^2}\, du = \frac{1}{4}u\sqrt{3 - u^2} + \frac{3}{4}\sin^{-1}(u/\sqrt{3}) + C = \frac{1}{2}e^x\sqrt{3 - 4e^{2x}} + \frac{3}{4}\sin^{-1}(2e^x/\sqrt{3}) + C$$

43. $u = 3x, du = 3dx$, Formula (112):

$$\frac{1}{3}\int \sqrt{\frac{5}{3}u - u^2}\, du = \frac{1}{6}\left(u - \frac{5}{6}\right)\sqrt{\frac{5}{3}u - u^2} + \frac{25}{216}\sin^{-1}\left(\frac{6u - 5}{5}\right) + C$$

$$= \frac{18x - 5}{36}\sqrt{5x - 9x^2} + \frac{25}{216}\sin^{-1}\left(\frac{18x - 5}{5}\right) + C$$

45. $u = 2x, du = 2dx$, Formula (44):

$$\int u \sin u \, du = (\sin u - u\cos u) + C = \sin 2x - 2x\cos 2x + C$$

47. $u = -\sqrt{x}, u^2 = x, 2u\,du = dx$, Formula (51): $\displaystyle 2\int ue^u du = -2(\sqrt{x} + 1)e^{-\sqrt{x}} + C$

49. $x^2 + 6x - 7 = (x + 3)^2 - 16; u = x + 3, du = dx$, Formula (70):

$$\int \frac{du}{u^2 - 16} = \frac{1}{8}\ln\left|\frac{u - 4}{u + 4}\right| + C = \frac{1}{8}\ln\left|\frac{x - 1}{x + 7}\right| + C$$

51. $x^2 - 4x - 5 = (x - 2)^2 - 9, u = x - 2, du = dx$, Formula (77):

$$\int \frac{u + 2}{\sqrt{9 - u^2}}\, du = \int \frac{u\,du}{\sqrt{9 - u^2}} + 2\int \frac{du}{\sqrt{9 - u^2}} = -\sqrt{9 - u^2} + 2\sin^{-1}\frac{u}{3} + C$$

$$= -\sqrt{5 + 4x - x^2} + 2\sin^{-1}\left(\frac{x - 2}{3}\right) + C$$

53. $u = \sqrt{x - 2},\ x = u^2 + 2,\ dx = 2u\,du$;

$$\int 2u^2(u^2 + 2)du = 2\int (u^4 + 2u^2)du = \frac{2}{5}u^5 + \frac{4}{3}u^3 + C = \frac{2}{5}(x - 2)^{5/2} + \frac{4}{3}(x - 2)^{3/2} + C$$

55. $u = \sqrt{x^3 + 1},\ x^3 = u^2 - 1,\ 3x^2 dx = 2u\,du$;

$$\frac{2}{3}\int u^2(u^2 - 1)du = \frac{2}{3}\int (u^4 - u^2)du = \frac{2}{15}u^5 - \frac{2}{9}u^3 + C = \frac{2}{15}(x^3 + 1)^{5/2} - \frac{2}{9}(x^3 + 1)^{3/2} + C$$

57. $u = x^{1/3}$, $x = u^3$, $dx = 3u^2\,du$;

$$\int \frac{3u^2}{u^3 - u}\,du = 3\int \frac{u}{u^2 - 1}\,du = 3\int \left[\frac{1}{2(u+1)} + \frac{1}{2(u-1)}\right]\,du$$
$$= \frac{3}{2}\ln|x^{1/3} + 1| + \frac{3}{2}\ln|x^{1/3} - 1| + C$$

59. $u = x^{1/4}$, $x = u^4$, $dx = 4u^3\,du$; $4\int \frac{1}{u(1-u)}\,du = 4\int \left[\frac{1}{u} + \frac{1}{1-u}\right]\,du = 4\ln\frac{x^{1/4}}{|1 - x^{1/4}|} + C$

61. $u = x^{1/6}$, $x = u^6$, $dx = 6u^5\,du$;

$$6\int \frac{u^3}{u - 1}\,du = 6\int \left[u^2 + u + 1 + \frac{1}{u-1}\right]\,du = 2x^{1/2} + 3x^{1/3} + 6x^{1/6} + 6\ln|x^{1/6} - 1| + C$$

63. $u = \sqrt{1 + x^2}$, $x^2 = u^2 - 1$, $2x\,dx = 2u\,du$, $x\,dx = u\,du$;

$$\int (u^2 - 1)\,du = \frac{1}{3}(1 + x^2)^{3/2} - (1 + x^2)^{1/2} + C$$

65. $\displaystyle\int \frac{1}{1 + \dfrac{2u}{1+u^2} + \dfrac{1-u^2}{1+u^2}}\frac{2}{1+u^2}\,du = \int \frac{1}{u+1}\,du = \ln|\tan(x/2) + 1| + C$

67. $u = \tan(\theta/2)$, $\displaystyle\int \frac{d\theta}{1 - \cos\theta} = \int \frac{1}{u^2}\,du = -\frac{1}{u} + C = -\cot(\theta/2) + C$

69. $u = \tan(x/2)$, $\dfrac{1}{2}\displaystyle\int \frac{1 - u^2}{u}\,du = \frac{1}{2}\int (1/u - u)\,du = \frac{1}{2}\ln|\tan(x/2)| - \frac{1}{4}\tan^2(x/2) + C$

71. $\displaystyle\int_2^x \frac{1}{t(4-t)}\,dt = \frac{1}{4}\ln\frac{t}{4-t}\Big]_2^x$ (Formula (65), $a = 4, b = -1$)

$$= \frac{1}{4}\left[\ln\frac{x}{4-x} - \ln 1\right] = \frac{1}{4}\ln\frac{x}{4-x}, \frac{1}{4}\ln\frac{x}{4-x} = 0.5, \ln\frac{x}{4-x} = 2,$$

$$\frac{x}{4-x} = e^2, x = 4e^2 - e^2 x, x(1 + e^2) = 4e^2, x = 4e^2/(1 + e^2) \approx 3.523188312$$

73. $A = \displaystyle\int_0^4 \sqrt{25 - x^2}\,dx = \left(\frac{1}{2}x\sqrt{25 - x^2} + \frac{25}{2}\sin^{-1}\frac{x}{5}\right)\Big]_0^4$ (Formula (74), $a = 5$)

$$= 6 + \frac{25}{2}\sin^{-1}\frac{4}{5} \approx 17.59119023$$

75. $A = \displaystyle\int_0^1 \frac{1}{25 - 16x^2}\,dx$; $u = 4x$,

$$A = \frac{1}{4}\int_0^4 \frac{1}{25 - u^2}\,du = \frac{1}{40}\ln\left|\frac{u+5}{u-5}\right|\Big]_0^4 = \frac{1}{40}\ln 9 \approx 0.054930614 \text{ (Formula (69), } a = 5)$$

77. $V = 2\pi \displaystyle\int_0^{\pi/2} x\cos x\,dx = 2\pi(\cos x + x\sin x)\Big]_0^{\pi/2} = \pi(\pi - 2) \approx 3.586419094$ (Formula (45))

79. $V = 2\pi \int_0^3 xe^{-x}dx; \; u = -x,$

$$V = 2\pi \int_0^{-3} ue^u du = 2\pi e^u (u-1) \Big]_0^{-3} = 2\pi(1 - 4e^{-3}) \approx 5.031899801 \quad \text{(Formula (51))}$$

81. $L = \int_0^2 \sqrt{1 + 16x^2}\, dx; \; u = 4x,$

$$L = \frac{1}{4} \int_0^8 \sqrt{1 + u^2}\, du = \frac{1}{4} \left(\frac{u}{2}\sqrt{1 + u^2} + \frac{1}{2}\ln\left(u + \sqrt{1 + u^2}\right) \right) \Big]_0^8 \quad \text{(Formula (72), } a^2 = 1\text{)}$$

$$= \sqrt{65} + \frac{1}{8}\ln(8 + \sqrt{65}) \approx 8.409316783$$

83. $S = 2\pi \int_0^\pi (\sin x)\sqrt{1 + \cos^2 x}\, dx; \; u = \cos x, a^2 = 1,$

$$S = -2\pi \int_1^{-1} \sqrt{1 + u^2}\, du = 4\pi \int_0^1 \sqrt{1 + u^2}\, du = 4\pi\left(\frac{u}{2}\sqrt{1 + u^2} + \frac{1}{2}\ln\left(u + \sqrt{1 + u^2}\right)\right)\Big]_0^1$$

$$= 2\pi\left[\sqrt{2} + \ln(1 + \sqrt{2})\right] \approx 14.42359945 \quad \text{(Formula (72))}$$

85. **(a)** $s(t) = 2 + \int_0^t 20\cos^6 u \sin^3 u\, du$

$$= -\frac{20}{9}\sin^2 t\cos^7 t - \frac{40}{63}\cos^7 t + \frac{166}{63}$$

(b)

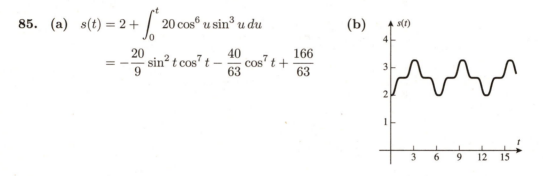

87. **(a)** $\displaystyle\int \sec x\, dx = \int \frac{1}{\cos x}dx = \int \frac{2}{1 - u^2}du = \ln\left|\frac{1 + u}{1 - u}\right| + C = \ln\left|\frac{1 + \tan(x/2)}{1 - \tan(x/2)}\right| + C$

$$= \ln\left\{ \left|\frac{\cos(x/2) + \sin(x/2)}{\cos(x/2) - \sin(x/2)}\right| \left|\frac{\cos(x/2) + \sin(x/2)}{\cos(x/2) + \sin(x/2)}\right| \right\} + C = \ln\left|\frac{1 + \sin x}{\cos x}\right| + C$$

$$= \ln|\sec x + \tan x| + C$$

(b) $\displaystyle\tan\left(\frac{\pi}{4} + \frac{x}{2}\right) = \frac{\tan\dfrac{\pi}{4} + \tan\dfrac{x}{2}}{1 - \tan\dfrac{\pi}{4}\tan\dfrac{x}{2}} = \frac{1 + \tan\dfrac{x}{2}}{1 - \tan\dfrac{x}{2}}$

89. Let $u = \tanh(x/2)$ then $\cosh(x/2) = 1/\operatorname{sech}(x/2) = 1/\sqrt{1 - \tanh^2(x/2)} = 1/\sqrt{1 - u^2},$

$\sinh(x/2) = \tanh(x/2)\cosh(x/2) = u/\sqrt{1 - u^2}$, so $\sinh x = 2\sinh(x/2)\cosh(x/2) = 2u/(1 - u^2),$

$\cosh x = \cosh^2(x/2) + \sinh^2(x/2) = (1 + u^2)/(1 - u^2)$, $x = 2\tanh^{-1} u, dx = [2/(1 - u^2)]du;$

$$\int \frac{dx}{2\cosh x + \sinh x} = \int \frac{1}{u^2 + u + 1}du = \frac{2}{\sqrt{3}}\tan^{-1}\frac{2u + 1}{\sqrt{3}} + C = \frac{2}{\sqrt{3}}\tan^{-1}\frac{2\tanh(x/2) + 1}{\sqrt{3}} + C.$$

91. $\displaystyle\int(\cos^{32}x\sin^{30}x-\cos^{30}x\sin^{32}x)dx=\int\cos^{30}x\sin^{30}x(\cos^2 x-\sin^2 x)dx$

$$=\frac{1}{2^{30}}\int\sin^{30}2x\cos 2x\,dx=\frac{\sin^{31}2x}{31(2^{31})}+C$$

93. $\displaystyle\int\frac{1}{x^{10}(1+x^{-9})}dx=-\frac{1}{9}\int\frac{1}{u}du=-\frac{1}{9}\ln|u|+C=-\frac{1}{9}\ln|1+x^{-9}|+C$

EXERCISE SET 7.7

1. exact value $=14/3\approx 4.666666667$

    (a) 4.667600662, $|E_M|\approx 0.000933995$

    (b) 4.664795676, $|E_T|\approx 0.001870991$

    (c) 4.666666602, $|E_S|\approx 9.9\cdot 10^{-7}$

3. exact value $=1$

    (a) 1.001028824, $|E_M|\approx 0.001028824$

    (b) 0.9979429864, $|E_T|\approx 0.0020570136$

    (c) 1.000000013, $|E_S|\approx 2.12\cdot 10^{-7}$

5. exact value $=\frac{1}{2}(e^{-2}-e^{-6})\approx 0.06642826551$

    (a) 0.06598746840, $|E_M|\approx 0.00044079711$

    (b) 0.06731162281, $|E_T|\approx 0.00088335730$

    (c) 0.06642830240, $|E_S|\approx 5.88\cdot 10^{-7}$

7. $f(x)=\sqrt{x+1}$, $f''(x)=-\frac{1}{4}(x+1)^{-3/2}$, $f^{(4)}(x)=-\frac{15}{16}(x+1)^{-7/2}$; $K_2=1/4$, $K_4=15/16$

    (a) $|E_M|\le\dfrac{27}{2400}(1/4)=0.002812500$

    (b) $|E_T|\le\dfrac{27}{1200}(1/4)=0.00562500$

    (c) $|E_S|\le\dfrac{81}{10240000}\approx 0.000007910156250$

9. $f(x)=\cos x$, $f''(x)=-\cos x$, $f^{(4)}(x)=\cos x$; $K_2=K_4=1$

    (a) $|E_M|\le\dfrac{\pi^3/8}{2400}(1)\approx 0.001028824$

    (b) $|E_T|\le\dfrac{\pi^3/8}{1200}(1)\approx 0.003229820488$

    (c) $|E_S|\le\dfrac{\pi^5}{180\times 10^4}(1)\approx 3.320526095\cdot 10^{-7}$

11. $f(x)=e^{-2x}$, $f''(x)=4e^{-2x}$; $f^{(4)}(x)=16e^{-2x}$; $K_2=4e^{-2}$; $K_4=16e^{-2}$

    (a) $|E_M|\le\dfrac{8}{2400}(4e^{-2})\approx 0.0018044704$

    (b) $|E_T|\le\dfrac{8}{1200}(4e^{-2})\approx 0.0036089409$

    (c) $|E_S|\le\dfrac{32}{180\times 10^5}(16e^{-2})\approx 0.000017778$

13. **(a)** $n>\left[\dfrac{(27)(1/4)}{(24)(5\times 10^{-4})}\right]^{1/2}\approx 23.7$; $n=24$

    (b) $n>\left[\dfrac{(27)(1/4)}{(12)(5\times 10^{-4})}\right]^{1/2}\approx 33.5$; $n=34$

    (c) $n>\left[\dfrac{(243)(15/16)}{(180)(5\times 10^{-4})}\right]^{1/4}\approx 7.1$; $n=8$

15. **(a)** $n>\left[\dfrac{(\pi^3/8)(1)}{(24)(10^{-3})}\right]^{1/2}\approx$; $n=13$

    (b) $n>\left[\dfrac{(\pi^3/8)(1)}{(12)(10^{-3})}\right]^{1/2}\approx 17.9$; $n=18$

    (c) $n>\left[\dfrac{(\pi^5/32)(1)}{(180)(10^{-3})}\right]^{1/4}\approx 2.7$; $n=4$

17. **(a)** $n > \left[\dfrac{(8)(4e^{-2})}{(24)(10^{-6})}\right]^{1/2} \approx 42.5;\; n = 43$ **(b)** $n > \left[\dfrac{(8)(4e^{-2})}{(12)(10^{-6})}\right]^{1/2} \approx 60.2;\; n = 61$

(c) $n > \left[\dfrac{(32)(16e^{-2})}{(180)(10^{-6})}\right]^{1/4} \approx 7.9;\; n = 8$

19. false; T_n is the average of L_n and R_n

21. false, it is the weighted average of M_{25} and T_{25}

23. $g(X_0) = aX_0^2 + bX_0 + c = 4a + 2b + c = f(X_0) = 1/X_0 = 1/2$; similarly
$9a + 3b + c = 1/3, 16a + 4b + c = 1/4$. Three equations in three unknowns, with solution
$a = 1/24, b = -3/8, c = 13/12, g(x) = x^2/24 - 3x/8 + 13/12$.

$$\int_2^4 g(x)\,dx = \int_2^4 \left(\frac{x^2}{24} - \frac{3x}{8} + \frac{13}{12}\right) dx = \frac{25}{36}$$

$$\frac{\Delta x}{3}[f(X_0) + 4f(X_1) + f(X_2)] = \frac{1}{3}\left[\frac{1}{2} + \frac{4}{3} + \frac{1}{4}\right] = \frac{25}{36}$$

25. 1.493648266,
1.493649897

27. 3.805639712,
3.805537256

29. 0.9045242448,
0.9045242380

31. exact value $= 4\tan^{-1}(x/2)\Big]_0^2 = \pi$

(a) $3.142425985, |E_M| \approx 0.000833331$
(b) $3.139925989, |E_T| \approx 0.001666665$
(c) $3.141592654, |E_S| \approx 0$

33. $S_{14} = 0.693147984, |E_S| \approx 0.000000803 = 8.03 \times 10^{-7}$; the method used in Example 6 results in a value of n which ensures that the magnitude of the error will be less than 10^{-6}, this is not necessarily the *smallest* value of n.

35. $f(x) = x\sin x, f''(x) = 2\cos x - x\sin x, |f''(x)| \leq 2|\cos x| + |x|\,|\sin x| \leq 2 + 2 = 4$ so $K_2 \leq 4$,

$$n > \left[\frac{(8)(4)}{(24)(10^{-4})}\right]^{1/2} \approx 115.5;\; n = 116 \text{ (a smaller } n \text{ might suffice)}$$

37. $f(x) = x\sqrt{x}, f''(x) = \dfrac{3}{4\sqrt{x}}, \lim\limits_{x \to 0^+} |f''(x)| = +\infty$

39. $s(x) = \displaystyle\int_0^x \sqrt{1 + (y'(t))^2}\,dt = \int_0^x \sqrt{1 + \cos^2 t}\,dt, \ell = \int_0^\pi \sqrt{1 + \cos^2 t}\,dt \approx 3.820197788$

41. $\displaystyle\int_0^{15} v\,dt \approx \frac{15}{(3)(6)}[0 + 4(37.1) + 2(60.1) + 4(90.9) + 2(98.3) + 4(104.1) + 114.4] \approx 772.5 \text{ ft}$

43. $\displaystyle\int_0^{180} v\,dt \approx \frac{180}{(3)(6)}[0.00 + 4(0.03) + 2(0.08) + 4(0.16) + 2(0.27) + 4(0.42) + 0.65] = 37.9 \text{ mi}$

45. $V = \int_0^{16} \pi r^2 dy = \pi \int_0^{16} r^2 dy \approx \pi \frac{16}{(3)(4)}[(8.5)^2 + 4(11.5)^2 + 2(13.8)^2 + 4(15.4)^2 + (16.8)^2]$

$$\approx 9270 \text{ cm}^3 \approx 9.3 \text{ L}$$

47. **(a)** The maximum value of $|f''(x)|$ is approximately 3.8442

    (b) $n = 18$

    (c) 0.9047406684

49. **(a)** $K_4 = \max_{0 \le x \le 1} |f^{(4)}(x)| \approx 12.40703740$ (calculator)

    (b) $\dfrac{(b-a)^5 K_4}{180 n^4} < 10^{-4}$ provided $n^4 > \dfrac{10^4 K_4}{180}, n > 5.12$, so $n \ge 6$

    (c) $\dfrac{K_4}{180} \cdot 6^4 \approx 0.0000531$ with $S_{12} = 0.9833961318$

51. **(a)** Left endpoint approximation $\approx \dfrac{b-a}{n}[y_0 + y_1 + \ldots + y_{n-2} + y_{n-1}]$

    Right endpoint approximation $\approx \dfrac{b-a}{n}[y_1 + y_2 + \ldots + y_{n-1} + y_n]$

    Average of the two $= \dfrac{b-a}{n}\dfrac{1}{2}[y_0 + 2y_1 + 2y_2 + \ldots + 2y_{n-2} + 2y_{n-1} + y_n]$

    (b) Area of trapezoid $= (x_{k+1} - x_k)\dfrac{y_k + y_{k+1}}{2}$. If we sum from $k = 0$ to $k = n - 1$ then we get the right hand side of (2).

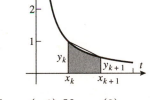

53. Given $g(x) = Ax^2 + Bx + C$, suppose $\Delta x = 1$ and $m = 0$. Then set $Y_0 = g(-1), Y_1 = g(0)$, $Y_2 = g(1)$. Also $Y_0 = g(-1) = A - B + C, Y_1 = g(0) = C, Y_2 = g(1) = A + B + C$, with solution $C = Y_1, B = \frac{1}{2}(Y_2 - Y_0)$, and $A = \frac{1}{2}(Y_0 + Y_2) - Y_1$.

Then $\int_{-1}^{1} g(x)\, dx = 2\int_0^1 (Ax^2 + C)\, dx = \frac{2}{3}A + 2C = \frac{1}{3}(Y_0 + Y_2) - \frac{2}{3}Y_1 + 2Y_1 = \frac{1}{3}(Y_0 + 4Y_1 + Y_2)$, which is exactly what one gets applying the Simpson's Rule.

The general case with the interval $(m - \Delta x, m + \Delta x)$ and values Y_0, Y_1, Y_2, can be converted by the change of variables $z = \dfrac{x - m}{\Delta x}$. Set $g(x) = h(z) = h((x - m)/\Delta x)$ to get $dx = \Delta x\, dz$ and

$\Delta x \int_{m-\Delta x}^{m+\Delta x} h(z)\, dz = \int_{-1}^{1} g(x)\, dx$. Finally, $Y_0 = g(m - \Delta x) = h(-1)$,

$Y_1 = g(m) = h(0), Y_2 = g(m + \Delta x) = h(1)$.

EXERCISE SET 7.8

1. **(a)** improper; infinite discontinuity at $x = 3$

    (b) continuous integrand, not improper

    (c) improper; infinite discontinuity at $x = 0$

    (d) improper; infinite interval of integration

    (e) improper; infinite interval of integration and infinite discontinuity at $x = 1$

    (f) continuous integrand, not improper

3. $\displaystyle \lim_{\ell \to +\infty} \left. \left(-\frac{1}{2}e^{-2x} \right) \right]_0^\ell = \frac{1}{2} \lim_{\ell \to +\infty} (-e^{-2\ell} + 1) = \frac{1}{2}$

5. $\displaystyle \lim_{\ell \to +\infty} \left. -2\coth^{-1} x \right]_3^\ell = \lim_{\ell \to +\infty} \left(2\coth^{-1} 3 - 2\coth^{-1} \ell \right) = 2\coth^{-1} 3$

7. $\displaystyle \lim_{\ell \to +\infty} \left. -\frac{1}{2\ln^2 x} \right]_e^\ell = \lim_{\ell \to +\infty} \left[-\frac{1}{2\ln^2 \ell} + \frac{1}{2} \right] = \frac{1}{2}$

9. $\displaystyle \lim_{\ell \to -\infty} \left. -\frac{1}{4(2x-1)^2} \right]_\ell^0 = \lim_{\ell \to -\infty} \frac{1}{4}[-1 + 1/(2\ell - 1)^2] = -1/4$

11. $\displaystyle \lim_{\ell \to -\infty} \left. \frac{1}{3}e^{3x} \right]_\ell^0 = \lim_{\ell \to -\infty} \left[\frac{1}{3} - \frac{1}{3}e^{3\ell} \right] = \frac{1}{3}$

13. $\displaystyle \int_{-\infty}^{+\infty} x\,dx$ converges if $\displaystyle \int_{-\infty}^0 x\,dx$ and $\displaystyle \int_0^{+\infty} x\,dx$ both converge; it diverges if either (or both)

diverges. $\displaystyle \int_0^{+\infty} x\,dx = \lim_{\ell \to +\infty} \left. \frac{1}{2}x^2 \right]_0^\ell = \lim_{\ell \to +\infty} \frac{1}{2}\ell^2 = +\infty$ so $\displaystyle \int_{-\infty}^{+\infty} x\,dx$ is divergent.

15. $\displaystyle \int_0^{+\infty} \frac{x}{(x^2+3)^2}dx = \lim_{\ell \to +\infty} \left. -\frac{1}{2(x^2+3)} \right]_0^\ell = \lim_{\ell \to +\infty} \frac{1}{2}[-1/(\ell^2+3) + 1/3] = \frac{1}{6}$,

similarly $\displaystyle \int_{-\infty}^0 \frac{x}{(x^2+3)^2}dx = -1/6$ so $\displaystyle \int_{-\infty}^\infty \frac{x}{(x^2+3)^2}dx = 1/6 + (-1/6) = 0$

17. $\displaystyle \lim_{\ell \to 4^-} \left. -\frac{1}{x-4} \right]_0^\ell = \lim_{\ell \to 4^-} \left[-\frac{1}{\ell-4} - \frac{1}{4} \right] = +\infty$, divergent

19. $\displaystyle \lim_{\ell \to \pi/2^-} \left. -\ln(\cos x) \right]_0^\ell = \lim_{\ell \to \pi/2^-} -\ln(\cos \ell) = +\infty$, divergent

21. $\displaystyle \lim_{\ell \to 1^-} \left. \sin^{-1} x \right]_0^\ell = \lim_{\ell \to 1^-} \sin^{-1} \ell = \pi/2$

23. $\displaystyle \lim_{\ell \to \pi/3^+} \left. \sqrt{1 - 2\cos x} \right]_\ell^{\pi/2} = \lim_{\ell \to \pi/3^+} (1 - \sqrt{1 - 2\cos \ell}) = 1$

25. $\displaystyle \int_0^2 \frac{dx}{x-2} = \lim_{\ell \to 2^-} \left. \ln|x-2| \right]_0^\ell = \lim_{\ell \to 2^-} (\ln|\ell - 2| - \ln 2) = -\infty$, divergent

27. $\displaystyle \int_0^8 x^{-1/3}dx = \lim_{\ell \to 0^+} \left. \frac{3}{2}x^{2/3} \right]_\ell^8 = \lim_{\ell \to 0^+} \frac{3}{2}(4 - \ell^{2/3}) = 6$,

$\displaystyle \int_{-1}^0 x^{-1/3}dx = \lim_{\ell \to 0^-} \left. \frac{3}{2}x^{2/3} \right]_{-1}^\ell = \lim_{\ell \to 0^-} \frac{3}{2}(\ell^{2/3} - 1) = -3/2$

so $\displaystyle \int_{-1}^8 x^{-1/3}dx = 6 + (-3/2) = 9/2$

29. Define $\displaystyle\int_0^{+\infty} \frac{1}{x^2}\,dx = \int_0^a \frac{1}{x^2}\,dx + \int_a^{+\infty} \frac{1}{x^2}\,dx$ where $a > 0$; take $a = 1$ for convenience,

$\displaystyle\int_0^1 \frac{1}{x^2}\,dx = \lim_{\ell \to 0^+}\left.(-1/x)\right]_\ell^1 = \lim_{\ell \to 0^+}(1/\ell - 1) = +\infty$ so $\displaystyle\int_0^{+\infty} \frac{1}{x^2}\,dx$ is divergent.

31. Let $u = \sqrt{x}, x = u^2, dx = 2u\,du$. Then

$\displaystyle\int \frac{dx}{\sqrt{x}(x+1)} = \int 2\frac{du}{u^2+1} = 2\tan^{-1}u + C = 2\tan^{-1}\sqrt{x} + C$ and

$\displaystyle\int_0^1 \frac{dx}{\sqrt{x}(x+1)} = 2\lim_{\epsilon \to 0^+}\left.\tan^{-1}\sqrt{x}\right]_\epsilon^1 = 2\lim_{\epsilon \to 0^+}(\pi/4 - \tan^{-1}\sqrt{\epsilon}) = \pi/2$

33. true, Theorem 7.8.2

35. false, neither 0 nor 3 lies in $[1, 2]$, so the integrand is continuous

37. $\displaystyle\int_0^{+\infty} \frac{e^{-\sqrt{x}}}{\sqrt{x}}\,dx = 2\int_0^{+\infty} e^{-u}\,du = 2\lim_{\ell \to +\infty}\left.(-e^{-u})\right]_0^\ell = 2\lim_{\ell \to +\infty}(1 - e^{-\ell}) = 2$

39. $\displaystyle\int_0^{+\infty} \frac{e^{-x}}{\sqrt{1-e^{-x}}}\,dx = \int_0^1 \frac{du}{\sqrt{u}} = \lim_{\ell \to 0^+}\left.2\sqrt{u}\right]_\ell^1 = \lim_{\ell \to 0^+}2(1 - \sqrt{\ell}) = 2$

41. $\displaystyle\lim_{\ell \to +\infty}\int_0^\ell e^{-x}\cos x\,dx = \lim_{\ell \to +\infty}\left.\frac{1}{2}e^{-x}(\sin x - \cos x)\right]_0^\ell = 1/2$

43. **(a)** 2.726585 **(b)** 2.804364 **(c)** 0.219384 **(d)** 0.504067

45. $\displaystyle 1 + \left(\frac{dy}{dx}\right)^2 = 1 + \frac{4 - x^{2/3}}{x^{2/3}} = \frac{4}{x^{2/3}}$; the arc length is $\displaystyle\int_0^8 \frac{2}{x^{1/3}}\,dx = \left.3x^{2/3}\right|_0^8 = 12$

47. $\displaystyle\int \ln x\,dx = x\ln x - x + C$,

$\displaystyle\int_0^1 \ln x\,dx = \lim_{\ell \to 0^+}\int_\ell^1 \ln x\,dx = \lim_{\ell \to 0^+}\left.(x\ln x - x)\right]_\ell^1 = \lim_{\ell \to 0^+}(-1 - \ell\ln\ell + \ell)$,

but $\displaystyle\lim_{\ell \to 0^+}\ell\ln\ell = \lim_{\ell \to 0^+}\frac{\ln\ell}{1/\ell} = \lim_{\ell \to 0^+}(-\ell) = 0$ so $\displaystyle\int_0^1 \ln x\,dx = -1$

49. **(a)** From the graph of $f^{(4)}(x)$ it is evident that $|f^{(4)}(x)| \leq 12.5$ on $[0, 1]$. We take $K_4 = 12.5$.

(b) Set $n = 2k$, then $\displaystyle\frac{12.5}{180 \cdot 2^4 k^4} \leq 10^{-4}$, or $\displaystyle k \geq \left[\frac{12.5}{180 \cdot 2^4 \cdot 10^{-4}}\right]^{1/4} = 2.57$, so
$k \geq 2.57, k \geq 3, n \geq 6$.

(c) $S_6 \approx 0.983347416$

51. **(a)** $\displaystyle V = \pi\int_0^{+\infty} e^{-2x}\,dx = -\frac{\pi}{2}\lim_{\ell \to +\infty}\left.e^{-2x}\right]_0^\ell = \pi/2$

(b) $\displaystyle S = 2\pi\int_0^{+\infty} e^{-x}\sqrt{1 + e^{-2x}}\,dx$, let $u = e^{-x}$ to get

$\displaystyle S = -2\pi\int_1^0 \sqrt{1 + u^2}\,du = 2\pi\left[\frac{u}{2}\sqrt{1+u^2} + \frac{1}{2}\ln\left|u + \sqrt{1+u^2}\right|\right]_0^1 = \pi\left[\sqrt{2} + \ln(1 + \sqrt{2})\right]$

53. (a) For $x \geq 1, x^2 \geq x, e^{-x^2} \leq e^{-x}$

(b) $\displaystyle\int_1^{+\infty} e^{-x}\,dx = \lim_{\ell \to +\infty}\int_1^\ell e^{-x}\,dx = \lim_{\ell \to +\infty} -e^{-x}\Big]_1^\ell = \lim_{\ell \to +\infty}(e^{-1} - e^{-\ell}) = 1/e$

(c) By Parts (a) and (b) and Exercise 52(b), $\displaystyle\int_1^{+\infty} e^{-x^2}\,dx$ is convergent and is $\leq 1/e$.

55. $V = \displaystyle\lim_{\ell \to +\infty}\int_1^\ell (\pi/x^2)\,dx = \lim_{\ell \to +\infty} -(\pi/x)\Big]_1^\ell = \lim_{\ell \to +\infty}(\pi - \pi/\ell) = \pi$

$A = \displaystyle\lim_{\ell \to +\infty}\int_1^\ell 2\pi(1/x)\sqrt{1 + 1/x^4}\,dx$; use Exercise 52(a) with $f(x) = 2\pi/x$, $g(x) = (2\pi/x)\sqrt{1 + 1/x^4}$

and $a = 1$ to see that the area is infinite.

57. The area under the curve $y = \dfrac{1}{1 + x^2}$, above the x-axis, and to the

right of the y-axis is given by $\displaystyle\int_0^\infty \dfrac{1}{1 + x^2}$. Solving for

$x = \sqrt{\dfrac{1 - y}{y}}$, the area is also given by the improper integral

$\displaystyle\int_0^1 \sqrt{\dfrac{1 - y}{y}}\,dy.$

59. Let $x = r\tan\theta$ to get $\displaystyle\int \dfrac{dx}{(r^2 + x^2)^{3/2}} = \dfrac{1}{r^2}\int \cos\theta\,d\theta = \dfrac{1}{r^2}\sin\theta + C = \dfrac{x}{r^2\sqrt{r^2 + x^2}} + C$

so $u = \dfrac{2\pi NIr}{k}\displaystyle\lim_{\ell \to +\infty}\dfrac{x}{r^2\sqrt{r^2 + x^2}}\Big]_a^\ell = \dfrac{2\pi NI}{kr}\lim_{\ell \to +\infty}(\ell/\sqrt{r^2 + \ell^2} - a/\sqrt{r^2 + a^2})$

$= \dfrac{2\pi NI}{kr}(1 - a/\sqrt{r^2 + a^2}).$

61. (a) Satellite's weight $= w(x) = k/x^2$ lb when $x =$ distance from center of Earth; $w(4000) = 6000$

so $k = 9.6 \times 10^{10}$ and $W = \displaystyle\int_{4000}^{4000+b} 9.6 \times 10^{10} x^{-2}\,dx$ mi·lb.

(b) $\displaystyle\int_{4000}^{+\infty} 9.6 \times 10^{10} x^{-2}\,dx = \lim_{\ell \to +\infty} -9.6 \times 10^{10}/x\Big]_{4000}^\ell = 2.4 \times 10^7$ mi·lb

63. (a) $\mathcal{L}\{f(t)\} = \displaystyle\int_0^{+\infty} te^{-st}\,dt = \lim_{\ell \to +\infty} -(t/s + 1/s^2)e^{-st}\Big]_0^\ell = \dfrac{1}{s^2}$

(b) $\mathcal{L}\{f(t)\} = \displaystyle\int_0^{+\infty} t^2 e^{-st}\,dt = \lim_{\ell \to +\infty} -(t^2/s + 2t/s^2 + 2/s^3)e^{-st}\Big]_0^\ell = \dfrac{2}{s^3}$

(c) $\mathcal{L}\{f(t)\} = \displaystyle\int_3^{+\infty} e^{-st}\,dt = \lim_{\ell \to +\infty} -\dfrac{1}{s}e^{-st}\Big]_3^\ell = \dfrac{e^{-3s}}{s}$

65. (a) $u = \sqrt{a}x, du = \sqrt{a}\,dx, 2\displaystyle\int_0^{+\infty} e^{-ax^2}\,dx = \dfrac{2}{\sqrt{a}}\int_0^{+\infty} e^{-u^2}\,du = \sqrt{\pi/a}$

(b) $x = \sqrt{2}\sigma u, dx = \sqrt{2}\sigma\,du, \dfrac{2}{\sqrt{2\pi}\sigma}\displaystyle\int_0^{+\infty} e^{-x^2/2\sigma^2}\,dx = \dfrac{2}{\sqrt{\pi}}\int_0^{+\infty} e^{-u^2}\,du = 1$

67. (a) $\displaystyle\int_0^4 \dfrac{1}{x^6 + 1}\,dx \approx 1.047; \pi/3 \approx 1.047$

(b) $\displaystyle\int_0^{+\infty}\frac{1}{x^6+1}dx = \int_0^4\frac{1}{x^6+1}dx + \int_4^{+\infty}\frac{1}{x^6+1}dx$ so

$$E = \int_4^{+\infty}\frac{1}{x^6+1}dx < \int_4^{+\infty}\frac{1}{x^6}dx = \frac{1}{5(4)^5} < 2\times 10^{-4}$$

69. If $p = 1$, then $\displaystyle\int_0^1\frac{dx}{x} = \lim_{\ell\to 0^+}\ln x\bigg]_\ell^1 = +\infty$;

if $p \neq 1$, then $\displaystyle\int_0^1\frac{dx}{x^p} = \lim_{\ell\to 0^+}\frac{x^{1-p}}{1-p}\bigg]_\ell^1 = \lim_{\ell\to 0^+}[(1-\ell^{1-p})/(1-p)] = \begin{cases} 1/(1-p), & p < 1 \\ +\infty, & p > 1 \end{cases}$.

71. $\displaystyle 2\int_0^1\cos(u^2)du \approx 1.809$

REVIEW EXERCISES, CHAPTER 7

1. $u = 4 + 9x, du = 9\,dx, \quad \dfrac{1}{9}\displaystyle\int u^{1/2}\,du = \dfrac{2}{27}(4+9x)^{3/2} + C$

3. $u = \cos\theta, -\displaystyle\int u^{1/2}du = -\dfrac{2}{3}\cos^{3/2}\theta + C$

5. $u = \tan(x^2), \dfrac{1}{2}\displaystyle\int u^2 du = \dfrac{1}{6}\tan^3(x^2) + C$

7. **(a)** With $u = \sqrt{x}$:

$$\int\frac{1}{\sqrt{x}\sqrt{2-x}}\,dx = 2\int\frac{1}{\sqrt{2-u^2}}\,du = 2\sin^{-1}(u/\sqrt{2}) + C = 2\sin^{-1}(\sqrt{x/2}) + C;$$

with $u = \sqrt{2-x}$:

$$\int\frac{1}{\sqrt{x}\sqrt{2-x}}\,dx = -2\int\frac{1}{\sqrt{2-u^2}}\,du = -2\sin^{-1}(u/\sqrt{2}) + C = -2\sin^{-1}(\sqrt{2-x}/\sqrt{2}) + C_1;$$

completing the square:

$$\int\frac{1}{\sqrt{1-(x-1)^2}}\,dx = \sin^{-1}(x-1) + C.$$

(b) In the three results in Part (a) the antiderivatives differ by a constant, in particular
$2\sin^{-1}(\sqrt{x/2}) = \pi - 2\sin^{-1}(\sqrt{2-x}/\sqrt{2}) = \pi/2 + \sin^{-1}(x-1)$.

9. $u = x, dv = e^{-x}dx, du = dx, v = -e^{-x}$;

$$\int xe^{-x}dx = -xe^{-x} + \int e^{-x}dx = -xe^{-x} - e^{-x} + C$$

11. $u = \ln(2x+3), dv = dx, du = \dfrac{2}{2x+3}dx, v = x; \displaystyle\int\ln(2x+3)dx = x\ln(2x+3) - \int\frac{2x}{2x+3}dx$

but $\displaystyle\int\frac{2x}{2x+3}dx = \int\left(1 - \frac{3}{2x+3}\right)dx = x - \frac{3}{2}\ln(2x+3) + C_1$ so

$$\int\ln(2x+3)dx = x\ln(2x+3) - x + \frac{3}{2}\ln(2x+3) + C$$

13.

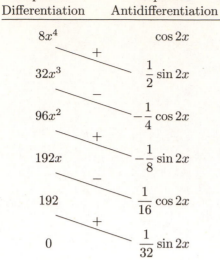

Repeated Differentiation	Repeated Antidifferentiation
$8x^4$	$\cos 2x$
$32x^3$	$\dfrac{1}{2}\sin 2x$
$96x^2$	$-\dfrac{1}{4}\cos 2x$
$192x$	$-\dfrac{1}{8}\sin 2x$
192	$\dfrac{1}{16}\cos 2x$
0	$\dfrac{1}{32}\sin 2x$

$$\int 8x^4 \cos 2x\, dx = (4x^4 - 12x^2 + 6)\sin 2x + (8x^3 - 12x)\cos 2x + C$$

15. $\displaystyle\int \sin^2 5\theta\, d\theta = \frac{1}{2}\int (1 - \cos 10\theta)d\theta = \frac{1}{2}\theta - \frac{1}{20}\sin 10\theta + C$

17. $\displaystyle\int \sin x \cos 2x\, dx = \frac{1}{2}\int (\sin 3x - \sin x)dx = -\frac{1}{6}\cos 3x + \frac{1}{2}\cos x + C$

19. $u = 2x,$

$$\int \sin^4 2x\, dx = \frac{1}{2}\int \sin^4 u\, du = \frac{1}{2}\left[-\frac{1}{4}\sin^3 u \cos u + \frac{3}{4}\int \sin^2 u\, du\right]$$

$$= -\frac{1}{8}\sin^3 u \cos u + \frac{3}{8}\left[-\frac{1}{2}\sin u \cos u + \frac{1}{2}\int du\right]$$

$$= -\frac{1}{8}\sin^3 u \cos u - \frac{3}{16}\sin u \cos u + \frac{3}{16}u + C$$

$$= -\frac{1}{8}\sin^3 2x \cos 2x - \frac{3}{16}\sin 2x \cos 2x + \frac{3}{8}x + C$$

21. $x = 3\sin\theta,\ dx = 3\cos\theta\, d\theta,$

$$9\int \sin^2 \theta\, d\theta = \frac{9}{2}\int (1 - \cos 2\theta)d\theta = \frac{9}{2}\theta - \frac{9}{4}\sin 2\theta + C = \frac{9}{2}\theta - \frac{9}{2}\sin\theta\cos\theta + C$$

$$= \frac{9}{2}\sin^{-1}(x/3) - \frac{1}{2}x\sqrt{9 - x^2} + C$$

23. $x = \sec\theta,\ dx = \sec\theta\tan\theta\, d\theta,\ \displaystyle\int \sec\theta\, d\theta = \ln|\sec\theta + \tan\theta| + C = \ln\left|x + \sqrt{x^2 - 1}\right| + C$

25. $x = 3\tan\theta,\ dx = 3\sec^2\theta\, d\theta,$

$$9\int \tan^2 \theta \sec\theta\, d\theta = 9\int \sec^3 \theta\, d\theta - 9\int \sec\theta\, d\theta$$

$$= \frac{9}{2}\sec\theta\tan\theta - \frac{9}{2}\ln|\sec\theta + \tan\theta| + C$$

$$= \frac{1}{2}x\sqrt{9 + x^2} - \frac{9}{2}\ln\left|\frac{1}{3}\sqrt{9 + x^2} + \frac{1}{3}x\right| + C$$

27. $\dfrac{1}{(x+4)(x-1)} = \dfrac{A}{x+4} + \dfrac{B}{x-1}$; $A = -\dfrac{1}{5}$, $B = \dfrac{1}{5}$ so

$$-\frac{1}{5}\int \frac{1}{x+4}dx + \frac{1}{5}\int \frac{1}{x-1}dx = -\frac{1}{5}\ln|x+4| + \frac{1}{5}\ln|x-1| + C = \frac{1}{5}\ln\left|\frac{x-1}{x+4}\right| + C$$

29. $\dfrac{x^2+2}{x+2} = x - 2 + \dfrac{6}{x+2}$, $\displaystyle\int\left(x - 2 + \dfrac{6}{x+2}\right)dx = \dfrac{1}{2}x^2 - 2x + 6\ \ln|x+2| + C$

31. $\dfrac{x^2}{(x+2)^3} = \dfrac{A}{x+2} + \dfrac{B}{(x+2)^2} + \dfrac{C}{(x+2)^3}$; $A = 1$, $B = -4$, $C = 4$ so

$$\int \frac{1}{x+2}dx - 4\int \frac{1}{(x+2)^2}dx + 4\int \frac{1}{(x+2)^3}dx = \ln|x+2| + \frac{4}{x+2} - \frac{2}{(x+2)^2} + C$$

33. (a) With $x = \sec\theta$:

$$\int \frac{1}{x^3-x}dx = \int \cot\theta\, d\theta = \ln|\sin\theta| + C = \ln\frac{\sqrt{x^2-1}}{|x|} + C;\ \text{valid for } |x| > 1.$$

(b) With $x = \sin\theta$:

$$\int \frac{1}{x^3-x}dx = -\int \frac{1}{\sin\theta\cos\theta}d\theta = -\int 2\csc 2\theta\, d\theta$$

$$= -\ln|\csc 2\theta - \cot 2\theta| + C = \ln|\cot\theta| + C = \ln\frac{\sqrt{1-x^2}}{|x|} + C,\ \ 0 < |x| < 1.$$

(c) $\dfrac{1}{x^3-x} = \dfrac{A}{x} + \dfrac{B}{x-1} + \dfrac{C}{x+1} = -\dfrac{1}{x} + \dfrac{1}{2(x-1)} + \dfrac{1}{2(x+1)}$

$\displaystyle\int \frac{1}{x^3-x}dx = -\ln|x| + \frac{1}{2}\ln|x-1| + \frac{1}{2}\ln|x+1| + C$, valid on any interval not containing the numbers $x = 0, \pm 1$

35. #40   **37.** #113   **39.** #28

41. exact value $= 4 - 2\sqrt{2} \approx 1.17157$

  (a) 1.17138, $|E_M| \approx 0.000190169$

  (b) 1.17195, $|E_T| \approx 0.000380588$

  (c) 1.17157, $|E_S| \approx 8.35 \times 10^{-8}$

43. $f(x) = \dfrac{1}{\sqrt{x+1}}$, $f''(x) = \dfrac{3}{4(x+1)^{5/2}}$, $f^{(4)}(x) = \dfrac{105}{16(x+1)^{9/2}}(x+1)^{-7/2}$; $K_2 = \dfrac{3}{2^4\sqrt{2}}$, $K_4 = \dfrac{105}{2^8\sqrt{2}}$

  (a) $|E_M| \leq \dfrac{2^3}{2400}\dfrac{3}{2^4\sqrt{2}} = \dfrac{1}{10^2 2^4\sqrt{2}} \approx 4.419417 \times 10^{-4}$

  (b) $|E_T| \leq \dfrac{2^3}{1200}\dfrac{3}{2^4\sqrt{2}} = 8.838834 \times 10^{-4}$

  (c) $|E_S| \leq \dfrac{2^5}{180 \times 20^4}\dfrac{105}{2^8\sqrt{2}} = \dfrac{7}{3 \cdot 10^4 \cdot 2^9\sqrt{2}} \approx 3.2224918 \times 10^{-7}$

45. (a) $n^2 \geq 10^4 \dfrac{8 \cdot 3}{24 \times 2^4\sqrt{2}}$, so $n \geq \dfrac{10^2}{2^2 2^{1/4}} \approx 21.02$, $n \geq 22$

  (b) $n^2 \geq \dfrac{10^4}{2^3\sqrt{2}}$, so $n \geq \dfrac{10^2}{2 \cdot 2^{3/4}} \approx 29.73$, $n \geq 30$

(c) Let $n = 2k$, then want $\dfrac{2^5 K_4}{180(2k)^4} \leq 10^{-4}$, or $k^4 \geq 10^4 \dfrac{2^5}{180} \dfrac{105}{2^4 \cdot 2^8 \sqrt{2}} = 10^4 \dfrac{7}{2^9 \cdot 3\sqrt{2}}$, so

$$k \geq 10 \left(\frac{7}{3 \cdot 2^9 \sqrt{2}} \right)^{1/4} \approx 2.38; \text{ so } k \geq 3, n \geq 6$$

47. $\displaystyle \lim_{\ell \to +\infty} (-e^{-x}) \Big]_0^{\ell} = \lim_{\ell \to +\infty} (-e^{-\ell} + 1) = 1$

49. $\displaystyle \lim_{\ell \to 9^-} -2\sqrt{9-x} \Big]_0^{\ell} = \lim_{\ell \to 9^-} 2(-\sqrt{9-\ell} + 3) = 6$

51. $A = \displaystyle \int_e^{+\infty} \frac{\ln x - 1}{x^2} dx = \lim_{\ell \to +\infty} c - \frac{\ln x}{x} \Big]_e^{\ell} = 1/e$

53. $\displaystyle \int_0^{+\infty} \frac{dx}{x^2 + a^2} = \lim_{\ell \to +\infty} \frac{1}{a} \tan^{-1}(x/a) \Big]_0^{\ell} = \lim_{\ell \to +\infty} \frac{1}{a} \tan^{-1}(\ell/a) = \frac{\pi}{2a} = 1, a = \pi/2$

55. $x = \sqrt{3} \tan \theta, dx = \sqrt{3} \sec^2 \theta \, d\theta,$

$$\frac{1}{3} \int \frac{1}{\sec \theta} d\theta = \frac{1}{3} \int \cos \theta \, d\theta = \frac{1}{3} \sin \theta + C = \frac{x}{3\sqrt{3 + x^2}} + C$$

57. Use Endpaper Formula (31) to get

$$\int_0^{\pi/4} \tan^7 \theta \, d\theta = \frac{1}{6} \tan^6 \theta \Big]_0^{\pi/4} - \frac{1}{4} \tan^4 \theta \Big]_0^{\pi/4} + \frac{1}{2} \tan^2 \theta \Big]_0^{\pi/4} + \ln|\cos \theta| \Big]_0^{\pi/4}$$

$$= \frac{1}{6} - \frac{1}{4} + \frac{1}{2} - \ln \sqrt{2} = \frac{5}{12} - \ln \sqrt{2}.$$

59. $\displaystyle \int \sin^2 2x \cos^3 2x \, dx = \int \sin^2 2x (1 - \sin^2 2x) \cos 2x \, dx = \int (\sin^2 2x - \sin^4 2x) \cos 2x \, dx$

$$= \frac{1}{6} \sin^3 2x - \frac{1}{10} \sin^5 2x + C$$

61. $u = e^{2x}, dv = \cos 3x \, dx, du = 2e^{2x} dx, v = \dfrac{1}{3} \sin 3x;$

$$\int e^{2x} \cos 3x \, dx = \frac{1}{3} e^{2x} \sin 3x - \frac{2}{3} \int e^{2x} \sin 3x \, dx. \text{ Use } u = e^{2x}, dv = \sin 3x \, dx \text{ to get}$$

$$\int e^{2x} \sin 3x \, dx = -\frac{1}{3} e^{2x} \cos 3x + \frac{2}{3} \int e^{2x} \cos 3x \, dx \text{ so}$$

$$\int e^{2x} \cos 3x \, dx = \frac{1}{3} e^{2x} \sin 3x + \frac{2}{9} e^{2x} \cos 3x - \frac{4}{9} \int e^{2x} \cos 3x \, dx,$$

$$\frac{13}{9} \int e^{2x} \cos 3x \, dx = \frac{1}{9} e^{2x} (3 \sin 3x + 2 \cos 3x) + C_1, \int e^{2x} \cos 3x \, dx = \frac{1}{13} e^{2x} (3 \sin 3x + 2 \cos 3x) + C$$

63. $\dfrac{1}{(x-1)(x+2)(x-3)} = \dfrac{A}{x-1} + \dfrac{B}{x+2} + \dfrac{C}{x-3}; A = -\dfrac{1}{6}, B = \dfrac{1}{15}, C = \dfrac{1}{10}$ so

$$-\frac{1}{6} \int \frac{1}{x-1} dx + \frac{1}{15} \int \frac{1}{x+2} dx + \frac{1}{10} \int \frac{1}{x-3} dx$$

$$= -\frac{1}{6} \ln|x-1| + \frac{1}{15} \ln|x+2| + \frac{1}{10} \ln|x-3| + C$$

65. $u = \sqrt{x-4}$, $x = u^2 + 4$, $dx = 2u\,du$,

$$\int_0^2 \frac{2u^2}{u^2+4}\,du = 2\int_0^2 \left[1 - \frac{4}{u^2+4}\right]du = \left[2u - 4\tan^{-1}(u/2)\right]_0^2 = 4 - \pi$$

67. $u = \sqrt{e^x + 1}$, $e^x = u^2 - 1$, $x = \ln(u^2 - 1)$, $dx = \frac{2u}{u^2-1}\,du$,

$$\int \frac{2}{u^2-1}\,du = \int\left[\frac{1}{u-1} - \frac{1}{u+1}\right]du = \ln|u-1| - \ln|u+1| + C = \ln\frac{\sqrt{e^x+1}-1}{\sqrt{e^x+1}+1} + C$$

69. $u = \sin^{-1} x$, $dv = dx$, $du = \frac{1}{\sqrt{1-x^2}}\,dx$, $v = x$;

$$\int_0^{1/2} \sin^{-1}x\,dx = x\sin^{-1}x\Big]_0^{1/2} - \int_0^{1/2}\frac{x}{\sqrt{1-x^2}}\,dx = \frac{1}{2}\sin^{-1}\frac{1}{2} + \sqrt{1-x^2}\Big]_0^{1/2}$$

$$= \frac{1}{2}\left(\frac{\pi}{6}\right) + \sqrt{\frac{3}{4}} - 1 = \frac{\pi}{12} + \frac{\sqrt{3}}{2} - 1$$

71. $\int \frac{x+3}{\sqrt{(x+1)^2+1}}\,dx$, let $u = x+1$,

$$\int \frac{u+2}{\sqrt{u^2+1}}\,du = \int\left[u(u^2+1)^{-1/2} + \frac{2}{\sqrt{u^2+1}}\right]du = \sqrt{u^2+1} + 2\sinh^{-1}u + C$$

$$= \sqrt{x^2+2x+2} + 2\sinh^{-1}(x+1) + C$$

Alternate solution: let $x+1 = \tan\theta$,

$$\int(\tan\theta + 2)\sec\theta\,d\theta = \int \sec\theta\tan\theta\,d\theta + 2\int\sec\theta\,d\theta = \sec\theta + 2\ln|\sec\theta + \tan\theta| + C$$

$$= \sqrt{x^2+2x+2} + 2\ln(\sqrt{x^2+2x+2} + x + 1) + C.$$

73. $\lim_{\ell\to+\infty} -\frac{1}{2(x^2+1)}\Big]_a^\ell = \lim_{\ell\to+\infty}\left[-\frac{1}{2(\ell^2+1)} + \frac{1}{2(a^2+1)}\right] = \frac{1}{2(a^2+1)}$

MAKING CONNECTIONS, CHAPTER 7

1. (a) $u = f(x)$, $dv = dx$, $du = f'(x)$, $v = x$;

$$\int_a^b f(x)\,dx = xf(x)\Big]_a^b - \int_a^b xf'(x)\,dx = bf(b) - af(a) - \int_a^b xf'(x)\,dx$$

(b) Substitute $y = f(x)$, $dy = f'(x)\,dx$, $x = a$ when $y = f(a)$, $x = b$ when $y = f(b)$,

$$\int_a^b xf'(x)\,dx = \int_{f(a)}^{f(b)} x\,dy = \int_{f(a)}^{f(b)} f^{-1}(y)\,dy$$

(c) From $a = f^{-1}(\alpha)$ and $b = f^{-1}(\beta)$ we get

$bf(b) - af(a) = \beta f^{-1}(\beta) - \alpha f^{-1}(\alpha);$ then

$$\int_\alpha^\beta f^{-1}(x)\,dx = \int_\alpha^\beta f^{-1}(y)\,dy = \int_{f(a)}^{f(b)} f^{-1}(y)\,dy,$$

which, by Part (b), yields

$$\int_\alpha^\beta f^{-1}(x)\,dx = bf(b) - af(a) - \int_a^b f(x)\,dx$$

$$= \beta f^{-1}(\beta) - \alpha f^{-1}(\alpha) - \int_{f^{-1}(\alpha)}^{f^{-1}(\beta)} f(x)\,dx$$

Note from the figure that $A_1 = \displaystyle\int_\alpha^\beta f^{-1}(x)\,dx,\ A_2 = \int_{f^{-1}(\alpha)}^{f^{-1}(\beta)} f(x)\,dx$, and

$A_1 + A_2 = \beta f^{-1}(\beta) - \alpha f^{-1}(\alpha)$, a "picture proof".

3. (a) $\Gamma(1) = \displaystyle\int_0^{+\infty} e^{-t}\,dt = \lim_{\ell\to+\infty} -e^{-t}\Big|_0^\ell = \lim_{\ell\to+\infty}(-e^{-\ell} + 1) = 1$

(b) $\Gamma(x+1) = \displaystyle\int_0^{+\infty} t^x e^{-t}\,dt$; let $u = t^x$, $dv = e^{-t}\,dt$ to get

$$\Gamma(x+1) = -t^x e^{-t}\Big|_0^{+\infty} + x\int_0^{+\infty} t^{x-1} e^{-t}\,dt = -t^x e^{-t}\Big|_0^{+\infty} + x\Gamma(x)$$

$\displaystyle\lim_{t\to+\infty} t^x e^{-t} = \lim_{t\to+\infty}\frac{t^x}{e^t} = 0$ (by multiple applications of L'Hôpital's rule)

so $\Gamma(x+1) = x\Gamma(x)$

(c) $\Gamma(2) = (1)\Gamma(1) = (1)(1) = 1$, $\Gamma(3) = 2\Gamma(2) = (2)(1) = 2$, $\Gamma(4) = 3\Gamma(3) = (3)(2) = 6$

Thus $\Gamma(n) = (n-1)!$ if n is a positive integer.

(d) $\Gamma\left(\dfrac{1}{2}\right) = \displaystyle\int_0^{+\infty} t^{-1/2} e^{-t}\,dt = 2\int_0^{+\infty} e^{-u^2}\,du$ (with $u = \sqrt{t}$) $= 2(\sqrt{\pi}/2) = \sqrt{\pi}$

(e) $\Gamma\left(\dfrac{3}{2}\right) = \dfrac{1}{2}\Gamma\left(\dfrac{1}{2}\right) = \dfrac{1}{2}\sqrt{\pi}$, $\Gamma\left(\dfrac{5}{2}\right) = \dfrac{3}{2}\Gamma\left(\dfrac{3}{2}\right) = \dfrac{3}{4}\sqrt{\pi}$

5. (a) $\sqrt{\cos\theta - \cos\theta_0} = \sqrt{2\left[\sin^2(\theta_0/2) - \sin^2(\theta/2)\right]} = \sqrt{2(k^2 - k^2\sin^2\phi)} = \sqrt{2k^2\cos^2\phi}$

$\qquad = \sqrt{2}\,k\cos\phi;\ k\sin\phi = \sin(\theta/2)$ so $k\cos\phi\,d\phi = \dfrac{1}{2}\cos(\theta/2)\,d\theta = \dfrac{1}{2}\sqrt{1 - \sin^2(\theta/2)}\,d\theta$

$\qquad = \dfrac{1}{2}\sqrt{1 - k^2\sin^2\phi}\,d\theta$, thus $d\theta = \dfrac{2k\cos\phi}{\sqrt{1 - k^2\sin^2\phi}}\,d\phi$ and hence

$$T = \sqrt{\frac{8L}{g}} \int_0^{\pi/2} \frac{1}{\sqrt{2}k\cos\phi} \cdot \frac{2k\cos\phi}{\sqrt{1 - k^2\sin^2\phi}}\,d\phi = 4\sqrt{\frac{L}{g}} \int_0^{\pi/2} \frac{1}{\sqrt{1 - k^2\sin^2\phi}}\,d\phi$$

(b) If $L = 1.5$ ft and $\theta_0 = (\pi/180)(20) = \pi/9$, then

$$T = \frac{\sqrt{3}}{2} \int_0^{\pi/2} \frac{d\phi}{\sqrt{1 - \sin^2(\pi/18)\sin^2\phi}} \approx 1.37\text{ s}.$$

CHAPTER 8
Mathematical Modeling with Differential Equations

EXERCISE SET 8.1

1. $y' = 9x^2 e^{x^3} = 3x^2 y$ and $y(0) = 3$ by inspection.

3. (a) first order; $\dfrac{dy}{dx} = c$; $(1+x)\dfrac{dy}{dx} = (1+x)c = y$

    (b) second order; $y' = c_1 \cos t - c_2 \sin t$, $y'' + y = -c_1 \sin t - c_2 \cos t + (c_1 \sin t + c_2 \cos t) = 0$

5. False. It is a first-order equation, because it involves y and dy/dx, but not $d^n y/dx^n$ for $n > 1$.

7. True. As mentioned in the marginal note after equation (2), the general solution of an n'th order differential equation usually involves n arbitrary constants.

9. (a) If $y = e^{-2x}$ then $y' = -2e^{-2x}$ and $y'' = 4e^{-2x}$, so
$y'' + y' - 2y = 4e^{-2x} + (-2e^{-2x}) - 2e^{-2x} = 0$.

    If $y = e^x$ then $y' = e^x$ and $y'' = e^x$, so $y'' + y' - 2y = e^x + e^x - 2e^x = 0$.

    (b) If $y = c_1 e^{-2x} + c_2 e^x$ then $y' = -2c_1 e^{-2x} + c_2 e^x$ and $y'' = 4c_1 e^{-2x} + c_2 e^x$, so
$y'' + y' - 2y = (4c_1 e^{-2x} + c_2 e^x) + (-2c_1 e^{-2x} + c_2 e^x) - 2(c_1 e^{-2x} + c_2 e^x) = 0$.

11. (a) If $y = e^{2x}$ then $y' = 2e^{2x}$ and $y'' = 4e^{2x}$, so $y'' - 4y' + 4y = 4e^{2x} - 4(2e^{2x}) + 4e^{2x} = 0$.
If $y = xe^{2x}$ then $y' = (2x+1)e^{2x}$ and $y'' = (4x+4)e^{2x}$, so
$y'' - 4y' + 4y = (4x+4)e^{2x} - 4(2x+1)e^{2x} + 4xe^{2x} = 0$.

    (b) If $y = c_1 e^{2x} + c_2 xe^{2x}$ then $y' = 2c_1 e^{2x} + c_2(2x+1)e^{2x}$ and $y'' = 4c_1 e^{2x} + c_2(4x+4)e^{2x}$, so
$y'' - 4y' + 4y = (4c_1 e^{2x} + c_2(4x+4)e^{2x}) - 4(2c_1 e^{2x} + c_2(2x+1)e^{2x}) + 4(c_1 e^{2x} + c_2 xe^{2x}) = 0$.

13. (a) If $y = \sin 2x$ then $y' = 2\cos 2x$ and $y'' = -4\sin 2x$, so $y'' + 4y = -4\sin 2x + 4\sin 2x = 0$.

    If $y = \cos 2x$ then $y' = -2\sin 2x$ and $y'' = -4\cos 2x$, so $y'' + 4y = -4\cos 2x + 4\cos 2x = 0$.

    (b) If $y = c_1 \sin 2x + c_2 \cos 2x$ then $y' = 2c_1 \cos 2x - 2c_2 \sin 2x$ and $y'' = -4c_1 \sin 2x - 4c_2 \cos 2x$,
so $y'' + 4y = (-4c_1 \sin 2x - 4c_2 \cos 2x) + 4(c_1 \sin 2x + c_2 \cos 2x) = 0$.

15. From Exercise 9, $y = c_1 e^{-2x} + c_2 e^x$ is a solution of the differential equation, with
$y' = -2c_1 e^{-2x} + c_2 e^x$. Setting $y(0) = -1$ and $y'(0) = -4$ gives $c_1 + c_2 = -1$ and $-2c_1 + c_2 = -4$.
So $c_1 = 1$, $c_2 = -2$, and $y = e^{-2x} - 2e^x$.

17. From Exercise 11, $y = c_1 e^{2x} + c_2 xe^{2x}$ is a solution of the differential equation, with
$y' = 2c_1 e^{2x} + c_2(2x+1)e^{2x}$. Setting $y(0) = 2$ and $y'(0) = 2$ gives $c_1 = 2$ and $2c_1 + c_2 = 2$,
so $c_2 = -2$ and $y = 2e^{2x} - 2xe^{2x}$.

19. From Exercise 13, $y = c_1 \sin 2x + c_2 \cos 2x$ is a solution of the differential equation, with
$y' = 2c_1 \cos 2x - 2c_2 \sin 2x$. Setting $y(0) = 1$ and $y'(0) = 2$ gives $c_2 = 1$ and $2c_1 = 2$,
so $c_1 = 1$ and $y = \sin 2x + \cos 2x$.

21. $y' = 2 - 4x$ so $y = \displaystyle\int (2-4x)\,dx = -2x^2 + 2x + C$. Setting $y(0) = 3$ gives $C = 3$, so $y = -2x^2 + 2x + 3$.

23. If the solution has an inverse function $x(y)$ then, by equation (3) of Section 3.3,
$\dfrac{dx}{dy} = \dfrac{1}{dy/dx} = y^{-2}$. So $x = \displaystyle\int y^{-2}\,dy = -y^{-1} + C$. When $x = 1$, $y = 2$, so $C = \dfrac{3}{2}$ and $x = \dfrac{3}{2} - y^{-1}$.

Solving for y gives $y = \dfrac{2}{3 - 2x}$. The solution is valid for $x < \dfrac{3}{2}$.

25. By the product rule, $\frac{d}{dx}(x^2 y) = x^2 y' + 2xy = 0$, so $x^2 y = C$ and $y = C/x^2$. Setting $y(1) = 2$ gives $C = 2$ so $y = 2/x^2$. The solution is valid for $x > 0$.

27. **(a)** $\frac{dy}{dt} = ky^2$, $y(0) = y_0$, $k > 0$ **(b)** $\frac{dy}{dt} = -ky^2$, $y(0) = y_0$, $k > 0$

29. **(a)** $\frac{ds}{dt} = \frac{1}{2}s$ **(b)** $\frac{d^2 s}{dt^2} = 2\frac{ds}{dt}$

31. **(a)** Since $k > 0$ and $y > 0$, equation (3) gives $\frac{dy}{dt} = ky > 0$, so y is increasing.

      (b) $\frac{d^2 y}{dt^2} = \frac{d}{dt}(ky) = k\frac{dy}{dt} = k^2 y > 0$, so y is concave upward.

33. **(a)** Both $y = 0$ and $y = L$ satisfy equation (6).

      (b) The rate of growth is $\frac{dy}{dt} = ky(L - y)$; we wish to find the value of y which maximizes this. Since $\frac{d}{dy}[ky(L - y)] = k(L - 2y)$, which is positive for $y < L/2$ and negative for $y > L/2$, the maximum growth rate occurs for $y = L/2$.

35. If $x = c_1 \cos\left(\sqrt{\frac{k}{m}}\, t\right) + c_2 \sin\left(\sqrt{\frac{k}{m}}\, t\right)$ then $\frac{dx}{dt} = c_2\sqrt{\frac{k}{m}}\cos\left(\sqrt{\frac{k}{m}}\, t\right) - c_1\sqrt{\frac{k}{m}}\sin\left(\sqrt{\frac{k}{m}}\, t\right)$

and $\frac{d^2 x}{dt^2} = -c_1\frac{k}{m}\cos\left(\sqrt{\frac{k}{m}}\, t\right) - c_2\frac{k}{m}\sin\left(\sqrt{\frac{k}{m}}\, t\right) = -\frac{k}{m}x$. So $m\frac{d^2 x}{dt^2} = -kx$; thus x satisfies the differential equation for the vibrating string.

EXERCISE SET 8.2

1. $\frac{1}{y}dy = \frac{1}{x}dx$, $\ln|y| = \ln|x| + C_1$, $\ln\left|\frac{y}{x}\right| = C_1$, $\frac{y}{x} = \pm e^{C_1} = C$, $y = Cx$
including $C = 0$ by inspection.

3. $\frac{dy}{1 + y} = -\frac{x}{\sqrt{1 + x^2}}dx$, $\ln|1 + y| = -\sqrt{1 + x^2} + C_1$, $1 + y = \pm e^{-\sqrt{1 + x^2}}e^{C_1} = Ce^{-\sqrt{1 + x^2}}$,
$y = Ce^{-\sqrt{1 + x^2}} - 1$, $C \neq 0$

5. $\frac{2(1 + y^2)}{y}dy = e^x dx$, $2\ln|y| + y^2 = e^x + C$; by inspection, $y = 0$ is also a solution.

7. $e^y dy = \frac{\sin x}{\cos^2 x}dx = \sec x \tan x\, dx$, $e^y = \sec x + C$, $y = \ln(\sec x + C)$

9. $\frac{dy}{y^2 - y} = \frac{dx}{\sin x}$, $\int\left[-\frac{1}{y} + \frac{1}{y - 1}\right]dy = \int \csc x\, dx$, $\ln\left|\frac{y - 1}{y}\right| = \ln|\csc x - \cot x| + C_1$,

$\frac{y - 1}{y} = \pm e^{C_1}(\csc x - \cot x) = C(\csc x - \cot x)$, $y = \frac{1}{1 - C(\csc x - \cot x)}$, $C \neq 0$;

by inspection, $y = 0$ is also a solution, as is $y = 1$.

11. $(2y + \cos y)\, dy = 3x^2\, dx, y^2 + \sin y = x^3 + C, \pi^2 + \sin \pi = C, C = \pi^2, y^2 + \sin y = x^3 + \pi^2$

13. $2(y-1)\, dy = (2t+1)\, dt, y^2 - 2y = t^2 + t + C, 1 + 2 = C, C = 3, y^2 - 2y = t^2 + t + 3$

15. (a) $\dfrac{dy}{y} = \dfrac{dx}{2x}, \ \ln|y| = \dfrac{1}{2}\ln|x| + C_1,$

$|y| = C_2 |x|^{1/2}, \ y^2 = Cx;$

by inspection $y = 0$ is also a solution.

(b) $1^2 = C \cdot 2, C = 1/2, y^2 = x/2$

17. $\dfrac{dy}{y} = -\dfrac{x\, dx}{x^2 + 4}$

$\ln|y| = -\dfrac{1}{2}\ln(x^2 + 4) + C_1$

$y = \dfrac{C}{\sqrt{x^2 + 4}}$

19. $(1 - y^2)\, dy = x^2\, dx,$

$y - \dfrac{y^3}{3} = \dfrac{x^3}{3} + C_1, x^3 + y^3 - 3y = C$

21. True. The equation can be rewritten as $\dfrac{1}{f(y)}\dfrac{dy}{dx} = 1$, which has the form (1).

23. True. After t minutes there will be $32 \cdot (1/2)^t$ grams left; when $t = 5$ there will be $32 \cdot (1/2)^5 = 1$ gram.

25. Of the solutions $y = \dfrac{1}{2x^2 - C}$, all pass through the point $\left(0, -\dfrac{1}{C}\right)$ and thus never through $(0, 0)$.

A solution of the initial value problem with $y(0) = 0$ is (by inspection) $y = 0$. The method of Example 1 fails in this case because it starts with a division by $y^2 = 0$.

27. $\dfrac{dy}{dx} = xe^{-y}, e^y\, dy = x\, dx, e^y = \dfrac{x^2}{2} + C, x = 2$ when $y = 0$ so $1 = 2 + C, C = -1, e^y = x^2/2 - 1$

29. (a) $\dfrac{dy}{dt} = 0.02y, y_0 = 10{,}000$ \qquad **(b)** $y = 10{,}000e^{t/50}$

(c) $T = \dfrac{1}{0.02}\ln 2 \approx 34.657$ h \qquad **(d)** $45{,}000 = 10{,}000e^{t/50},$

$$t = 50\ln\dfrac{45{,}000}{10{,}000} \approx 75.20 \text{ h}$$

31. (a) $\dfrac{dy}{dt} = -ky$, $y(0) = 5.0 \times 10^7$; $3.83 = T = \dfrac{1}{k}\ln 2$, so $k = \dfrac{\ln 2}{3.83} \approx 0.1810$

   (b) $y = 5.0 \times 10^7 e^{-0.181t}$

   (c) $y(30) = 5.0 \times 10^7 e^{-0.1810(30)} \approx 219{,}000$

   (d) $y(t) = (0.1)y_0 = y_0 e^{-kt}$, $-kt = \ln 0.1$, $t = -\dfrac{\ln 0.1}{0.1810} = 12.72$ days

33. $100e^{0.02t} = 10{,}000$, $e^{0.02t} = 100$, $t = \dfrac{1}{0.02}\ln 100 \approx 230$ days

35. $y(t) = y_0 e^{-kt} = 10.0 e^{-kt}$, $3.5 = 10.0 e^{-k(5)}$, $k = -\dfrac{1}{5}\ln\dfrac{3.5}{10.0} \approx 0.2100$, $T = \dfrac{1}{k}\ln 2 \approx 3.30$ days

39. (a) $T = \dfrac{\ln 2}{k}$; and $\ln 2 \approx 0.6931$. If k is measured in percent, $k' = 100k$,

   then $T = \dfrac{\ln 2}{k} \approx \dfrac{69.31}{k'} \approx \dfrac{70}{k'}$.

   (b) 70 yr   **(c)** 20 yr   **(d)** 7%

41. From (19), $y(t) = y_0 e^{-0.000121t}$. If $0.27 = \dfrac{y(t)}{y_0} = e^{-0.000121t}$ then $t = -\dfrac{\ln 0.27}{0.000121} \approx 10{,}820$ yr, and

   if $0.30 = \dfrac{y(t)}{y_0}$ then $t = -\dfrac{\ln 0.30}{0.000121} \approx 9950$, or roughly between 9000 B.C. and 8000 B.C.

43. (a) Let $T_1 = 5730 - 40 = 5690$, $k_1 = \dfrac{\ln 2}{T_1} \approx 0.00012182$; $T_2 = 5730 + 40 = 5770$, $k_2 \approx 0.00012013$.

   With $y/y_0 = 0.92, 0.93$, $t_1 = -\dfrac{1}{k_1}\ln\dfrac{y}{y_0} = 684.5, 595.7$; $t_2 = -\dfrac{1}{k_2}\ln(y/y_0) = 694.1, 604.1$; in

   1988 the shroud was at most 695 years old, which places its creation in or after the year 1293.

   (b) Suppose T is the true half-life of carbon-14 and $T_1 = T(1 + r/100)$ is the false half-life. Then

   with $k = \dfrac{\ln 2}{T}$, $k_1 = \dfrac{\ln 2}{T_1}$ we have the formulae $y(t) = y_0 e^{-kt}$, $y_1(t) = y_0 e^{-k_1 t}$. At a certain

   point in time a reading of the carbon-14 is taken resulting in a certain value y, which in the

   case of the true formula is given by $y = y(t)$ for some t, and in the case of the false formula

   is given by $y = y_1(t_1)$ for some t_1.

   If the true formula is used then the time t since the beginning is given by $t = -\dfrac{1}{k}\ln\dfrac{y}{y_0}$. If

   the false formula is used we get a false value $t_1 = -\dfrac{1}{k_1}\ln\dfrac{y}{y_0}$; note that in both cases the

   value y/y_0 is the same. Thus $t_1/t = k/k_1 = T_1/T = 1 + r/100$, so the percentage error in

   the time to be measured is the same as the percentage error in the half-life.

45. (a) If $y = y_0 e^{kt}$, then $y_1 = y_0 e^{kt_1}$, $y_2 = y_0 e^{kt_2}$, divide: $y_2/y_1 = e^{k(t_2 - t_1)}$, $k = \dfrac{1}{t_2 - t_1}\ln(y_2/y_1)$,

   $T = \dfrac{\ln 2}{k} = \dfrac{(t_2 - t_1)\ln 2}{\ln(y_2/y_1)}$. If $y = y_0 e^{-kt}$, then $y_1 = y_0 e^{-kt_1}$, $y_2 = y_0 e^{-kt_2}$,

   $y_2/y_1 = e^{-k(t_2 - t_1)}$, $k = -\dfrac{1}{t_2 - t_1}\ln(y_2/y_1)$, $T = \dfrac{\ln 2}{k} = -\dfrac{(t_2 - t_1)\ln 2}{\ln(y_2/y_1)}$.

   In either case, T is positive, so $T = \left|\dfrac{(t_2 - t_1)\ln 2}{\ln(y_2/y_1)}\right|$.

(b) In part (a) assume $t_2 = t_1 + 1$ and $y_2 = 1.25y_1$. Then $T = \dfrac{\ln 2}{\ln 1.25} \approx 3.1$ h.

47. **(a)** $A = 1000e^{(0.08)(5)} = 1000e^{0.4} \approx \$1,491.82$

(b) $Pe^{(0.08)(10)} = 10,000$, $Pe^{0.8} = 10,000$, $P = 10,000e^{-0.8} \approx \$4,493.29$

(c) From (11), with $k = r = 0.08$, $T = (\ln 2)/0.08 \approx 8.7$ years.

49. **(a)** Given $\dfrac{dy}{dt} = k\left(1 - \dfrac{y}{L}\right)y$, separation of variables yields $\left(\dfrac{1}{y} + \dfrac{1}{L-y}\right)dy = k\,dt$ so that

$\ln\dfrac{y}{L-y} = \ln y - \ln(L-y) = kt + C$. The initial condition gives $C = \ln\dfrac{y_0}{L-y_0}$ so

$\ln\dfrac{y}{L-y} = kt + \ln\dfrac{y_0}{L-y_0}$, $\dfrac{y}{L-y} = e^{kt}\dfrac{y_0}{L-y_0}$, and $y(t) = \dfrac{y_0 L}{y_0 + (L-y_0)e^{-kt}}$.

(b) If $y_0 > 0$ then $y_0 + (L-y_0)e^{-kt} = Le^{-kt} + y_0(1 - e^{-kt}) > 0$ for all $t \geq 0$, so $y(t)$ exists for all such t. Since $\lim\limits_{t \to +\infty} e^{-kt} = 0$, $\lim\limits_{t \to +\infty} y(t) = \dfrac{y_0 L}{y_0 + (L-y_0) \cdot 0} = L$.

(Note that for $y_0 < 0$ the solution "blows up" at $t = -\dfrac{1}{k}\ln\dfrac{-y_0}{L-y_0}$, so $\lim\limits_{t \to +\infty} y(t)$ is undefined.)

51. **(a)** $k = L = 1, y_0 = 2$ 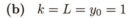 **(b)** $k = L = y_0 = 1$

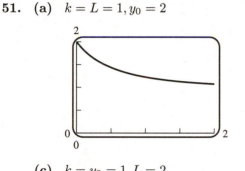

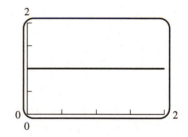

(c) $k = y_0 = 1, L = 2$ 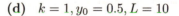 **(d)** $k = 1, y_0 = 0.5, L = 10$

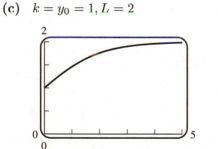

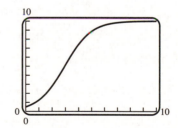

53. $y_0 \approx 2$, $L \approx 8$; since the curve $y = \dfrac{2 \cdot 8}{2 + 6e^{-kt}}$ passes through the point $(2, 4)$, $4 = \dfrac{16}{2 + 6e^{-2k}}$,

$6e^{-2k} = 2$, $k = \dfrac{1}{2}\ln 3 \approx 0.5493$.

55. **(a)** $y_0 = 5$ **(b)** $L = 12$ **(c)** $k = 1$

(d) $L/2 = 6 = \dfrac{60}{5 + 7e^{-t}}$, $5 + 7e^{-t} = 10$, $t = -\ln(5/7) \approx 0.3365$

(e) $\dfrac{dy}{dt} = \dfrac{1}{12}y(12 - y)$, $y(0) = 5$

57. (a) Assume that $y(t)$ students have had the flu t days after the break. If the disease spreads as predicted by equation (6) of Section 8.1 and if nobody is immune, then Exercise 50 gives

$$y(t) = \frac{y_0 L}{y_0 + (L - y_0)e^{-kLt}}, \text{ where } y_0 = 20 \text{ and } L = 1000. \text{ So } y(t) = \frac{20000}{20 + 980e^{-1000kt}} =$$

$$\frac{1000}{1 + 49e^{-1000kt}}. \text{ Using } y(5) = 35 \text{ we find that } k = -\frac{\ln(193/343)}{5000}. \text{ Hence } y = \frac{1000}{1 + 49(193/343)^{t/5}}.$$

(b)

t	0	1	2	3	4	5	6	7	8	9	10	11	12	13	14
$y(t)$	20	22	25	28	31	35	39	44	49	54	61	67	75	83	93

(c)

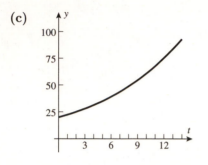

59. (a) From Exercise 58 with $T_0 = 95$ and $T_a = 21$, we have $T = 21 + 74e^{-kt}$ for some $k > 0$.

(b) $85 = T(1) = 21 + 74e^{-k}, \ k = -\ln\frac{64}{74} = -\ln\frac{32}{37}, \ T = 21 + 74e^{t\ln(32/37)} = 21 + 74\left(\frac{32}{37}\right)^t$,

$T = 51$ when $\frac{30}{74} = \left(\frac{32}{37}\right)^t, \ t = \frac{\ln(30/74)}{\ln(32/37)} \approx 6.22$ min

61. (a) $\frac{dv}{dt} = \frac{ck}{m_0 - kt} - g, v = -c\ln(m_0 - kt) - gt + C; v = 0$ when $t = 0$ so $0 = -c\ln m_0 + C,$

$C = c\ln m_0, v = c\ln m_0 - c\ln(m_0 - kt) - gt = c\ln\frac{m_0}{m_0 - kt} - gt.$

(b) $m_0 - kt = 0.2m_0$ when $t = 100$ so

$v = 2500\ln\frac{m_0}{0.2m_0} - 9.8(100) = 2500\ln 5 - 980 \approx 3044 \text{ m/s}.$

63. (a) $A(h) = \pi(1)^2 = \pi, \pi\frac{dh}{dt} = -0.025\sqrt{h}, \frac{\pi}{\sqrt{h}}dh = -0.025dt, 2\pi\sqrt{h} = -0.025t + C; h = 4$ when

$t = 0$, so $4\pi = C, 2\pi\sqrt{h} = -0.025t + 4\pi, \sqrt{h} = 2 - \frac{0.025}{2\pi}t, h \approx (2 - 0.003979\,t)^2.$

(b) $h = 0$ when $t \approx 2/0.003979 \approx 502.6$ s ≈ 8.4 min.

65. $\frac{dv}{dt} = -\frac{1}{32}v^2, \frac{1}{v^2}dv = -\frac{1}{32}dt, -\frac{1}{v} = -\frac{1}{32}t + C; v = 128$ when $t = 0$ so $-\frac{1}{128} = C,$

$-\frac{1}{v} = -\frac{1}{32}t - \frac{1}{128}, v = \frac{128}{4t+1}$ cm/s. But $v = \frac{dx}{dt}$ so $\frac{dx}{dt} = \frac{128}{4t+1}, x = 32\ln(4t+1) + C_1;$

$x = 0$ when $t = 0$ so $C_1 = 0, x = 32\ln(4t+1)$ cm.

67. Suppose that $H(y) = G(x) + C$. Then $\frac{dH}{dy}\frac{dy}{dx} = G'(x)$. But $\frac{dH}{dy} = h(y)$ and $\frac{dG}{dx} = g(x)$, hence $y(x)$ is a solution of (1).

69. If $h(y) = 0$ then (1) implies that $g(x) = 0$, so $h(y)\,dy = 0 = g(x)\,dx$. Otherwise the slope of L is $\dfrac{dy}{dx} = \dfrac{g(x)}{h(y)}$. Since (x_1, y_1) and (x_2, y_2) lie on L, we have $\dfrac{y_2 - y_1}{x_2 - x_1} = \dfrac{g(x)}{h(y)}$. So $h(y)(y_2 - y_1) = g(x)(x_2 - x_1)$; i.e. $h(y)\,dy = g(x)\,dx$.

EXERCISE SET 8.3

1.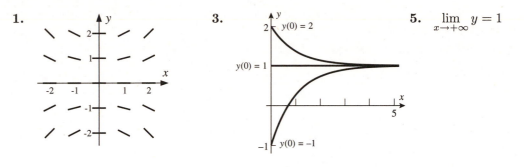

3.

5. $\displaystyle\lim_{x \to +\infty} y = 1$

7. $y_0 = 1, y_{n+1} = y_n + \frac{1}{2}y_n^{1/3}$

n	0	1	2	3	4	5	6	7	8
x_n	0	0.5	1	1.5	2	2.5	3	3.5	4
y_n	1	1.50	2.07	2.71	3.41	4.16	4.96	5.81	6.71

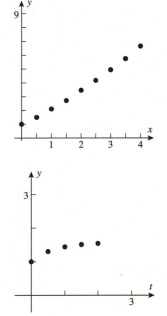

9. $y_0 = 1, y_{n+1} = y_n + \dfrac{1}{2}\cos y_n$

n	0	1	2	3	4
t_n	0	0.5	1	1.5	2
y_n	1	1.27	1.42	1.49	1.53

11. $h = 1/5, \ y_0 = 1, \ y_{n+1} = y_n + \dfrac{1}{5}\sin(\pi n/5)$

n	0	1	2	3	4	5
t_n	0	0.2	0.4	0.6	0.8	1.0
y_n	0	0.12	0.31	0.50	0.62	0.62

13. True. $\dfrac{dy}{dx} = e^{xy} > 0$ for all x and y. So, for any integral curve, y is an increasing function of x.

15. True. Every cubic polynomial has at least one real root. If $p(y_0) = 0$ then $y = y_0$ is an integral curve that is a horizontal line.

17. (b) $y\,dy = -x\,dx$, $y^2/2 = -x^2/2 + C_1$, $x^2 + y^2 = C$; if $y(0) = 1$ then $C = 1$ so $y(1/2) = \sqrt{3}/2$.

19. (b) The equation $y' = 1 - y$ is separable: $\dfrac{dy}{1-y} = dx$, so $\displaystyle\int \dfrac{dy}{1-y} = \int dx$, $-\ln|1-y| = x + C$.

Substituting $x = 0$ and $y = -1$ gives $C = -\ln 2$, so $x = \ln 2 - \ln|1-y| = \ln\left|\dfrac{2}{1-y}\right|$. Since the

integral curve stays below the line $y = 1$, we can drop the absolute value signs: $x = \ln\dfrac{2}{1-y}$

and $y = 1 - 2e^{-x}$. Solving $y = 0$ shows that the x-intercept is $\ln 2 \approx 0.693$.

21. (a) The slope field does not vary with x, hence along a given parallel line all values are equal since they only depend on the height y.

(b) As in part (a), the slope field does not vary with x; it is independent of x.

(c) From $G(y) - x = C$ we obtain $\dfrac{d}{dx}(G(y) - x) = \dfrac{1}{f(y)}\dfrac{dy}{dx} - 1 = \dfrac{d}{dx}C = 0$, i.e. $\dfrac{dy}{dx} = f(y)$.

23. (a) By implicit differentiation, $y^3 + 3xy^2\dfrac{dy}{dx} - 2xy - x^2\dfrac{dy}{dx} = 0$, $\dfrac{dy}{dx} = \dfrac{2xy - y^3}{3xy^2 - x^2}$.

(b) If $y(x)$ is an integral curve of the slope field in part (a), then

$\dfrac{d}{dx}\{x[y(x)]^3 - x^2y(x)\} = [y(x)]^3 + 3xy(x)^2y'(x) - 2xy(x) - x^2y'(x) = 0$, so the integral curve

must be of the form $x[y(x)]^3 - x^2y(x) = C$.

(c) $x[y(x)]^3 - x^2y(x) = 2$

25. (a) For any n, y_n is the value of the discrete approximation at the right endpoint, that, is an approximation of $y(1)$. By increasing the number of subdivisions of the interval $[0,1]$ one might expect more accuracy, and hence in the limit $y(1)$.

(b) For a fixed value of n we have, for $k = 1, 2, \ldots, n$, $y_k = y_{k-1} + y_{k-1}\dfrac{1}{n} = \dfrac{n+1}{n}y_{k-1}$. In partic-

ular $y_n = \dfrac{n+1}{n}y_{n-1} = \left(\dfrac{n+1}{n}\right)^2 y_{n-2} = \ldots = \left(\dfrac{n+1}{n}\right)^n y_0 = \left(\dfrac{n+1}{n}\right)^n$. Consequently,

$\displaystyle\lim_{n\to+\infty} y_n = \lim_{n\to+\infty}\left(\dfrac{n+1}{n}\right)^n = e$, which is the (correct) value $y = e^x\big|_{x=1}$.

EXERCISE SET 8.4

1. $\mu = e^{\int 4\,dx} = e^{4x}$, $e^{4x}y = \displaystyle\int e^x\,dx = e^x + C$, $y = e^{-3x} + Ce^{-4x}$

3. $\mu = e^{\int dx} = e^x$, $e^x y = \displaystyle\int e^x \cos(e^x)\,dx = \sin(e^x) + C$, $y = e^{-x}\sin(e^x) + Ce^{-x}$

5. $\dfrac{dy}{dx} + \dfrac{x}{x^2+1}y = 0$, $\mu = e^{\int (x/(x^2+1))\,dx} = e^{\frac{1}{2}\ln(x^2+1)} = \sqrt{x^2+1}$,

$\dfrac{d}{dx}\left[y\sqrt{x^2+1}\right] = 0$, $y\sqrt{x^2+1} = C$, $y = \dfrac{C}{\sqrt{x^2+1}}$

7. $\dfrac{dy}{dx} + \dfrac{1}{x}y = 1$, $\mu = e^{\int(1/x)\,dx} = e^{\ln x} = x$, $\dfrac{d}{dx}[xy] = x$, $xy = \dfrac{1}{2}x^2 + C$, $y = \dfrac{x}{2} + \dfrac{C}{x}$,

$2 = y(1) = \dfrac{1}{2} + C$, $C = \dfrac{3}{2}$, $y = \dfrac{x}{2} + \dfrac{3}{2x}$

9. $\mu = e^{-2\int x\,dx} = e^{-x^2}$, $e^{-x^2}y = \int 2xe^{-x^2}\,dx = -e^{-x^2} + C$,

$y = -1 + Ce^{x^2}$, $3 = -1 + C$, $C = 4$, $y = -1 + 4e^{x^2}$

11. False. If y_1 and y_2 both satisfy $\dfrac{dy}{dx} + p(x)y = q(x)$ then $\dfrac{d}{dx}(y_1 + y_2) + p(x)(y_1 + y_2) = 2q(x)$. Unless $q(x) = 0$ for all x, $y_1 + y_2$ is not a solution of the original differential equation.

13. True. The concentration in the tank will approach the concentration in the solution flowing into the tank.

15.

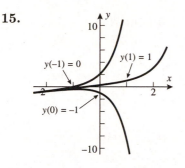

17. It appears that $\displaystyle\lim_{x\to+\infty} y = \begin{cases} +\infty, & \text{if } y_0 \geq 1/4; \\ -\infty, & \text{if } y_0 < 1/4. \end{cases}$

To confirm this, we solve the equation using the method of integrating factors:

$\dfrac{dy}{dx} - 2y = -x$, $\quad \mu = e^{-2\int dx} = e^{-2x}$, $\quad \dfrac{d}{dx}\left[ye^{-2x}\right] = -xe^{-2x}$, $\quad ye^{-2x} = \dfrac{1}{4}(2x+1)e^{-2x} + C$,

$y = \dfrac{1}{4}(2x+1) + Ce^{2x}$. Setting $y(0) = y_0$ gives $C = y_0 - \dfrac{1}{4}$, so $y = \dfrac{1}{4}(2x+1) + \left(y_0 - \dfrac{1}{4}\right)e^{2x}$.

If $y_0 = 1/4$, then $y = \dfrac{1}{4}(2x+1) \to +\infty$ as $x \to +\infty$. Otherwise, we rewrite the solution as

$y = e^{2x}\left(y_0 - \dfrac{1}{4} + \dfrac{2x+1}{4e^{2x}}\right)$; since $\displaystyle\lim_{x\to+\infty} \dfrac{2x+1}{4e^{2x}} = 0$, we obtain the conjectured limit.

19. (a) $y_0 = 1$,

$y_{n+1} = y_n + (x_n + y_n)(0.2) = (x_n + 6y_n)/5$

n	0	1	2	3	4	5
x_n	0	0.2	0.4	0.6	0.8	1.0
y_n	1	1.20	1.48	1.86	2.35	2.98

(b) $y' - y = x$, $\mu = e^{-x}$, $\dfrac{d}{dx}\left[ye^{-x}\right] = xe^{-x}$,

$ye^{-x} = -(x+1)e^{-x} + C$, $1 = -1 + C$,

$C = 2$, $y = -(x+1) + 2e^x$

x_n	0	0.2	0.4	0.6	0.8	1.0
$y(x_n)$	1	1.24	1.58	2.04	2.65	3.44
abs. error	0	0.04	0.10	0.19	0.30	0.46
perc. error	0	3	6	9	11	13

(c)

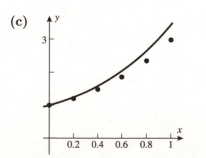

21. $\dfrac{dy}{dt}$ = rate in − rate out, where y is the amount of salt at time t,

$\dfrac{dy}{dt} = (4)(2) - \left(\dfrac{y}{50}\right)(2) = 8 - \dfrac{1}{25}y$, so $\dfrac{dy}{dt} + \dfrac{1}{25}y = 8$ and $y(0) = 25$.

$\mu = e^{\int (1/25)dt} = e^{t/25}$, $e^{t/25}y = \displaystyle\int 8e^{t/25}\,dt = 200e^{t/25} + C$,

$y = 200 + Ce^{-t/25}$, $25 = 200 + C$, $C = -175$,

(a) $y = 200 - 175e^{-t/25}$ oz $\qquad\qquad$ **(b)** when $t = 25$, $y = 200 - 175e^{-1} \approx 136$ oz

23. The volume V of the (polluted) water is $V(t) = 500 + (20 - 10)t = 500 + 10t$; if $y(t)$ is the number of pounds of particulate matter in the water, then $y(0) = 50$ and $\dfrac{dy}{dt} = 0 - 10\dfrac{y}{V} = -\dfrac{y}{50 + t}$. Using the method of integrating factors, we have $\dfrac{dy}{dt} + \dfrac{1}{50 + t}y = 0$; $\mu = e^{\int \frac{dt}{50+t}} = 50 + t$; $\dfrac{d}{dt}[(50+t)y] = 0$, $(50 + t)y = C$, $2500 = 50y(0) = C$, $y(t) = 2500/(50 + t)$. (The differential equation may also be solved by separation of variables.)

The tank reaches the point of overflowing when $V = 500 + 10t = 1000$, $t = 50$ min, so $y = 2500/(50 + 50) = 25$ lb.

25. (a) $\dfrac{dv}{dt} + \dfrac{c}{m}v = -g$, $\mu = e^{(c/m)\int dt} = e^{ct/m}$, $\dfrac{d}{dt}\left[ve^{ct/m}\right] = -ge^{ct/m}$, $ve^{ct/m} = -\dfrac{gm}{c}e^{ct/m} + C$,

$v = -\dfrac{gm}{c} + Ce^{-ct/m}$, but $v_0 = v(0) = -\dfrac{gm}{c} + C$, $C = v_0 + \dfrac{gm}{c}$, $v = -\dfrac{gm}{c} + \left(v_0 + \dfrac{gm}{c}\right)e^{-ct/m}$

(b) Replace $\dfrac{mg}{c}$ with v_τ and $-ct/m$ with $-gt/v_\tau$ in (16).

(c) From part (b), $s(t) = C - v_\tau t - (v_0 + v_\tau)\dfrac{v_\tau}{g}e^{-gt/v_\tau}$; $s_0 = s(0) = C - (v_0 + v_\tau)\dfrac{v_\tau}{g}$,

$C = s_0 + (v_0 + v_\tau)\dfrac{v_\tau}{g}$, $s(t) = s_0 - v_\tau t + \dfrac{v_\tau}{g}(v_0 + v_\tau)\left(1 - e^{-gt/v_\tau}\right)$

27. $\dfrac{dI}{dt} + \dfrac{R}{L}I = \dfrac{V(t)}{L}$, $\mu = e^{(R/L)\int dt} = e^{Rt/L}$, $\dfrac{d}{dt}(e^{Rt/L}I) = \dfrac{V(t)}{L}e^{Rt/L}$,

$Ie^{Rt/L} = I(0) + \dfrac{1}{L}\displaystyle\int_0^t V(u)e^{Ru/L}\,du$, $I(t) = I(0)e^{-Rt/L} + \dfrac{1}{L}e^{-Rt/L}\displaystyle\int_0^t V(u)e^{Ru/L}\,du$.

(a) $I(t) = \dfrac{1}{5}e^{-2t}\displaystyle\int_0^t 20e^{2u}\,du = 2e^{-2t}e^{2u}\Big]_0^t = 2\left(1 - e^{-2t}\right)$ A.

(b) $\displaystyle\lim_{t \to +\infty} I(t) = 2$ A

29. (a) Let $y = \dfrac{1}{\mu}[H(x) + C]$ where $\mu = e^{P(x)}$, $\dfrac{dP}{dx} = p(x)$, $\dfrac{d}{dx}H(x) = \mu q$, and C is an arbitrary constant. Then

$\dfrac{dy}{dx} + p(x)y = \dfrac{1}{\mu}H'(x) - \dfrac{\mu'}{\mu^2}[H(x) + C] + p(x)y = q - \dfrac{p}{\mu}[H(x) + C] + p(x)y = q$

(b) Given the initial value problem, let $C = \mu(x_0)y_0 - H(x_0)$. Then $y = \dfrac{1}{\mu}[H(x) + C]$ is a solution of the initial value problem with $y(x_0) = y_0$. This shows that the initial value problem has a solution.

To show uniqueness, suppose $u(x)$ also satisfies (3) together with $u(x_0) = y_0$. Following the arguments in the text we arrive at $u(x) = \dfrac{1}{\mu}[H(x) + C]$ for some constant C. The initial condition requires $C = \mu(x_0)y_0 - H(x_0)$, and thus $u(x)$ is identical with $y(x)$.

REVIEW EXERCISES, CHAPTER 8

3. $\dfrac{dy}{1+y^2} = x^2\, dx$, $\tan^{-1} y = \dfrac{1}{3}x^3 + C$, $y = \tan\left(\dfrac{1}{3}x^3 + C\right)$

5. $\left(\dfrac{1}{y} + y\right) dy = e^x dx$, $\ln|y| + y^2/2 = e^x + C$; by inspection, $y = 0$ is also a solution

7. $\left(\dfrac{1}{y^5} + \dfrac{1}{y}\right) dy = \dfrac{dx}{x}$, $-\dfrac{1}{4}y^{-4} + \ln|y| = \ln|x| + C$; $-\dfrac{1}{4} = C$, $y^{-4} + 4\ln(x/y) = 1$

9. $\dfrac{dy}{y^2} = -2x\, dx$, $-\dfrac{1}{y} = -x^2 + C$, $-1 = C$, $y = 1/(x^2 + 1)$

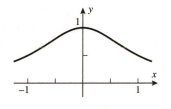

11.

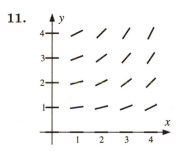

13. $y_0 = 1, y_{n+1} = y_n + \sqrt{y_n}/2$

n	0	1	2	3	4	5	6	7	8
x_n	0	0.5	1	1.5	2	2.5	3	3.5	4
y_n	1	1.50	2.11	2.84	3.68	4.64	5.72	6.91	8.23

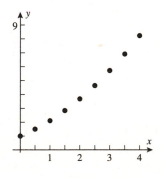

15. $h = 1/5$, $y_0 = 1$, $y_{n+1} = y_n + \dfrac{1}{5}\cos(2\pi n/5)$

n	0	1	2	3	4	5
t_n	0	0.2	0.4	0.6	0.8	1.0
y_n	1.00	1.20	1.26	1.10	0.94	1.00

17. From formula (19) of Section 8.2, $y(t) = y_0 e^{-0.000121t}$, so $0.785y_0 = y_0 e^{-0.000121t}$, $t = -\ln 0.785/0.000121 \approx$ 2000.6 yr

19. $\mu = e^{\int 3\,dx} = e^{3x}$, $e^{3x}y = \int e^x\,dx = e^x + C$, $y = e^{-2x} + Ce^{-3x}$

21. $\mu = e^{-\int x\,dx} = e^{-x^2/2}$, $e^{-x^2/2}y = \int xe^{-x^2/2}dx = -e^{-x^2/2} + C$,

$y = -1 + Ce^{x^2/2}$, $3 = -1 + C$, $C = 4$, $y = -1 + 4e^{x^2/2}$

23. By inspection, the left side of the equation is $\dfrac{d}{dx}(y\cosh x)$, so $\dfrac{d}{dx}(y\cosh x) = \cosh^2 x = $

$\dfrac{1}{2}(1 + \cosh 2x)$ and $y\cosh x = \dfrac{1}{2}x + \dfrac{1}{4}\sinh 2x + C = \dfrac{1}{2}(x + \sinh x\cosh x) + C$.

When $x = 0$, $y = 2$ so $2 = C$, and $y = 2\operatorname{sech} x + \dfrac{1}{2}(x\operatorname{sech} x + \sinh x)$.

25. (a) linear **(b)** both **(c)** separable **(d)** neither

27. Assume the tank contains $y(t)$ oz of salt at time t. Then $y_0 = 0$ and for $0 < t < 15$,

$\dfrac{dy}{dt} = 5\cdot 10 - \dfrac{y}{1000}10 = (50 - y/100)$ oz/min, with solution $y = 5000 + Ce^{-t/100}$. But $y(0) = 0$ so

$C = -5000$, $y = 5000(1 - e^{-t/100})$ for $0 \le t \le 15$, and $y(15) = 5000(1 - e^{-0.15})$. For $15 < t < 30$,

$\dfrac{dy}{dt} = 0 - \dfrac{y}{1000}5$, $y = C_1 e^{-t/200}$, $C_1 e^{-0.075} = y(15) = 5000(1 - e^{-0.15})$, $C_1 = 5000(e^{0.075} - e^{-0.075})$,

$y = 5000(e^{0.075} - e^{-0.075})e^{-t/200}$, $y(30) = 5000(e^{0.075} - e^{-0.075})e^{-0.15} \approx 646.14$ oz.

MAKING CONNECTIONS, CHAPTER 8

1. (a) $u(x) = q - py(x)$ so $\dfrac{du}{dx} = -p\dfrac{dy}{dx} = -p(q - py(x)) = (-p)u(x)$.

If $p < 0$ then $-p > 0$ so $u(x)$ grows exponentially.

If $p > 0$ then $-p < 0$ so $u(x)$ decays exponentially.

(b) From (a), $u(x) = 4 - 2y(x)$ satisfies $\dfrac{du}{dx} = -2u(x)$, so equation (14) of Section 8.2 gives

$u(x) = u_0 e^{-2x}$ for some constant u_0. Since $u(0) = 4 - 2y(0) = 6$, we have $u(x) = 6e^{-2x}$;

hence $y(x) = 2 - 3e^{-2x}$.

3. (a) $\dfrac{du}{dx} = \dfrac{d}{dx}\left(\dfrac{y}{x}\right) = \dfrac{x\dfrac{dy}{dx} - y}{x^2} = \dfrac{xf\left(\dfrac{y}{x}\right) - y}{x^2}$. Since $y = ux$, $\dfrac{du}{dx} = \dfrac{xf(u) - ux}{x^2} = \dfrac{f(u) - u}{x}$

and $\dfrac{1}{f(u) - u}\dfrac{du}{dx} = \dfrac{1}{x}$.

(b) $\dfrac{dy}{dx} = \dfrac{x - y}{x + y} = \dfrac{1 - y/x}{1 + y/x}$ has the form given in (a), with $f(t) = \dfrac{1 - t}{1 + t}$. So $\dfrac{1}{\dfrac{1 - u}{1 + u} - u}\dfrac{du}{dx} = \dfrac{1}{x}$,

$\dfrac{1 + u}{1 - 2u - u^2}du = \dfrac{dx}{x}$, $\displaystyle\int \dfrac{1 + u}{1 - 2u - u^2}du = \int \dfrac{dx}{x}$, $-\dfrac{1}{2}\ln|1 - 2u - u^2| = \ln|x| + C_1$, and

$|1 - 2u - u^2| = e^{-2C_1}x^{-2}$. Hence $1 - 2u - u^2 = Cx^{-2}$ where C is either e^{-2C_1} or $-e^{-2C_1}$.

Substituting $u = \dfrac{y}{x}$ gives $1 - \dfrac{2y}{x} - \dfrac{y^2}{x^2} = Cx^{-2}$, and $x^2 - 2xy - y^2 = C$.

CHAPTER 9
Infinite Series

EXERCISE SET 9.1

1. (a) $\dfrac{1}{3^{n-1}}$ (b) $\dfrac{(-1)^{n-1}}{3^{n-1}}$ (c) $\dfrac{2n-1}{2n}$ (d) $\dfrac{n^2}{\pi^{1/(n+1)}}$

3. (a) $2, 0, 2, 0$ (b) $1, -1, 1, -1$ (c) $2(1+(-1)^n); 2 + 2\cos n\pi$

5. (a) no; $f(n)$ oscillates between the values ± 1 and 0
    (b) $-1, +1, -1, +1, -1$
    (c) no, it oscillates between $+1$ and -1

7. $1/3, 2/4, 3/5, 4/6, 5/7, \ldots$; $\displaystyle\lim_{n\to+\infty}\frac{n}{n+2} = 1$, converges

9. $2, 2, 2, 2, 2, \ldots$; $\displaystyle\lim_{n\to+\infty} 2 = 2$, converges

11. $\dfrac{\ln 1}{1}, \dfrac{\ln 2}{2}, \dfrac{\ln 3}{3}, \dfrac{\ln 4}{4}, \dfrac{\ln 5}{5}, \ldots$;

$\displaystyle\lim_{n\to+\infty}\frac{\ln n}{n} = \lim_{n\to+\infty}\frac{1}{n} = 0 \left(\text{apply L'Hôpital's Rule to } \frac{\ln x}{x}\right)$, converges

13. $0, 2, 0, 2, 0, \ldots$; diverges

15. $-1, 16/9, -54/28, 128/65, -250/126, \ldots$; diverges because odd-numbered terms approach -2, even-numbered terms approach 2.

17. $6/2, 12/8, 20/18, 30/32, 42/50, \ldots$; $\displaystyle\lim_{n\to+\infty}\frac{1}{2}(1+1/n)(1+2/n) = 1/2$, converges

19. $e^{-1}, 4e^{-2}, 9e^{-3}, 16e^{-4}, 25e^{-5}, \ldots$; $\displaystyle\lim_{x\to+\infty} x^2 e^{-x} = \lim_{x\to+\infty}\frac{x^2}{e^x} = 0$, so $\displaystyle\lim_{n\to+\infty} n^2 e^{-n} = 0$, converges

21. $2, (5/3)^2, (6/4)^3, (7/5)^4, (8/6)^5, \ldots$; let $y = \left[\dfrac{x+3}{x+1}\right]^x$, converges because

$\displaystyle\lim_{x\to+\infty}\ln y = \lim_{x\to+\infty}\frac{\ln\dfrac{x+3}{x+1}}{1/x} = \lim_{x\to+\infty}\frac{2x^2}{(x+1)(x+3)} = 2$, so $\displaystyle\lim_{n\to+\infty}\left[\frac{n+3}{n+1}\right]^n = e^2$

23. $\left\{\dfrac{2n-1}{2n}\right\}_{n=1}^{+\infty}$; $\displaystyle\lim_{n\to+\infty}\frac{2n-1}{2n} = 1$, converges

25. $\left\{(-1)^{n-1}\dfrac{1}{3^n}\right\}_{n=1}^{+\infty}$; $\displaystyle\lim_{n\to+\infty}\frac{(-1)^{n-1}}{3^n} = 0$, converges

27. $\left(\dfrac{1}{n} - \dfrac{1}{n+1}\right)_{n=1}^{+\infty}$; the sequence converges to 0

29. $\{\sqrt{n+1} - \sqrt{n+2}\}_{n=1}^{+\infty}$; converges because

$$\lim_{n \to +\infty} (\sqrt{n+1} - \sqrt{n+2}) = \lim_{n \to +\infty} \frac{(n+1) - (n+2)}{\sqrt{n+1} + \sqrt{n+2}} = \lim_{n \to +\infty} \frac{-1}{\sqrt{n+1} + \sqrt{n+2}} = 0$$

31. true; a function whose domain is a set of integers

33. false, e.g. $a_n = (-1)^n$

35. Let $a_n = 0, b_n = \dfrac{\sin^2 n}{n}, c_n = \dfrac{1}{n}$; then $a_n \le b_n \le c_n$, $\lim\limits_{n \to +\infty} a_n = \lim\limits_{n \to +\infty} c_n = 0$, so $\lim\limits_{n \to +\infty} b_n = 0$.

37. $a_n = \begin{cases} +1 & k \text{ even} \\ -1 & k \text{ odd} \end{cases}$ oscillates; there is no limit point which attracts all of the a_n.

$b_n = \cos n$; the terms lie all over the interval $[-1, 1]$ without any limit.

39. **(a)** $1, 2, 1, 4, 1, 6$ **(b)** $a_n = \begin{cases} n, & n \text{ odd} \\ 1/2^n, & n \text{ even} \end{cases}$ **(c)** $a_n = \begin{cases} 1/n, & n \text{ odd} \\ 1/(n+1), & n \text{ even} \end{cases}$

(d) In Part (a) the sequence diverges, since the even terms diverge to $+\infty$ and the odd terms equal 1; in Part (b) the sequence diverges, since the odd terms diverge to $+\infty$ and the even terms tend to zero; in Part (c) $\lim\limits_{n \to +\infty} a_n = 0$.

41. $\lim\limits_{n \to +\infty} x_{n+1} = \dfrac{1}{2} \lim\limits_{n \to +\infty} \left(x_n + \dfrac{a}{x_n} \right)$ or $L = \dfrac{1}{2}\left(L + \dfrac{a}{L} \right), 2L^2 - L^2 - a = 0, L = \sqrt{a}$ (we reject $-\sqrt{a}$

because $x_n > 0$, thus $L \ge 0$.)

43. **(a)** $a_1 = (0.5)^2, a_2 = a_1^2 = (0.5)^4, \ldots, a_n = (0.5)^{2^n}$

(c) $\lim\limits_{n \to +\infty} a_n = \lim\limits_{n \to +\infty} e^{2^n \ln(0.5)} = 0$, since $\ln(0.5) < 0$.

(d) Replace 0.5 in Part (a) with a_0; then the sequence converges for $-1 \le a_0 \le 1$, because if $a_0 = \pm 1$, then $a_n = 1$ for $n \ge 1$; if $a_0 = 0$ then $a_n = 0$ for $n \ge 1$; and if $0 < |a_0| < 1$ then $a_1 = a_0^2 > 0$ and $\lim\limits_{n \to +\infty} a_n = \lim\limits_{n \to +\infty} e^{2^{n-1} \ln a_1} = 0$ since $0 < a_1 < 1$. This same argument proves divergence to $+\infty$ for $|a| > 1$ since then $\ln a_1 > 0$.

45. **(a)**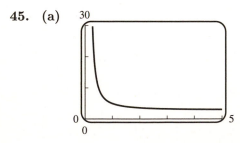

(b) Let $y = (2^x + 3^x)^{1/x}$, $\lim\limits_{x \to +\infty} \ln y = \lim\limits_{x \to +\infty} \dfrac{\ln(2^x + 3^x)}{x} = \lim\limits_{x \to +\infty} \dfrac{2^x \ln 2 + 3^x \ln 3}{2^x + 3^x}$

$= \lim\limits_{x \to +\infty} \dfrac{(2/3)^x \ln 2 + \ln 3}{(2/3)^x + 1} = \ln 3$, so $\lim\limits_{n \to +\infty} (2^n + 3^n)^{1/n} = e^{\ln 3} = 3$

Alternate proof: $3 = (3^n)^{1/n} < (2^n + 3^n)^{1/n} < (2 \cdot 3^n)^{1/n} = 3 \cdot 2^{1/n}$. Then apply the Squeezing Theorem.

47. (a) If $n \geq 1$, then $a_{n+2} = a_{n+1} + a_n$, so $\dfrac{a_{n+2}}{a_{n+1}} = 1 + \dfrac{a_n}{a_{n+1}}$.

(c) With $L = \lim\limits_{n \to +\infty} (a_{n+2}/a_{n+1}) = \lim\limits_{n \to +\infty} (a_{n+1}/a_n)$, $L = 1 + 1/L$, $L^2 - L - 1 = 0$,

$L = (1 \pm \sqrt{5})/2$, so $L = (1 + \sqrt{5})/2$ because the limit cannot be negative.

49. $\left| \dfrac{n}{n+1} - 1 \right| = \dfrac{1}{n+1} < \epsilon$ if $n + 1 > 1/\epsilon$, $n > 1/\epsilon - 1$

(a) $1/\epsilon - 1 = 1/0.25 - 1 = 3$, $N = 4$ (b) $1/\epsilon - 1 = 1/0.1 - 1 = 9$, $N = 10$

(c) $1/\epsilon - 1 = 1/0.001 - 1 = 999$, $N = 1000$

EXERCISE SET 9.2

1. $a_{n+1} - a_n = \dfrac{1}{n+1} - \dfrac{1}{n} = -\dfrac{1}{n(n+1)} < 0$ for $n \geq 1$, so strictly decreasing.

3. $a_{n+1} - a_n = \dfrac{n+1}{2n+3} - \dfrac{n}{2n+1} = \dfrac{1}{(2n+1)(2n+3)} > 0$ for $n \geq 1$, so strictly increasing.

5. $a_{n+1} - a_n = (n + 1 - 2^{n+1}) - (n - 2^n) = 1 - 2^n < 0$ for $n \geq 1$, so strictly decreasing.

7. $\dfrac{a_{n+1}}{a_n} = \dfrac{(n+1)/(2n+3)}{n/(2n+1)} = \dfrac{(n+1)(2n+1)}{n(2n+3)} = \dfrac{2n^2 + 3n + 1}{2n^2 + 3n} > 1$ for $n \geq 1$, so strictly increasing.

9. $\dfrac{a_{n+1}}{a_n} = \dfrac{(n+1)e^{-(n+1)}}{ne^{-n}} = (1 + 1/n)e^{-1} < 1$ for $n \geq 1$, so strictly decreasing.

11. $\dfrac{a_{n+1}}{a_n} = \dfrac{(n+1)^{n+1}}{(n+1)!} \cdot \dfrac{n!}{n^n} = \dfrac{(n+1)^n}{n^n} = (1 + 1/n)^n > 1$ for $n \geq 1$, so strictly increasing.

13. true by definition   15. false, e.g. $a_n = (-1)^n$

17. $f(x) = x/(2x + 1)$, $f'(x) = 1/(2x + 1)^2 > 0$ for $x \geq 1$, so strictly increasing.

19. $f(x) = \tan^{-1} x$, $f'(x) = 1/(1 + x^2) > 0$ for $x \geq 1$, so strictly increasing.

21. $f(x) = 2x^2 - 7x$, $f'(x) = 4x - 7 > 0$ for $x \geq 2$, so eventually strictly increasing.

23. $\dfrac{a_{n+1}}{a_n} = \dfrac{(n+1)!}{3^{n+1}} \cdot \dfrac{3^n}{n!} = \dfrac{n+1}{3} > 1$ for $n \geq 3$, so eventually strictly increasing.

25. Yes: a monotone sequence is increasing or decreasing; if it is increasing, then it is increasing and bounded above, so by Theorem 9.2.3 it converges; if decreasing, then use Theorem 9.2.4. The limit lies in the interval $[1, 2]$.

27. (a) $\sqrt{2}, \sqrt{2 + \sqrt{2}}, \sqrt{2 + \sqrt{2 + \sqrt{2}}}$

(b) $a_1 = \sqrt{2} < 2$ so $a_2 = \sqrt{2 + a_1} < \sqrt{2 + 2} = 2$, $a_3 = \sqrt{2 + a_2} < \sqrt{2 + 2} = 2$, and so on indefinitely.

(c) $a_{n+1}^2 - a_n^2 = (2 + a_n) - a_n^2 = 2 + a_n - a_n^2 = (2 - a_n)(1 + a_n)$

(d) $a_n > 0$ and, from Part (b), $a_n < 2$ so $2 - a_n > 0$ and $1 + a_n > 0$ thus, from Part (c), $a_{n+1}^2 - a_n^2 > 0$, $a_{n+1} - a_n > 0$, $a_{n+1} > a_n$; $\{a_n\}$ is a strictly increasing sequence.

(e) The sequence is increasing and has 2 as an upper bound so it must converge to a limit L,
$$\lim_{n \to +\infty} a_{n+1} = \lim_{n \to +\infty} \sqrt{2 + a_n}, \quad L = \sqrt{2 + L}, \quad L^2 - L - 2 = 0, \quad (L - 2)(L + 1) = 0$$
thus $\lim_{n \to +\infty} a_n = 2$.

29. (a) $a_{n+1} = \dfrac{|x|^{n+1}}{(n + 1)!} = \dfrac{|x|}{n + 1} \dfrac{|x|^n}{n!} = \dfrac{|x|}{n + 1} a_n$

(b) $a_{n+1}/a_n = |x|/(n + 1) < 1$ if $n > |x| - 1$.

(c) From Part (b) the sequence is eventually decreasing, and it is bounded below by 0, so by Theorem 9.2.4 it converges.

31. $n! > \dfrac{n^n}{e^{n-1}}$, $\sqrt[n]{n!} > \dfrac{n}{e^{1-1/n}}$, $\lim\limits_{n \to +\infty} \dfrac{n}{e^{1-1/n}} = +\infty$ so $\lim\limits_{n \to +\infty} \sqrt[n]{n!} = +\infty$.

EXERCISE SET 9.3

1. (a) $s_1 = 2$, $s_2 = 12/5$, $s_3 = \dfrac{62}{25}$, $s_4 = \dfrac{312}{125}$ $s_n = \dfrac{2 - 2(1/5)^n}{1 - 1/5} = \dfrac{5}{2} - \dfrac{5}{2}(1/5)^n$,

$\lim\limits_{n \to +\infty} s_n = \dfrac{5}{2}$, converges

(b) $s_1 = \dfrac{1}{4}$, $s_2 = \dfrac{3}{4}$, $s_3 = \dfrac{7}{4}$, $s_4 = \dfrac{15}{4}$ $s_n = \dfrac{(1/4) - (1/4)2^n}{1 - 2} = -\dfrac{1}{4} + \dfrac{1}{4}(2^n)$,

$\lim\limits_{n \to +\infty} s_n = +\infty$, diverges

(c) $\dfrac{1}{(k + 1)(k + 2)} = \dfrac{1}{k + 1} - \dfrac{1}{k + 2}$, $s_1 = \dfrac{1}{6}$, $s_2 = \dfrac{1}{4}$, $s_3 = \dfrac{3}{10}$, $s_4 = \dfrac{1}{3}$;

$s_n = \dfrac{1}{2} - \dfrac{1}{n + 2}$, $\lim\limits_{n \to +\infty} s_n = \dfrac{1}{2}$, converges

3. geometric, $a = 1$, $r = -3/4$, sum $= \dfrac{1}{1 - (-3/4)} = 4/7$

5. geometric, $a = 7$, $r = -1/6$, sum $= \dfrac{7}{1 + 1/6} = 6$

7. $s_n = \sum\limits_{k=1}^{n} \left(\dfrac{1}{k + 2} - \dfrac{1}{k + 3} \right) = \dfrac{1}{3} - \dfrac{1}{n + 3}$, $\lim\limits_{n \to +\infty} s_n = 1/3$

9. $s_n = \sum\limits_{k=1}^{n} \left(\dfrac{1/3}{3k - 1} - \dfrac{1/3}{3k + 2} \right) = \dfrac{1}{6} - \dfrac{1/3}{3n + 2}$, $\lim\limits_{n \to +\infty} s_n = 1/6$

11. $\sum\limits_{k=3}^{\infty} \dfrac{1}{k - 2} = \sum\limits_{k=1}^{\infty} 1/k$, the harmonic series, so the series diverges.

13. $\displaystyle\sum_{k=1}^{\infty}\frac{4^{k+2}}{7^{k-1}}=\sum_{k=1}^{\infty}64\left(\frac{4}{7}\right)^{k-1}$; geometric, $a=64$, $r=4/7$, sum $=\dfrac{64}{1-4/7}=448/3$

15. **(a)** Exercise 5   **(b)** Exercise 3   **(c)** Exercise 7   **(d)** Exercise 9

17. false; e.g. $a_n=1/n$ **19.** true

21. $0.9999\cdots=0.9+0.09+0.009+\cdots=\dfrac{0.9}{1-0.1}=1$

23. $5.373737\cdots=5+0.37+0.0037+0.000037+\cdots=5+\dfrac{0.37}{1-0.01}=5+37/99=532/99$

25. $0.a_1a_2\cdots a_n9999\cdots=0.a_1a_2\cdots a_n+0.9\left(10^{-n}\right)+0.09\left(10^{-n}\right)+\cdots$

$$=0.a_1a_2\cdots a_n+\frac{0.9\left(10^{-n}\right)}{1-0.1}=0.a_1a_2\cdots a_n+10^{-n}$$

$$=0.a_1a_2\cdots(a_n+1)=0.a_1a_2\cdots(a_n+1)0000\cdots$$

27. $d=10+2\cdot\dfrac{3}{4}\cdot10+2\cdot\dfrac{3}{4}\cdot\dfrac{3}{4}\cdot10+2\cdot\dfrac{3}{4}\cdot\dfrac{3}{4}\cdot\dfrac{3}{4}\cdot10+\cdots$

$$=10+20\left(\frac{3}{4}\right)+20\left(\frac{3}{4}\right)^2+20\left(\frac{3}{4}\right)^3+\cdots=10+\frac{20(3/4)}{1-3/4}=10+60=70\text{ meters}$$

29. **(a)** $s_n=\ln\dfrac{1}{2}+\ln\dfrac{2}{3}+\ln\dfrac{3}{4}+\cdots+\ln\dfrac{n}{n+1}=\ln\left(\dfrac{1}{2}\cdot\dfrac{2}{3}\cdot\dfrac{3}{4}\cdots\dfrac{n}{n+1}\right)=\ln\dfrac{1}{n+1}=-\ln(n+1)$,

$\displaystyle\lim_{n\to+\infty}s_n=-\infty$, series diverges.

(b) $\ln(1-1/k^2)=\ln\dfrac{k^2-1}{k^2}=\ln\dfrac{(k-1)(k+1)}{k^2}=\ln\dfrac{k-1}{k}+\ln\dfrac{k+1}{k}=\ln\dfrac{k-1}{k}-\ln\dfrac{k}{k+1}$,

$$s_n=\sum_{k=2}^{n+1}\left[\ln\frac{k-1}{k}-\ln\frac{k}{k+1}\right]$$

$$=\left(\ln\frac{1}{2}-\ln\frac{2}{3}\right)+\left(\ln\frac{2}{3}-\ln\frac{3}{4}\right)+\left(\ln\frac{3}{4}-\ln\frac{4}{5}\right)+\cdots+\left(\ln\frac{n}{n+1}-\ln\frac{n+1}{n+2}\right)$$

$$=\ln\frac{1}{2}-\ln\frac{n+1}{n+2},\ \lim_{n\to+\infty}s_n=\ln\frac{1}{2}=-\ln2$$

31. **(a)** Geometric series, $a=x$, $r=-x^2$. Converges for $|-x^2|<1$, $|x|<1$;

$S=\dfrac{x}{1-(-x^2)}=\dfrac{x}{1+x^2}$.

(b) Geometric series, $a=1/x^2$, $r=2/x$. Converges for $|2/x|<1$, $|x|>2$;

$S=\dfrac{1/x^2}{1-2/x}=\dfrac{1}{x^2-2x}$.

(c) Geometric series, $a=e^{-x}$, $r=e^{-x}$. Converges for $|e^{-x}|<1$, $e^{-x}<1$, $e^x>1$, $x>0$;

$S=\dfrac{e^{-x}}{1-e^{-x}}=\dfrac{1}{e^x-1}$.

33. $a_2=\dfrac{1}{2}a_1+\dfrac{1}{2}$, $a_3=\dfrac{1}{2}a_2+\dfrac{1}{2}=\dfrac{1}{2^2}a_1+\dfrac{1}{2^2}+\dfrac{1}{2}$, $a_4=\dfrac{1}{2}a_3+\dfrac{1}{2}=\dfrac{1}{2^3}a_1+\dfrac{1}{2^3}+\dfrac{1}{2^2}+\dfrac{1}{2}$,

$$a_5 = \frac{1}{2}a_4 + \frac{1}{2} = \frac{1}{2^4}a_1 + \frac{1}{2^4} + \frac{1}{2^3} + \frac{1}{2^2} + \frac{1}{2}, \ldots, a_n = \frac{1}{2^{n-1}}a_1 + \frac{1}{2^{n-1}} + \frac{1}{2^{n-2}} + \cdots + \frac{1}{2},$$

$$\lim_{n \to +\infty} a_n = \lim_{n \to +\infty} \frac{a_1}{2^{n-1}} + \sum_{n=1}^{\infty}\left(\frac{1}{2}\right)^n = 0 + \frac{1/2}{1 - 1/2} = 1$$

35. $s_n = (1 - 1/3) + (1/2 - 1/4) + (1/3 - 1/5) + (1/4 - 1/6) + \cdots + [1/n - 1/(n+2)]$

$= (1 + 1/2 + 1/3 + \cdots + 1/n) - (1/3 + 1/4 + 1/5 + \cdots + 1/(n+2))$

$= 3/2 - 1/(n+1) - 1/(n+2), \displaystyle\lim_{n \to +\infty} s_n = 3/2$

37. $s_n = \displaystyle\sum_{k=1}^{n} \frac{1}{(2k-1)(2k+1)} = \sum_{k=1}^{n}\left[\frac{1/2}{2k-1} - \frac{1/2}{2k+1}\right] = \frac{1}{2}\left[\sum_{k=1}^{n}\frac{1}{2k-1} - \sum_{k=1}^{n}\frac{1}{2k+1}\right]$

$= \dfrac{1}{2}\left[\displaystyle\sum_{k=1}^{n}\frac{1}{2k-1} - \sum_{k=2}^{n+1}\frac{1}{2k-1}\right] = \frac{1}{2}\left[1 - \frac{1}{2n+1}\right]; \displaystyle\lim_{n \to +\infty} s_n = \frac{1}{2}$

39. By inspection, $\dfrac{\theta}{2} - \dfrac{\theta}{4} + \dfrac{\theta}{8} - \dfrac{\theta}{16} + \cdots = \dfrac{\theta/2}{1 - (-1/2)} = \theta/3$

EXERCISE SET 9.4

1. (a) $\displaystyle\sum_{k=1}^{\infty} \frac{1}{2^k} = \frac{1/2}{1 - 1/2} = 1;$ $\displaystyle\sum_{k=1}^{\infty} \frac{1}{4^k} = \frac{1/4}{1 - 1/4} =$

$1/3;$ $\displaystyle\sum_{k=1}^{\infty}\left(\frac{1}{2^k} + \frac{1}{4^k}\right) = 1 + 1/3 = 4/3$

(b) $\displaystyle\sum_{k=1}^{\infty} \frac{1}{5^k} = \frac{1/5}{1 - 1/5} = 1/4;$ $\displaystyle\sum_{k=1}^{\infty} \frac{1}{k(k+1)} = 1$ (Example 5, Section 9.3);

$\displaystyle\sum_{k=1}^{\infty}\left[\frac{1}{5^k} - \frac{1}{k(k+1)}\right] = 1/4 - 1 = -3/4$

3. (a) $p = 3$, converges   **(b)** $p = 1/2$, diverges   **(c)** $p = 1$, diverges   **(d)** $p = 2/3$, diverges

5. (a) $\displaystyle\lim_{k \to +\infty} \frac{k^2 + k + 3}{2k^2 + 1} = \frac{1}{2}$; the series diverges.   **(b)** $\displaystyle\lim_{k \to +\infty}\left(1 + \frac{1}{k}\right)^k = e$; the series diverges.

(c) $\displaystyle\lim_{k \to +\infty} \cos k\pi$ does not exist; the series diverges.   **(d)** $\displaystyle\lim_{k \to +\infty} \frac{1}{k!} = 0$; no information

7. (a) $\displaystyle\int_{1}^{+\infty} \frac{1}{5x+2} = \lim_{\ell \to +\infty} \frac{1}{5}\ln(5x+2)\Big]_{1}^{\ell} = +\infty$, the series diverges by the Integral Test.

(b) $\displaystyle\int_{1}^{+\infty}\frac{1}{1+9x^2}dx = \lim_{\ell\to+\infty}\frac{1}{3}\tan^{-1}3x\Big]_{1}^{\ell} = \frac{1}{3}\left(\pi/2 - \tan^{-1}3\right),$

the series converges by the Integral Test.

9. $\displaystyle\sum_{k=1}^{\infty}\frac{1}{k+6} = \sum_{k=7}^{\infty}\frac{1}{k}$, diverges because the harmonic series diverges.

11. $\displaystyle\sum_{k=1}^{\infty}\frac{1}{\sqrt{k+5}} = \sum_{k=6}^{\infty}\frac{1}{\sqrt{k}}$, diverges because the p-series with $p = 1/2 \le 1$ diverges.

13. $\displaystyle\int_{1}^{+\infty}(2x-1)^{-1/3}dx = \lim_{\ell\to+\infty}\frac{3}{4}(2x-1)^{2/3}\Big]_{1}^{\ell} = +\infty$, the series diverges by the Integral Test.

15. $\displaystyle\lim_{k\to+\infty}\frac{k}{\ln(k+1)} = \lim_{k\to+\infty}\frac{1}{1/(k+1)} = +\infty$, the series diverges because $\displaystyle\lim_{k\to+\infty}u_k \neq 0$.

17. $\displaystyle\lim_{k\to+\infty}(1+1/k)^{-k} = 1/e \neq 0$, the series diverges.

19. $\displaystyle\int_{1}^{+\infty}\frac{\tan^{-1}x}{1+x^2}dx = \lim_{\ell\to+\infty}\frac{1}{2}\left(\tan^{-1}x\right)^2\Big]_{1}^{\ell} = 3\pi^2/32$, the series converges by the Integral Test, since

$\displaystyle\frac{d}{dx}\frac{\tan^{-1}x}{1+x^2} = \frac{1-2x\tan^{-1}x}{(1+x^2)^2} < 0$ for $x \ge 1$.

21. $\displaystyle\lim_{k\to+\infty}k^2\sin^2(1/k) = 1 \neq 0$, the series diverges.

23. $\displaystyle 7\sum_{k=5}^{\infty}k^{-1.01}$, p-series with $p > 1$, converges

25. $\displaystyle\frac{1}{x(\ln x)^p}$ is decreasing for $x \ge e^{-p}$, so use the Integral Test with $a = e^\alpha$, i.e. $\displaystyle\int_{e^\alpha}^{+\infty}\frac{dx}{x(\ln x)^p}$ to get

$\displaystyle\lim_{\ell\to+\infty}\ln(\ln x)\Big]_{e^\alpha}^{\ell} = +\infty$ if $p = 1$, $\qquad\displaystyle\lim_{\ell\to+\infty}\frac{(\ln x)^{1-p}}{1-p}\Big]_{e^\alpha}^{\ell} = \begin{cases} +\infty & \text{if } p < 1 \\[2mm] \dfrac{\alpha^{1-p}}{p-1} & \text{if } p > 1 \end{cases}$

Thus the series converges for $p > 1$, assuming α was chosen large enough.

27. Suppose $\Sigma(u_k + v_k)$ converges; then so does $\Sigma[(u_k + v_k) - u_k]$, but $\Sigma[(u_k + v_k) - u_k] = \Sigma v_k$, so Σv_k converges which contradicts the assumption that Σv_k diverges. Suppose $\Sigma(u_k - v_k)$ converges; then so does $\Sigma[u_k - (u_k - v_k)] = \Sigma v_k$ which leads to the same contradiction as before.

29. **(a)** diverges because $\displaystyle\sum_{k=1}^{\infty}(2/3)^{k-1}$ converges and $\displaystyle\sum_{k=1}^{\infty}1/k$ diverges.

(b) diverges because $\displaystyle\sum_{k=1}^{\infty}1/(3k+2)$ diverges and $\displaystyle\sum_{k=1}^{\infty}1/k^{3/2}$ converges.

31. false; if $\sum u_k$ converges then $\lim u_k = 0$, so $\displaystyle\lim\frac{1}{u_k}$ diverges, so $\displaystyle\sum\frac{1}{u_k}$ cannot converge

33. true, see Theorem 9.4.4

35. (a) $3\sum_{k=1}^{\infty}\dfrac{1}{k^2}-\sum_{k=1}^{\infty}\dfrac{1}{k^4}=\pi^2/2-\pi^4/90$

(b) $\sum_{k=1}^{\infty}\dfrac{1}{k^2}-1-\dfrac{1}{2^2}=\pi^2/6-5/4$

(c) $\sum_{k=2}^{\infty}\dfrac{1}{(k-1)^4}=\sum_{k=1}^{\infty}\dfrac{1}{k^4}=\pi^4/90$

37. (a) In Exercise 36 above let $f(x)=\dfrac{1}{x^2}$. Then $\displaystyle\int_{n}^{+\infty}f(x)\,dx=-\dfrac{1}{x}\Big]_{n}^{+\infty}=\dfrac{1}{n}$;

use this result and the same result with $n+1$ replacing n to obtain the desired result.

(b) $s_3=1+1/4+1/9=49/36$; $58/36=s_3+\dfrac{1}{4}<\dfrac{1}{6}\pi^2<s_3+\dfrac{1}{3}=61/36$

(d) $1/11<\dfrac{1}{6}\pi^2-s_{10}<1/10$

39. (a) Let $S_n=\displaystyle\sum_{k=1}^{n}\dfrac{1}{k^4}$ By Exercise 36(a), with $f(x)=\dfrac{1}{x^4}$, the result follows.

(b) $h(x)=\dfrac{1}{3x^3}-\dfrac{1}{3(x+1)^3}$ is a decreasing function, and the smallest n such that

$\left|\dfrac{1}{3n^3}-\dfrac{1}{3(n+1)^3}\right|\le 0.001$ is $n=7$.

(c) The midpoint of the interval indicated in Part c is $S_7+\dfrac{\dfrac{1}{3\cdot 7^3}+\dfrac{1}{3\cdot 8^3}}{2}\approx 1.08235$. A calculator

gives $\pi^4/90\approx 1.08232$.

41. $x^2 e^{-x}$ is decreasing and positive for $x>2$ so the Integral Test applies:

$\displaystyle\int_{1}^{\infty}x^2 e^{-x}\,dx=-(x^2+2x+2)e^{-x}\Big]_{1}^{\infty}=5e^{-1}$ so the series converges.

EXERCISE SET 9.5

1. (a) $\dfrac{1}{5k^2-k}\le\dfrac{1}{5k^2-k^2}=\dfrac{1}{4k^2}$, $\displaystyle\sum_{k=1}^{\infty}\dfrac{1}{4k^2}$ converges

(b) $\dfrac{3}{k-1/4}>\dfrac{3}{k}$, $\displaystyle\sum_{k=1}^{\infty}3/k$ diverges

3. (a) $\dfrac{1}{3^k+5}<\dfrac{1}{3^k}$, $\displaystyle\sum_{k=1}^{\infty}\dfrac{1}{3^k}$ converges

(b) $\dfrac{5\sin^2 k}{k!}<\dfrac{5}{k!}$, $\displaystyle\sum_{k=1}^{\infty}\dfrac{5}{k!}$ converges

5. compare with the convergent series $\displaystyle\sum_{k=1}^{\infty}1/k^5$, $\rho=\lim_{k\to+\infty}\dfrac{4k^7-2k^6+6k^5}{8k^7+k-8}=1/2$, converges

7. compare with the convergent series $\displaystyle\sum_{k=1}^{\infty}5/3^k$, $\rho=\lim_{k\to+\infty}\dfrac{3^k}{3^k+1}=1$, converges

9. compare with the divergent series $\displaystyle\sum_{k=1}^{\infty} \frac{1}{k^{2/3}}$,

$$\rho = \lim_{k\to+\infty} \frac{k^{2/3}}{(8k^2 - 3k)^{1/3}} = \lim_{k\to+\infty} \frac{1}{(8 - 3/k)^{1/3}} = 1/2, \text{ diverges}$$

11. $\displaystyle\rho = \lim_{k\to+\infty} \frac{3^{k+1}/(k+1)!}{3^k/k!} = \lim_{k\to+\infty} \frac{3}{k+1} = 0$, the series converges

13. $\displaystyle\rho = \lim_{k\to+\infty} \frac{k}{k+1} = 1$, the result is inconclusive

15. $\displaystyle\rho = \lim_{k\to+\infty} \frac{(k+1)!/(k+1)^3}{k!/k^3} = \lim_{k\to+\infty} \frac{k^3}{(k+1)^2} = +\infty$, the series diverges

17. $\displaystyle\rho = \lim_{k\to+\infty} \frac{3k+2}{2k-1} = 3/2$, the series diverges

19. $\displaystyle\rho = \lim_{k\to+\infty} \frac{k^{1/k}}{5} = 1/5$, the series converges

21. false; it uses terms from two different sequences

23. true, Limit Comparison Test with $u_k, 1/k^2$

25. Ratio Test, $\displaystyle\rho = \lim_{k\to+\infty} 7/(k+1) = 0$, converges

27. Ratio Test, $\displaystyle\rho = \lim_{k\to+\infty} \frac{(k+1)^2}{5k^2} = 1/5$, converges

29. Ratio Test, $\displaystyle\rho = \lim_{k\to+\infty} e^{-1}(k+1)^{50}/k^{50} = e^{-1} < 1$, converges

31. Limit Comparison Test, compare with the convergent series $\displaystyle\sum_{k=1}^{\infty} 1/k^{5/2}$, $\rho = \lim_{k\to+\infty} \frac{k^3}{k^3 + 1} = 1$, converges

33. Limit Comparison Test, compare with the divergent series $\displaystyle\sum_{k=1}^{\infty} 1/k$, $\rho = \lim_{k\to+\infty} \frac{k}{\sqrt{k^2 + k}} = 1$, diverges

35. Limit Comparison Test, compare with the divergent series $\displaystyle\sum_{k=1}^{\infty} 1/\sqrt{k}$

37. Ratio Test, $\displaystyle\rho = \lim_{k\to+\infty} \frac{\ln(k+1)}{e \ln k} = \lim_{k\to+\infty} \frac{k}{e(k+1)} = 1/e < 1$, converges

39. Ratio Test, $\displaystyle\rho = \lim_{k\to+\infty} \frac{k+5}{4(k+1)} = 1/4$, converges

41. diverges because $\displaystyle\lim_{k\to+\infty} \frac{1}{4 + 2^{-k}} = 1/4 \neq 0$

43. $\displaystyle\frac{\tan^{-1} k}{k^2} < \frac{\pi/2}{k^2}, \sum_{k=1}^{\infty} \frac{\pi/2}{k^2}$ converges so $\displaystyle\sum_{k=1}^{\infty} \frac{\tan^{-1} k}{k^2}$ converges

45. Ratio Test, $\rho = \lim\limits_{k \to +\infty} \dfrac{(k+1)^2}{(2k+2)(2k+1)} = 1/4$, converges

47. $a_k = \dfrac{\ln k}{3^k}, \dfrac{a_{k+1}}{a_k} = \dfrac{\ln(k+1)}{\ln k} \dfrac{3^k}{3^{k+1}} \to \dfrac{1}{3}$, converges

49. $u_k = \dfrac{k!}{1 \cdot 3 \cdot 5 \cdots (2k-1)}$, by the Ratio Test $\rho = \lim\limits_{k \to +\infty} \dfrac{k+1}{2k+1} = 1/2$; converges

51. Set $g(x) = \sqrt{x} - \ln x;$ $\dfrac{d}{dx} g(x) = \dfrac{1}{2\sqrt{x}} - \dfrac{1}{x} = 0$ only at $x = 4$. Since $\lim\limits_{x \to 0+} g(x) = \lim\limits_{x \to +\infty} g(x) = +\infty$
it follows that $g(x)$ has its absolute minimum at $x = 4$, $g(4) = \sqrt{4} - \ln 4 > 0$, and thus $\sqrt{x} - \ln x > 0$
for $x > 0$.

    (a) $\dfrac{\ln k}{k^2} < \dfrac{\sqrt{k}}{k^2} = \dfrac{1}{k^{3/2}}, \sum\limits_{k=1}^{\infty} \dfrac{1}{k^{3/2}}$ converges so $\sum\limits_{k=1}^{\infty} \dfrac{\ln k}{k^2}$ converges.

    (b) $\dfrac{1}{(\ln k)^2} > \dfrac{1}{k}, \sum\limits_{k=2}^{\infty} \dfrac{1}{k}$ diverges so $\sum\limits_{k=2}^{\infty} \dfrac{1}{(\ln k)^2}$ diverges.

53. **(a)** $\cos x \approx 1 - x^2/2, 1 - \cos\left(\dfrac{1}{k}\right) \approx \dfrac{1}{2k^2}$ **(b)** $\rho = \lim\limits_{k \to +\infty} \dfrac{1 - \cos(1/k)}{1/k^2} = 1/2$, converges

55. **(a)** If $\sum b_k$ converges, then set $M = \sum b_k$. Then $a_1 + a_2 + \cdots + a_n \le b_1 + b_2 + \cdots + b_n \le M$;
apply Theorem 9.4.6 to get convergence of $\sum a_k$.

    (b) Assume the contrary, that $\sum b_k$ converges; then use Part (a) of the Theorem to show that
$\sum a_k$ converges, a contradiction.

EXERCISE SET 9.6

1. $a_{k+1} < a_k, \lim\limits_{k \to +\infty} a_k = 0, a_k > 0$

3. diverges because $\lim\limits_{k \to +\infty} a_k = \lim\limits_{k \to +\infty} \dfrac{k+1}{3k+1} = 1/3 \ne 0$

5. $\{e^{-k}\}$ is decreasing and $\lim\limits_{k \to +\infty} e^{-k} = 0$, converges

7. $\rho = \lim\limits_{k \to +\infty} \dfrac{(3/5)^{k+1}}{(3/5)^k} = 3/5$, converges absolutely

9. $\rho = \lim\limits_{k \to +\infty} \dfrac{3k^2}{(k+1)^2} = 3$, diverges

11. $\rho = \lim\limits_{k \to +\infty} \dfrac{(k+1)^3}{ek^3} = 1/e$, converges absolutely

13. conditionally convergent, $\sum\limits_{k=1}^{\infty} \dfrac{(-1)^{k+1}}{3k}$ converges by the Alternating Series Test but $\sum\limits_{k=1}^{\infty} \dfrac{1}{3k}$ diverges

15. divergent, $\lim\limits_{k \to +\infty} a_k \ne 0$

17. $\sum_{k=1}^{\infty} \dfrac{\cos k\pi}{k} = \sum_{k=1}^{\infty} \dfrac{(-1)^k}{k}$ is conditionally convergent, $\sum_{k=1}^{\infty} \dfrac{(-1)^k}{k}$ converges by the Alternating Series

Test but $\sum_{k=1}^{\infty} 1/k$ diverges.

19. conditionally convergent, $\sum_{k=1}^{\infty} (-1)^{k+1} \dfrac{k+2}{k(k+3)}$ converges by the

Alternating Series Test but $\sum_{k=1}^{\infty} \dfrac{k+2}{k(k+3)}$ diverges (Limit Comparison Test with $\sum 1/k$)

21. $\sum_{k=1}^{\infty} \sin(k\pi/2) = 1 + 0 - 1 + 0 + 1 + 0 - 1 + 0 + \cdots$, divergent ($\lim_{k \to +\infty} \sin(k\pi/2)$ does not exist)

23. conditionally convergent, $\sum_{k=2}^{\infty} \dfrac{(-1)^k}{k \ln k}$ converges by the Alternating Series Test but $\sum_{k=2}^{\infty} \dfrac{1}{k \ln k}$ diverges

(Integral Test)

25. absolutely convergent, $\sum_{k=2}^{\infty} (1/\ln k)^k$ converges by the Root Test

27. absolutely convergent by the Ratio Test, $\rho = \lim_{k \to +\infty} \dfrac{k+1}{(2k+1)(2k)} = 0$

29. false **31.** true

33. $|\text{error}| < a_8 = 1/8 = 0.125$ **35.** $|\text{error}| < a_{100} = 1/\sqrt{100} = 0.1$

37. $|\text{error}| < 0.0001$ if $a_{n+1} \leq 0.0001$, $1/(n+1) \leq 0.0001$, $n+1 \geq 10,000$, $n \geq 9,999$, $n = 9,999$

39. $|\text{error}| < 0.005$ if $a_{n+1} \leq 0.005$, $1/\sqrt{n+1} \leq 0.005$, $\sqrt{n+1} \geq 200$, $n+1 \geq 40,000$, $n \geq 39,999$, $n = 39,999$

41. $a_k = \dfrac{3}{2^{k+1}}$, $|\text{error}| < a_{11} = \dfrac{3}{2^{12}} < 0.00074$; $s_{10} \approx 0.4995$; $S = \dfrac{3/4}{1-(-1/2)} = 0.5$

43. $a_k = \dfrac{1}{(2k-1)!}$, $a_{n+1} = \dfrac{1}{(2n+1)!} \leq 0.005$, $(2n+1)! \geq 200$, $2n+1 \geq 6$, $n \geq 2.5$; $n = 3$,

$s_3 = 1 - 1/6 + 1/120 \approx 0.84$

45. $a_k = \dfrac{1}{k2^k}$, $a_{n+1} = \dfrac{1}{(n+1)2^{n+1}} \leq 0.005$, $(n+1)2^{n+1} \geq 200$, $n+1 \geq 6$, $n \geq 5$; $n = 5$, $s_5 \approx 0.41$

47. (c) $a_k = \dfrac{1}{2k-1}$, $a_{n+1} = \dfrac{1}{2n+1} \leq 10^{-2}$, $2n+1 \geq 100$, $n \geq 49.5$; $n = 50$

49. (a) $\sum \dfrac{(-1)^k}{\sqrt{k}}$ converges but $\sum \dfrac{1}{k}$ diverges; $\sum \dfrac{(-1)^k}{k}$ converges and $\sum \dfrac{1}{k^2}$ converges.

(b) Let $a_k = \dfrac{(-1)^k}{k}$, then $\sum a_k^2$ converges but $\sum |a_k|$ diverges, $\sum a_k$ converges.

51. Every positive integer can be written in exactly one of the three forms $2k - 1$ or $4k - 2$ or $4k$, so a rearrangement is

$$\left(1 - \frac{1}{2} - \frac{1}{4}\right) + \left(\frac{1}{3} - \frac{1}{6} - \frac{1}{8}\right) + \left(\frac{1}{5} - \frac{1}{10} - \frac{1}{12}\right) + \cdots + \left(\frac{1}{2k - 1} - \frac{1}{4k - 2} - \frac{1}{4k}\right) + \cdots$$

$$= \left(\frac{1}{2} - \frac{1}{4}\right) + \left(\frac{1}{6} - \frac{1}{8}\right) + \left(\frac{1}{10} - \frac{1}{12}\right) + \cdots + \left(\frac{1}{4k - 2} - \frac{1}{4k}\right) + \cdots = \frac{1}{2}\ln 2$$

53. Let $A = 1 - \frac{1}{2^2} + \frac{1}{3^2} - \frac{1}{4^2} + \cdots$; since the series all converge absolutely,

$$\frac{\pi^2}{6} - A = 2\frac{1}{2^2} + 2\frac{1}{4^2} + 2\frac{1}{6^2} + \cdots = \frac{1}{2}\left(1 + \frac{1}{2^2} + \frac{1}{3^2} + \cdots\right) = \frac{1}{2}\frac{\pi^2}{6}, \text{ so } A = \frac{1}{2}\frac{\pi^2}{6} = \frac{\pi^2}{12}.$$

EXERCISE SET 9.7

1. (a) $f^{(k)}(x) = (-1)^k e^{-x}$, $f^{(k)}(0) = (-1)^k$; $e^{-x} \approx 1 - x + x^2/2$ (quadratic), $e^{-x} \approx 1 - x$ (linear)

(b) $f'(x) = -\sin x$, $f''(x) = -\cos x$, $f(0) = 1$, $f'(0) = 0$, $f''(0) = -1$,

$\cos x \approx 1 - x^2/2$ (quadratic), $\cos x \approx 1$ (linear)

3. (a) $f'(x) = \frac{1}{2}x^{-1/2}$, $f''(x) = -\frac{1}{4}x^{-3/2}$; $f(1) = 1$, $f'(1) = \frac{1}{2}$, $f''(1) = -\frac{1}{4}$;

$\sqrt{x} \approx 1 + \frac{1}{2}(x - 1) - \frac{1}{8}(x - 1)^2$

(b) $x = 1.1, x_0 = 1, \sqrt{1.1} \approx 1 + \frac{1}{2}(0.1) - \frac{1}{8}(0.1)^2 = 1.04875$, calculator value ≈ 1.0488088

5. $f(x) = \tan x$, $61° = \pi/3 + \pi/180$ rad; $x_0 = \pi/3$, $f'(x) = \sec^2 x$, $f''(x) = 2\sec^2 x \tan x$;

$f(\pi/3) = \sqrt{3}$, $f'(\pi/3) = 4$, $f''(\pi/3) = 8\sqrt{3}$; $\tan x \approx \sqrt{3} + 4(x - \pi/3) + 4\sqrt{3}(x - \pi/3)^2$,

$\tan 61° = \tan(\pi/3 + \pi/180) \approx \sqrt{3} + 4\pi/180 + 4\sqrt{3}(\pi/180)^2 \approx 1.80397443$,

calculator value ≈ 1.80404776

7. $f^{(k)}(x) = (-1)^k e^{-x}$, $f^{(k)}(0) = (-1)^k$; $p_0(x) = 1$, $p_1(x) = 1 - x$, $p_2(x) = 1 - x + \frac{1}{2}x^2$,

$p_3(x) = 1 - x + \frac{1}{2}x^2 - \frac{1}{3!}x^3$, $p_4(x) = 1 - x + \frac{1}{2}x^2 - \frac{1}{3!}x^3 + \frac{1}{4!}x^4$; $\sum_{k=0}^{n} \frac{(-1)^k}{k!}x^k$

9. $f^{(k)}(0) = 0$ if k is odd, $f^{(k)}(0)$ is alternately π^k and $-\pi^k$ if k is even; $p_0(x) = 1$, $p_1(x) = 1$,

$p_2(x) = 1 - \frac{\pi^2}{2!}x^2$; $p_3(x) = 1 - \frac{\pi^2}{2!}x^2$, $p_4(x) = 1 - \frac{\pi^2}{2!}x^2 + \frac{\pi^4}{4!}x^4$; $\sum_{k=0}^{\lfloor \frac{n}{2} \rfloor} \frac{(-1)^k \pi^{2k}}{(2k)!}x^{2k}$

NB: The function $\lfloor x \rfloor$ defined for real x indicates the greatest integer which is $\leq x$.

11. $f^{(0)}(0) = 0$; for $k \geq 1$, $f^{(k)}(x) = \frac{(-1)^{k+1}(k-1)!}{(1+x)^k}$, $f^{(k)}(0) = (-1)^{k+1}(k-1)!$; $p_0(x) = 0$,

$p_1(x) = x$, $p_2(x) = x - \frac{1}{2}x^2$, $p_3(x) = x - \frac{1}{2}x^2 + \frac{1}{3}x^3$, $p_4(x) = x - \frac{1}{2}x^2 + \frac{1}{3}x^3 - \frac{1}{4}x^4$; $\sum_{k=1}^{n} \frac{(-1)^{k+1}}{k}x^k$

13. $f^{(k)}(0) = 0$ if k is odd, $f^{(k)}(0) = 1$ if k is even; $p_0(x) = 1, p_1(x) = 1$,

$p_2(x) = 1 + x^2/2$, $p_3(x) = 1 + x^2/2$, $p_4(x) = 1 + x^2/2 + x^4/4!$; $\sum_{k=0}^{\lfloor \frac{n}{2} \rfloor} \frac{1}{(2k)!}x^{2k}$

15. $f^{(k)}(x) = \begin{cases} (-1)^{k/2}(x\sin x - k\cos x) & k \text{ even} \\ (-1)^{(k-1)/2}(x\cos x + k\sin x) & k \text{ odd} \end{cases}$, $\quad f^{(k)}(0) = \begin{cases} (-1)^{1+k/2}k & k \text{ even} \\ 0 & k \text{ odd} \end{cases}$

$p_0(x) = 0$, $p_1(x) = 0$, $p_2(x) = x^2$, $p_3(x) = x^2$, $p_4(x) = x^2 - \dfrac{1}{6}x^4$; $\quad \displaystyle\sum_{k=0}^{\lfloor\frac{n}{2}\rfloor-1} \dfrac{(-1)^k}{(2k+1)!}x^{2k+2}$

17. $f^{(k)}(x_0) = e$; $p_0(x) = e$, $p_1(x) = e + e(x-1)$,

$p_2(x) = e + e(x-1) + \dfrac{e}{2}(x-1)^2$, $p_3(x) = e + e(x-1) + \dfrac{e}{2}(x-1)^2 + \dfrac{e}{3!}(x-1)^3$,

$p_4(x) = e + e(x-1) + \dfrac{e}{2}(x-1)^2 + \dfrac{e}{3!}(x-1)^3 + \dfrac{e}{4!}(x-1)^4$; $\quad \displaystyle\sum_{k=0}^{n} \dfrac{e}{k!}(x-1)^k$

19. $f^{(k)}(x) = \dfrac{(-1)^k k!}{x^{k+1}}$, $f^{(k)}(-1) = -k!$; $p_0(x) = -1$; $p_1(x) = -1 - (x+1)$;

$p_2(x) = -1 - (x+1) - (x+1)^2$; $p_3(x) = -1 - (x+1) - (x+1)^2 - (x+1)^3$;

$p_4(x) = -1 - (x+1) - (x+1)^2 - (x+1)^3 - (x+1)^4$; $\quad \displaystyle\sum_{k=0}^{n}(-1)(x+1)^k$

21. $f^{(k)}(1/2) = 0$ if k is odd, $f^{(k)}(1/2)$ is alternately π^k and $-\pi^k$ if k is even;

$p_0(x) = p_1(x) = 1$, $p_2(x) = p_3(x) = 1 - \dfrac{\pi^2}{2}(x-1/2)^2$,

$p_4(x) = 1 - \dfrac{\pi^2}{2}(x-1/2)^2 + \dfrac{\pi^4}{4!}(x-1/2)^4$; $\quad \displaystyle\sum_{k=0}^{\lfloor\frac{n}{2}\rfloor} \dfrac{(-1)^k\pi^{2k}}{(2k)!}(x-1/2)^{2k}$

23. $f(1) = 0$, for $k \geq 1$, $f^{(k)}(x) = \dfrac{(-1)^{k-1}(k-1)!}{x^k}$; $f^{(k)}(1) = (-1)^{k-1}(k-1)!$;

$p_0(x) = 0$, $p_1(x) = (x-1)$; $p_2(x) = (x-1) - \dfrac{1}{2}(x-1)^2$; $p_3(x) = (x-1) - \dfrac{1}{2}(x-1)^2 + \dfrac{1}{3}(x-1)^3$,

$p_4(x) = (x-1) - \dfrac{1}{2}(x-1)^2 + \dfrac{1}{3}(x-1)^3 - \dfrac{1}{4}(x-1)^4$; $\quad \displaystyle\sum_{k=1}^{n} \dfrac{(-1)^{k-1}}{k}(x-1)^k$

25. (a) $f(0) = 1$, $f'(0) = 2$, $f''(0) = -2$, $f'''(0) = 6$, the third MacLaurin polynomial for $f(x)$ is $f(x)$.

(b) $f(1) = 1$, $f'(1) = 2$, $f''(1) = -2$, $f'''(1) = 6$, the third Taylor polynomial for $f(x)$ is $f(x)$.

27. $f^{(k)}(0) = (-2)^k$; $p_0(x) = 1$, $p_1(x) = 1 - 2x$,

$p_2(x) = 1 - 2x + 2x^2$, $p_3(x) = 1 - 2x + 2x^2 - \dfrac{4}{3}x^3$

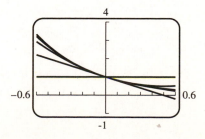

29. $f^{(k)}(\pi) = 0$ if k is odd, $f^{(k)}(\pi)$ is alternately -1 and 1 if k is even; $p_0(x) = -1$, $p_2(x) = -1 + \frac{1}{2}(x - \pi)^2$,

$$p_4(x) = -1 + \frac{1}{2}(x - \pi)^2 - \frac{1}{24}(x - \pi)^4,$$

$$p_6(x) = -1 + \frac{1}{2}(x - \pi)^2 - \frac{1}{24}(x - \pi)^4 + \frac{1}{720}(x - \pi)^6$$

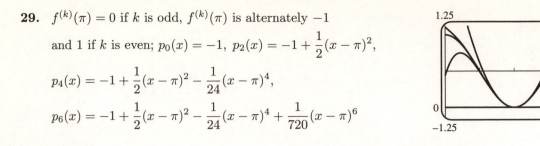

31. true

33. false, $p_6^{(4)}(x_0) = f^{(4)}(x_0)$

35. $\sqrt{e} = e^{1/2}$, $f(x) = e^x$, $M = e^{1/2}$, $|e^{1/2} - p_n(1/2)| \le M\dfrac{|x - 1/2|^{n+1}}{(n+1)!} \le 0.00005$, by experimentation take $n = 5$, $\sqrt{e} \approx p_5(1/2) \approx 1.648698$, calculator value ≈ 1.648721, difference ≈ 0.000023

37. $p(0) = 1$, $p(x)$ has slope -1 at $x = 0$, and $p(x)$ is concave up at $x = 0$, eliminating I, II and III respectively and leaving IV.

39. From Exercise 2(a), $p_1(x) = 1 + x$, $p_2(x) = 1 + x + x^2/2$

(a)

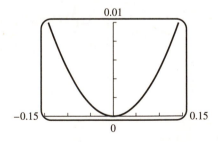

(b)

x	-1.000	-0.750	-0.500	-0.250	0.000	0.250	0.500	0.750	1.000
$f(x)$	0.431	0.506	0.619	0.781	1.000	1.281	1.615	1.977	2.320
$p_1(x)$	0.000	0.250	0.500	0.750	1.000	1.250	1.500	1.750	2.000
$p_2(x)$	0.500	0.531	0.625	0.781	1.000	1.281	1.625	2.031	2.500

(c) $|e^{\sin x} - (1 + x)| < 0.01$
for $-0.14 < x < 0.14$

(d) $|e^{\sin x} - (1 + x + x^2/2)| < 0.01$
for $-0.50 < x < 0.50$

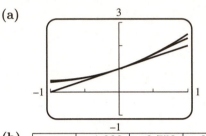

41. (a) $f^{(k)}(x) = e^x \le e^b$,

$$|R_2(x)| \le \frac{e^b b^3}{3!} < 0.0005,$$

$e^b b^3 < 0.003$ if $b \le 0.137$ (by trial and error with a hand calculator), so $[0, 0.137]$.

(b)

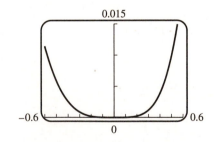

43. $\sin x = x - \dfrac{x^3}{3!} + 0 \cdot x^4 + R_4(x),$

$|R_4(x)| \leq \dfrac{|x|^5}{5!} < 0.5 \times 10^{-3}$ if $|x|^5 < 0.06,$

$|x| < (0.06)^{1/5} \approx 0.569, (-0.569, 0.569)$

45. $f^{(5)}(x) = -\dfrac{3840x^5}{(1+x^2)^6} + \dfrac{3840x^3}{(1+x^2)^5} - \dfrac{720x}{(1+x^2)^4},$
Assume first that $|x| < 1/2,$ then
$|f^{(5)}(x)| < 3840|x|^5 + 3840|x|^3 + 720|x|,$
so let $M = 960,$
$R_4(x) \leq \dfrac{960}{4!}|x|^4 < 0.0005$ if $x < 0.059$

EXERCISE SET 9.8

1. $f^{(k)}(x) = (-1)^k e^{-x},\ f^{(k)}(0) = (-1)^k;\quad \displaystyle\sum_{k=0}^{\infty} \dfrac{(-1)^k}{k!} x^k$

3. $f^{(k)}(0) = 0$ if k is odd, $f^{(k)}(0)$ is alternately π^k and $-\pi^k$ if k is even; $\quad \displaystyle\sum_{k=0}^{\infty} \dfrac{(-1)^k \pi^{2k}}{(2k)!} x^{2k}$

5. $f^{(0)}(0) = 0;$ for $k \geq 1,\ f^{(k)}(x) = \dfrac{(-1)^{k+1}(k-1)!}{(1+x)^k},\ f^{(k)}(0) = (-1)^{k+1}(k-1)!;\quad \displaystyle\sum_{k=1}^{\infty} \dfrac{(-1)^{k+1}}{k} x^k$

7. $f^{(k)}(0) = 0$ if k is odd, $f^{(k)}(0) = 1$ if k is even; $\quad \displaystyle\sum_{k=0}^{\infty} \dfrac{1}{(2k)!} x^{2k}$

9. $f^{(k)}(x) = \begin{cases} (-1)^{k/2}(x\sin x - k\cos x) & k \text{ even} \\ (-1)^{(k-1)/2}(x\cos x + k\sin x) & k \text{ odd} \end{cases},\qquad f^{(k)}(0) = \begin{cases} (-1)^{1+k/2}k & k \text{ even} \\ 0 & k \text{ odd} \end{cases}$

$\displaystyle\sum_{k=0}^{\infty} \dfrac{(-1)^k}{(2k+1)!} x^{2k+2}$

11. $f^{(k)}(x_0) = e;\quad \displaystyle\sum_{k=0}^{\infty} \dfrac{e}{k!}(x-1)^k$

13. $f^{(k)}(x) = \dfrac{(-1)^k k!}{x^{k+1}},\ f^{(k)}(-1) = -k!;\quad \displaystyle\sum_{k=0}^{\infty}(-1)(x+1)^k$

15. $f^{(k)}(1/2) = 0$ if k is odd, $f^{(k)}(1/2)$ is alternatively π^k and $-\pi^k$ if k is even;

$\displaystyle\sum_{k=0}^{\infty} \dfrac{(-1)^k \pi^{2k}}{(2k)!}(x-1/2)^{2k}$

17. $f(1) = 0$, for $k \geq 1$, $f^{(k)}(x) = \dfrac{(-1)^{k-1}(k-1)!}{x^k}$; $f^{(k)}(1) = (-1)^{k-1}(k-1)!$;

$$\sum_{k=1}^{\infty} \frac{(-1)^{k-1}}{k}(x-1)^k$$

19. geometric series, $\rho = \lim\limits_{k \to +\infty} \left| \dfrac{u_{k+1}}{u_k} \right| = |x|$, so the interval of convergence is $-1 < x < 1$, converges

there to $\dfrac{1}{1+x}$ (the series diverges for $x = \pm 1$)

21. geometric series, $\rho = \lim\limits_{k \to +\infty} \left| \dfrac{u_{k+1}}{u_k} \right| = |x-2|$, so the interval of convergence is $1 < x < 3$, converges

there to $\dfrac{1}{1-(x-2)} = \dfrac{1}{3-x}$ (the series diverges for $x = 1, 3$)

23. **(a)** geometric series, $\rho = \lim\limits_{k \to +\infty} \left| \dfrac{u_{k+1}}{u_k} \right| = |x/2|$, so the interval of convergence is $-2 < x < 2$,

converges there to $\dfrac{1}{1+x/2} = \dfrac{2}{2+x}$; (the series diverges for $x = -2, 2$)

(b) $f(0) = 1$; $f(1) = 2/3$

25. true $\qquad\qquad\qquad\qquad\qquad$ **27.** true

29. $\rho = \lim\limits_{k \to +\infty} \dfrac{k+1}{k+2}|x| = |x|$, the series converges if $|x| < 1$ and diverges if $|x| > 1$. If $x = -1$,

$\sum\limits_{k=0}^{\infty} \dfrac{(-1)^k}{k+1}$ converges by the Alternating Series Test; if $x = 1$, $\sum\limits_{k=0}^{\infty} \dfrac{1}{k+1}$ diverges. The radius of

convergence is 1, the interval of convergence is $[-1, 1)$.

31. $\rho = \lim\limits_{k \to +\infty} \dfrac{|x|}{k+1} = 0$, the radius of convergence is $+\infty$, the interval is $(-\infty, +\infty)$.

33. $\rho = \lim\limits_{k \to +\infty} \dfrac{5k^2|x|}{(k+1)^2} = 5|x|$, converges if $|x| < 1/5$ and diverges if $|x| > 1/5$. If $x = -1/5$, $\sum\limits_{k=1}^{\infty} \dfrac{(-1)^k}{k^2}$

converges; if $x = 1/5$, $\sum\limits_{k=1}^{\infty} 1/k^2$ converges. Radius of convergence is $1/5$, interval of convergence is

$[-1/5, 1/5]$.

35. $\rho = \lim\limits_{k \to +\infty} \dfrac{k|x|}{k+2} = |x|$, converges if $|x| < 1$, diverges if $|x| > 1$. If $x = -1$, $\sum\limits_{k=1}^{\infty} \dfrac{(-1)^k}{k(k+1)}$ converges;

if $x = 1$, $\sum\limits_{k=1}^{\infty} \dfrac{1}{k(k+1)}$ converges. Radius of convergence is 1, interval of convergence is $[-1, 1]$.

37. $\rho = \lim\limits_{k \to +\infty} \dfrac{\sqrt{k}}{\sqrt{k+1}}|x| = |x|$, converges if $|x| < 1$, diverges if $|x| > 1$. If $x = -1$, $\sum\limits_{k=1}^{\infty} \dfrac{-1}{\sqrt{k}}$ diverges; if

$x = 1$, $\sum\limits_{k=1}^{\infty} \dfrac{(-1)^{k-1}}{\sqrt{k}}$ converges. Radius of convergence is 1, interval of convergence is $(-1, 1]$.

39. $\rho = \lim\limits_{k \to +\infty} \dfrac{3|x|}{k+1} = 0$, radius of convergence is $+\infty$, interval of convergence is $(-\infty, +\infty)$.

41. $\rho = \lim\limits_{k \to +\infty} \dfrac{1+k^2}{1+(k+1)^2}|x| = |x|$, converges if $|x| < 1$, diverges if $|x| > 1$. If $x = -1$, $\sum\limits_{k=0}^{\infty} \dfrac{(-1)^k}{1+k^2}$ converges; if $x = 1$, $\sum\limits_{k=0}^{\infty} \dfrac{1}{1+k^2}$ converges. Radius of convergence is 1, interval of convergence is $[-1, 1]$.

43. $\rho = \lim\limits_{k \to +\infty} \dfrac{k|x+1|}{k+1} = |x+1|$, converges if $|x+1| < 1$, diverges if $|x+1| > 1$. If $x = -2$, $\sum\limits_{k=1}^{\infty} \dfrac{-1}{k}$ diverges; if $x = 0$, $\sum\limits_{k=1}^{\infty} \dfrac{(-1)^{k+1}}{k}$ converges. Radius of convergence is 1, interval of convergence is $(-2, 0]$.

45. $\rho = \lim\limits_{k \to +\infty} (3/4)|x+5| = \dfrac{3}{4}|x+5|$, converges if $|x+5| < 4/3$, diverges if $|x+5| > 4/3$. If $x = -19/3$, $\sum\limits_{k=0}^{\infty}(-1)^k$ diverges; if $x = -11/3$, $\sum\limits_{k=0}^{\infty} 1$ diverges. Radius of convergence is 4/3, interval of convergence is $(-19/3, -11/3)$.

47. $\rho = \lim\limits_{k \to +\infty} \dfrac{\pi|x-1|^2}{(2k+3)(2k+2)} = 0$, radius of convergence $+\infty$, interval of convergence $(-\infty, +\infty)$.

49. $\rho = \lim\limits_{k \to +\infty} \sqrt[k]{|u_k|} = \lim\limits_{k \to +\infty} \dfrac{|x|}{\ln k} = 0$, the series converges absolutely for all x so the interval of convergence is $(-\infty, +\infty)$.

51. If $x \geq 0$, then $\cos\sqrt{x} = 1 - \dfrac{(\sqrt{x})^2}{2!} + \dfrac{(\sqrt{x})^4}{4!} - \dfrac{(\sqrt{x})^6}{6!} + \cdots = 1 - \dfrac{x}{2!} + \dfrac{x^2}{4!} - \dfrac{x^3}{6!} + \cdots$; if $x \leq 0$, then

$\cosh(\sqrt{-x}) = 1 + \dfrac{(\sqrt{-x})^2}{2!} + \dfrac{(\sqrt{-x})^4}{4!} + \dfrac{(\sqrt{-x})^6}{6!} + \cdots = 1 - \dfrac{x}{2!} + \dfrac{x^2}{4!} - \dfrac{x^3}{6!} + \cdots$.

53. By Exercise 76 of Section 3.6, the derivative of an odd (even) function is even (odd); hence all odd-numbered derivatives of an odd function are even, all even-numbered derivatives of an odd function are odd; a similar statement holds for an even function.

    (a) If $f(x)$ is an even function, then $f^{(2k-1)}(x)$ is an odd function, so $f^{(2k-1)}(0) = 0$, and thus the MacLaurin series coefficients $a_{2k-1} = 0$, $k = 1, 2, \cdots$.

    (b) If $f(x)$ is an odd function, then $f^{(2k)}(x)$ is an odd function, so $f^{(2k)}(0) = 0$, and thus the MacLaurin series coefficients $a_{2k} = 0$, $k = 1, 2, \cdots$.

55. By Theorem 10.4.3(a) the series $\sum(c_k + d_k)(x - x_0)^k$ converges if $|x - x_0| < R$; if $|x - x_0| > R$ then $\sum(c_k + d_k)(x - x_0)^k$ cannot converge, as otherwise $\sum c_k(x - x_0)^k$ would converge by the same Theorem. Hence the radius of convergence of $\sum(c_k + d_k)(x - x_0)^k$ is R.

57. By the Ratio Test for absolute convergence,

$$\rho = \lim_{k \to +\infty} \frac{(pk+p)!(k!)^p}{(pk)![(k+1)!]^p}|x| = \lim_{k \to +\infty} \frac{(pk+p)(pk+p-1)(pk+p-2)\cdots(pk+p-[p-1])}{(k+1)^p}|x|$$

$$= \lim_{k \to +\infty} p\left(p - \frac{1}{k+1}\right)\left(p - \frac{2}{k+1}\right)\cdots\left(p - \frac{p-1}{k+1}\right)|x| = p^p|x|,$$

converges if $|x| < 1/p^p$, diverges if $|x| > 1/p^p$. Radius of convergence is $1/p^p$.

59. Ratio Test: $\rho = \lim\limits_{k \to +\infty} \dfrac{|x|^2}{4(k+1)(k+2)} = 0$, $R = +\infty$

61. (a) $\displaystyle\int_n^{+\infty} \frac{1}{x^{3.7}}dx < 0.005$ if $n > 4.93$; let $n = 5$.

(b) $s_n \approx 1.1062$; $s_n : 1.10628824$

63. By assumption $\displaystyle\sum_{k=0}^{\infty} c_k x^k$ converges if $|x| < R$ so $\displaystyle\sum_{k=0}^{\infty} c_k x^{2k} = \sum_{k=0}^{\infty} c_k(x^2)^k$ converges if $|x^2| < R$,

$|x| < \sqrt{R}$. Moreover, $\displaystyle\sum_{k=0}^{\infty} c_k x^{2k} = \sum_{k=0}^{\infty} c_k(x^2)^k$ diverges if $|x^2| > R$, $|x| > \sqrt{R}$. Thus $\displaystyle\sum_{k=0}^{\infty} c_k x^{2k}$

has radius of convergence $\sqrt{R}$.

EXERCISE SET 9.9

1. $f(x) = \sin x$, $f^{(n+1)}(x) = \pm \sin x$ or $\pm \cos x$, $|f^{(n+1)}(x)| \le 1$, $|R_n(x)| \le \dfrac{|x - \pi/4|^{n+1}}{(n+1)!}$,

$\lim\limits_{n \to +\infty} \dfrac{|x - \pi/4|^{n+1}}{(n+1)!} = 0$; by the Squeezing Theorem, $\lim\limits_{n \to +\infty} |R_n(x)| = 0$

so $\lim\limits_{n \to +\infty} R_n(x) = 0$ for all x.

3. $\sin 4° = \sin\left(\dfrac{\pi}{45}\right) = \dfrac{\pi}{45} - \dfrac{(\pi/45)^3}{3!} + \dfrac{(\pi/45)^5}{5!} - \cdots$

(a) Method 1: $|R_n(\pi/45)| \le \dfrac{(\pi/45)^{n+1}}{(n+1)!} < 0.000005$ for $n + 1 = 4, n = 3$;

$\sin 4° \approx \dfrac{\pi}{45} - \dfrac{(\pi/45)^3}{3!} \approx 0.069756$

(b) Method 2: The first term in the alternating series that is less than 0.000005 is $\dfrac{(\pi/45)^5}{5!}$, so the result is the same as in Part (a).

5. $|R_n(0.1)| \le \dfrac{(0.1)^{n+1}}{(n+1)!} \le 0.000005$ for $n = 3$; $\cos 0.1 \approx 1 - (0.1)^2/2 = 0.99500$, calculator value $0.995004\ldots$

7. Expand about $\pi/2$ to get $\sin x = 1 - \dfrac{1}{2!}(x - \pi/2)^2 + \dfrac{1}{4!}(x - \pi/2)^4 - \cdots$, $85° = 17\pi/36$ radians,

$|R_n(x)| \le \dfrac{|x - \pi/2|^{n+1}}{(n+1)!}$, $|R_n(17\pi/36)| \le \dfrac{|17\pi/36 - \pi/2|^{n+1}}{(n+1)!} = \dfrac{(\pi/36)^{n+1}}{(n+1)!} < 0.5 \times 10^{-4}$

if $n = 3$, $\sin 85° \approx 1 - \dfrac{1}{2}(-\pi/36)^2 \approx 0.99619$, calculator value $0.99619\ldots$

9. $f^{(k)}(x) = \cosh x$ or $\sinh x, |f^{(k)}(x)| \leq \cosh x \leq \cosh 0.5 = \dfrac{1}{2}\left(e^{0.5} + e^{-0.5}\right) < \dfrac{1}{2}(2 + 1) = 1.5$

so $|R_n(x)| < \dfrac{1.5(0.5)^{n+1}}{(n+1)!} \leq 0.5 \times 10^{-3}$ if $n = 4$, $\sinh 0.5 \approx 0.5 + \dfrac{(0.5)^3}{3!} \approx 0.5208$, calculator

value $0.52109\ldots$

11. (a) Let $x = 1/9$ in series (12).

(b) $\ln 1.25 \approx 2\left(1/9 + \dfrac{(1/9)^3}{3}\right) = 2(1/9 + 1/3^7) \approx 0.223$, which agrees with the calculator value

$0.22314\ldots$ to three decimal places.

13. (a) $(1/2)^9/9 < 0.5 \times 10^{-3}$ and $(1/3)^7/7 < 0.5 \times 10^{-3}$ so

$\tan^{-1}(1/2) \approx 1/2 - \dfrac{(1/2)^3}{3} + \dfrac{(1/2)^5}{5} - \dfrac{(1/2)^7}{7} \approx 0.4635$

$\tan^{-1}(1/3) \approx 1/3 - \dfrac{(1/3)^3}{3} + \dfrac{(1/3)^5}{5} \approx 0.3218$

(b) From Formula (16), $\pi \approx 4(0.4635 + 0.3218) = 3.1412$

(c) Let $a = \tan^{-1}\dfrac{1}{2}, b = \tan^{-1}\dfrac{1}{3}$; then $|a - 0.4635| < 0.0005$ and $|b - 0.3218| < 0.0005$, so

$|4(a + b) - 3.1412| \leq 4|a - 0.4635| + 4|b - 0.3218| < 0.004$, so two decimal-place accuracy is

guaranteed, but not three.

15. (a) $\cos x = 1 - \dfrac{x^2}{2!} + \dfrac{x^4}{4!} + (0)x^5 + R_5(x),$ **(b)**

$|R_5(x)| \leq \dfrac{|x|^6}{6!} \leq \dfrac{(0.2)^6}{6!} < 9 \times 10^{-8}$

17. (a) $(1 + x)^{-1} = 1 - x + \dfrac{-1(-2)}{2!}x^2 + \dfrac{-1(-2)(-3)}{3!}x^3 + \cdots + \dfrac{-1(-2)(-3)\cdots(-k)}{k!}x^k + \cdots$

$= \displaystyle\sum_{k=0}^{\infty}(-1)^k x^k$

(b) $(1 + x)^{1/3} = 1 + (1/3)x + \dfrac{(1/3)(-2/3)}{2!}x^2 + \dfrac{(1/3)(-2/3)(-5/3)}{3!}x^3 + \cdots$

$+ \dfrac{(1/3)(-2/3)\cdots(4 - 3k)/3}{k!}x^k + \cdots = 1 + x/3 + \displaystyle\sum_{k=2}^{\infty}(-1)^{k-1}\dfrac{2\cdot 5\cdots(3k - 4)}{3^k k!}x^k$

(c) $(1 + x)^{-3} = 1 - 3x + \dfrac{(-3)(-4)}{2!}x^2 + \dfrac{(-3)(-4)(-5)}{3!}x^3 + \cdots + \dfrac{(-3)(-4)\cdots(-2 - k)}{k!}x^k + \cdots$

$= \displaystyle\sum_{k=0}^{\infty}(-1)^k\dfrac{(k + 2)!}{2\cdot k!}x^k = \sum_{k=0}^{\infty}(-1)^k\dfrac{(k + 2)(k + 1)}{2}x^k$

19. **(a)** $\frac{d}{dx}\ln(1+x) = \frac{1}{1+x}, \frac{d^k}{dx^k}\ln(1+x) = (-1)^{k-1}\frac{(k-1)!}{(1+x)^k}$; similarly $\frac{d}{dx}\ln(1-x) = -\frac{(k-1)!}{(1-x)^k}$,

so $f^{(n+1)}(x) = n!\left[\frac{(-1)^n}{(1+x)^{n+1}} + \frac{1}{(1-x)^{n+1}}\right]$.

(b) $\left|f^{(n+1)}(x)\right| \le n!\left|\frac{(-1)^n}{(1+x)^{n+1}}\right| + n!\left|\frac{1}{(1-x)^{n+1}}\right| = n!\left[\frac{1}{(1+x)^{n+1}} + \frac{1}{(1-x)^{n+1}}\right]$

(c) If $\left|f^{(n+1)}(x)\right| \le M$ on the interval $[0, 1/3]$ then $|R_n(1/3)| \le \frac{M}{(n+1)!}\left(\frac{1}{3}\right)^{n+1}$.

(d) If $0 \le x \le 1/3$ then $1+x \ge 1, 1-x \ge 2/3, \left|f^{(n+1)}(x)\right| \le M = n!\left[1 + \frac{1}{(2/3)^{n+1}}\right]$.

(e) $0.000005 \ge \frac{M}{(n+1)!}\left(\frac{1}{3}\right)^{n+1} = \frac{1}{n+1}\left[\left(\frac{1}{3}\right)^{n+1} + \frac{(1/3)^{n+1}}{(2/3)^{n+1}}\right] = \frac{1}{n+1}\left[\left(\frac{1}{3}\right)^{n+1} + \left(\frac{1}{2}\right)^{n+1}\right]$

By inspection the inequality holds for $n = 13$ but for no smaller n.

21. $f(x) = \cos x, f^{(n+1)}(x) = \pm\sin x$ or $\pm\cos x, \left|f^{(n+1)}(x)\right| \le 1$, set $M = 1$,

$|R_n(x)| \le \frac{1}{(n+1)!}|x - x_0|^{n+1}$, $\lim\limits_{n\to+\infty}\frac{|x - x_0|^{n+1}}{(n+1)!} = 0$ so $\lim\limits_{n\to+\infty} R_n(x) = 0$ for all x.

23. $e^{-x} = 1 - x + x^2/2! + \cdots$. Replace x with $\left(\frac{x-100}{16}\right)^2/2$ to obtain

$e^{-\left(\frac{x-100}{16}\right)^2/2} = 1 - \frac{(x-100)^2}{2\cdot16^2} + \frac{(x-100)^4}{8\cdot16^4} + \cdots$, thus

$p \approx \frac{1}{16\sqrt{2\pi}}\int_{100}^{110}\left[1 - \frac{(x-100)^2}{2\cdot16^2} + \frac{(x-100)^4}{8\cdot16^4}\right]dx \approx 0.23406$ or 23.406%.

EXERCISE SET 9.10

1. **(a)** Replace x with $-x$: $\frac{1}{1+x} = 1 - x + x^2 - \cdots + (-1)^k x^k + \cdots$; $R = 1$.

(b) Replace x with x^2 : $\frac{1}{1-x^2} = 1 + x^2 + x^4 + \cdots + x^{2k} + \cdots$; $R = 1$.

(c) Replace x with $2x$: $\frac{1}{1-2x} = 1 + 2x + 4x^2 + \cdots + 2^k x^k + \cdots$; $R = 1/2$.

(d) $\frac{1}{2-x} = \frac{1/2}{1-x/2}$; replace x with $x/2$: $\frac{1}{2-x} = \frac{1}{2} + \frac{1}{2^2}x + \frac{1}{2^3}x^2 + \cdots + \frac{1}{2^{k+1}}x^k + \cdots$; $R = 2$.

3. **(a)** From Section 9.9, Example 4(b), $\frac{1}{\sqrt{1+x}} = 1 - \frac{1}{2}x + \frac{1\cdot3}{2^2\cdot2!}x^2 - \frac{1\cdot3\cdot5}{2^3\cdot3!}x^3 + \cdots$, so

$(2+x)^{-1/2} = \frac{1}{\sqrt{2}\sqrt{1+x/2}} = \frac{1}{2^{1/2}} - \frac{1}{2^{5/2}}x + \frac{1\cdot3}{2^{9/2}\cdot2!}x^2 - \frac{1\cdot3\cdot5}{2^{13/2}\cdot3!}x^3 + \cdots$

(b) Example 4(a): $\frac{1}{(1+x)^2} = 1 - 2x + 3x^2 - 4x^3 + \cdots$, so $\frac{1}{(1-x^2)^2} = 1 + 2x^2 + 3x^4 + 4x^6 + \cdots$

5. **(a)** $2x - \frac{2^3}{3!}x^3 + \frac{2^5}{5!}x^5 - \frac{2^7}{7!}x^7 + \cdots$; $R = +\infty$

(b) $1 - 2x + 2x^2 - \frac{4}{3}x^3 + \cdots; \; R = +\infty$

(c) $1 + x^2 + \frac{1}{2!}x^4 + \frac{1}{3!}x^6 + \cdots; \; R = +\infty$

(d) $x^2 - \frac{\pi^2}{2}x^4 + \frac{\pi^4}{4!}x^6 - \frac{\pi^6}{6!}x^8 + \cdots; \; R = +\infty$

7. **(a)** $x^2 \left(1 - 3x + 9x^2 - 27x^3 + \cdots\right) = x^2 - 3x^3 + 9x^4 - 27x^5 + \cdots; \; R = 1/3$

(b) $x \left(2x + \frac{2^3}{3!}x^3 + \frac{2^5}{5!}x^5 + \frac{2^7}{7!}x^7 + \cdots\right) = 2x^2 + \frac{2^3}{3!}x^4 + \frac{2^5}{5!}x^6 + \frac{2^7}{7!}x^8 + \cdots; \; R = +\infty$

(c) Substitute $3/2$ for m and $-x^2$ for x in Equation (17) of Section 9.9, then multiply by x:

$x - \frac{3}{2}x^3 + \frac{3}{8}x^5 + \frac{1}{16}x^7 + \cdots; \; R = 1$

9. **(a)** $\sin^2 x = \frac{1}{2}(1 - \cos 2x) = \frac{1}{2}\left[1 - \left(1 - \frac{2^2}{2!}x^2 + \frac{2^4}{4!}x^4 - \frac{2^6}{6!}x^6 + \cdots\right)\right]$

$= x^2 - \frac{2^3}{4!}x^4 + \frac{2^5}{6!}x^6 - \frac{2^7}{8!}x^8 + \cdots$

(b) $\ln\left[(1 + x^3)^{12}\right] = 12\ln(1 + x^3) = 12x^3 - 6x^6 + 4x^9 - 3x^{12} + \cdots$

11. **(a)** $\frac{1}{x} = \frac{1}{1 - (1 - x)} = 1 + (1 - x) + (1 - x)^2 + \cdots + (1 - x)^k + \cdots$

$= 1 - (x - 1) + (x - 1)^2 - \cdots + (-1)^k(x - 1)^k + \cdots$

(b) $(0, 2)$

13. **(a)** $(1 + x + x^2/2 + x^3/3! + x^4/4! + \cdots)(x - x^3/3! + x^5/5! - \cdots) = x + x^2 + x^3/3 - x^5/30 + \cdots$

(b) $(1 + x/2 - x^2/8 + x^3/16 - (5/128)x^4 + \cdots)(x - x^2/2 + x^3/3 - x^4/4 + x^5/5 - \cdots)$

$= x - x^3/24 + x^4/24 - (71/1920)x^5 + \cdots$

15. **(a)** $\frac{1}{\cos x} = 1 \Big/ \left(1 - \frac{1}{2!}x^2 + \frac{1}{4!}x^4 - \frac{1}{6!}x^6 + \cdots\right) = 1 + \frac{1}{2}x^2 + \frac{5}{24}x^4 + \frac{61}{720}x^6 + \cdots$

(b) $\frac{\sin x}{e^x} = \left(x - \frac{x^3}{3!} + \frac{x^5}{5!} - \cdots\right) \Big/ \left(1 + x + \frac{x^2}{2!} + \frac{x^3}{3!} + \frac{x^4}{4!} + \cdots\right) = x - x^2 + \frac{1}{3}x^3 - \frac{1}{30}x^5 + \cdots$

17. $e^x = 1 + x + x^2/2 + x^3/3! + \cdots + x^k/k! + \cdots, \; e^{-x} = 1 - x + x^2/2 - x^3/3! + \cdots + (-1)^k x^k/k! + \cdots;$

$\sinh x = \frac{1}{2}\left(e^x - e^{-x}\right) = x + x^3/3! + x^5/5! + \cdots + x^{2k+1}/(2k + 1)! + \cdots, R = +\infty$

$\cosh x = \frac{1}{2}\left(e^x + e^{-x}\right) = 1 + x^2/2 + x^4/4! + \cdots + x^{2k}/(2k)! + \cdots, R = +\infty$

19. $\frac{4x - 2}{x^2 - 1} = \frac{-1}{1 - x} + \frac{3}{1 + x} = -\left(1 + x + x^2 + x^3 + x^4 + \cdots\right) + 3\left(1 - x + x^2 - x^3 + x^4 + \cdots\right)$

$= 2 - 4x + 2x^2 - 4x^3 + 2x^4 + \cdots$

21. (a) $\dfrac{d}{dx}\left(1 - x^2/2! + x^4/4! - x^6/6! + \cdots\right) = -x + x^3/3! - x^5/5! + \cdots = -\sin x$

(b) $\dfrac{d}{dx}\left(x - x^2/2 + x^3/3 - \cdots\right) = 1 - x + x^2 - \cdots = 1/(1+x)$

23. (a) $\displaystyle\int \left(1 + x + x^2/2! + \cdots\right) dx = \left(x + x^2/2! + x^3/3! + \cdots\right) + C_1$

$$= \left(1 + x + x^2/2! + x^3/3! + \cdots\right) + C_1 - 1 = e^x + C$$

(b) $\displaystyle\int \left(x + x^3/3! + x^5/5! + \cdots\right) = x^2/2! + x^4/4! + \cdots + C_1$

$$= 1 + x^2/2! + x^4/4! + \cdots + C_1 - 1 = \cosh x + C$$

25. $\dfrac{d}{dx}\displaystyle\sum_{k=0}^{\infty} \dfrac{x^{k+1}}{(k+1)(k+2)} = \sum_{k=0}^{\infty} \dfrac{x^k}{k+2}$. Each series has radius of convergence $\rho = 1$, as can be seen from the Ratio Test.

27. (a) Substitute x^2 for x in the Maclaurin Series for $1/(1-x)$ (Table 9.9.1)

and then multiply by x: $\dfrac{x}{1-x^2} = x\displaystyle\sum_{k=0}^{\infty}(x^2)^k = \sum_{k=0}^{\infty} x^{2k+1}$

(b) $f^{(5)}(0) = 5!c_5 = 5!$, $f^{(6)}(0) = 6!c_6 = 0$ **(c)** $f^{(n)}(0) = n!c_n = \begin{cases} n! & \text{if } n \text{ odd} \\ 0 & \text{if } n \text{ even} \end{cases}$

29. (a) $\displaystyle\lim_{x \to 0} \dfrac{\sin x}{x} = \lim_{x \to 0}\left(1 - x^2/3! + x^4/5! - \cdots\right) = 1$

(b) $\displaystyle\lim_{x \to 0} \dfrac{\tan^{-1} x - x}{x^3} = \lim_{x \to 0} \dfrac{\left(x - x^3/3 + x^5/5 - x^7/7 + \cdots\right) - x}{x^3} = -1/3$

31. $\displaystyle\int_0^1 \sin\left(x^2\right) dx = \int_0^1 \left(x^2 - \dfrac{1}{3!}x^6 + \dfrac{1}{5!}x^{10} - \dfrac{1}{7!}x^{14} + \cdots\right) dx$

$$= \dfrac{1}{3}x^3 - \dfrac{1}{7 \cdot 3!}x^7 + \dfrac{1}{11 \cdot 5!}x^{11} - \dfrac{1}{15 \cdot 7!}x^{15} + \cdots \Bigg]_0^1$$

$$= \dfrac{1}{3} - \dfrac{1}{7 \cdot 3!} + \dfrac{1}{11 \cdot 5!} - \dfrac{1}{15 \cdot 7!} + \cdots,$$

but $\dfrac{1}{15 \cdot 7!} < 0.5 \times 10^{-3}$ so $\displaystyle\int_0^1 \sin(x^2)dx \approx \dfrac{1}{3} - \dfrac{1}{7 \cdot 3!} + \dfrac{1}{11 \cdot 5!} \approx 0.3103$

33. $\displaystyle\int_0^{0.2} \left(1 + x^4\right)^{1/3} dx = \int_0^{0.2} \left(1 + \dfrac{1}{3}x^4 - \dfrac{1}{9}x^8 + \cdots\right) dx$

$$= x + \dfrac{1}{15}x^5 - \dfrac{1}{81}x^9 + \cdots \Bigg]_0^{0.2} = 0.2 + \dfrac{1}{15}(0.2)^5 - \dfrac{1}{81}(0.2)^9 + \cdots,$$

but $\dfrac{1}{15}(0.2)^5 < 0.5 \times 10^{-3}$ so $\displaystyle\int_0^{0.2} \left(1 + x^4\right)^{1/3}dx \approx 0.200$

35. (a) Substitute x^4 for x in the MacLaurin Series for e^x to obtain $\displaystyle\sum_{k=0}^{+\infty}\frac{x^{4k}}{k!}$. The radius of convergence is $R=+\infty$.

(b) The first method is to multiply the MacLaurin Series for e^{x^4} by x^3: $\displaystyle x^3 e^{x^4}=\sum_{k=0}^{+\infty}\frac{x^{4k+3}}{k!}$. The second method involves differentiation: $\dfrac{d}{dx}e^{x^4}=4x^3 e^{x^4}$, so

$$x^3 e^{x^4}=\frac{1}{4}\frac{d}{dx}e^{x^4}=\frac{1}{4}\frac{d}{dx}\sum_{k=0}^{+\infty}\frac{x^{4k}}{k!}=\frac{1}{4}\sum_{k=0}^{+\infty}\frac{4kx^{4k-1}}{k!}=\sum_{k=0}^{+\infty}\frac{x^{4k-1}}{(k-1)!}.$$ Use the change of variable $j=k-1$ to show equality of the two series.

37. (a) In Exercise 36(a), set $x=\dfrac{1}{3}$, $S=\dfrac{1/3}{(1-1/3)^2}=\dfrac{3}{4}$

(b) In Part (b) set $x=1/4, S=\ln(4/3)$

39. (a) $\displaystyle\sinh^{-1}x=\int\left(1+x^2\right)^{-1/2}dx-C=\int\left(1-\frac{1}{2}x^2+\frac{3}{8}x^4-\frac{5}{16}x^6+\cdots\right)dx-C$

$$=\left(x-\frac{1}{6}x^3+\frac{3}{40}x^5-\frac{5}{112}x^7+\cdots\right)-C;\ \sinh^{-1}0=0\text{ so }C=0.$$

(b) $\displaystyle\left(1+x^2\right)^{-1/2}=1+\sum_{k=1}^{\infty}\frac{(-1/2)(-3/2)(-5/2)\cdots(-1/2-k+1)}{k!}(x^2)^k$

$$=1+\sum_{k=1}^{\infty}(-1)^k\frac{1\cdot3\cdot5\cdots(2k-1)}{2^k k!}x^{2k},$$

$$\sinh^{-1}x=x+\sum_{k=1}^{\infty}(-1)^k\frac{1\cdot3\cdot5\cdots(2k-1)}{2^k k!(2k+1)}x^{2k+1}$$

(c) $R=1$

41. (a) $\displaystyle y(t)=y_0\sum_{k=0}^{\infty}\frac{(-1)^k(0.000121)^k t^k}{k!}$

(b) $y(1)\approx y_0\left(1-0.000121t\right)\Big]_{t=1}=0.999879y_0$

(c) $y_0 e^{-0.000121}\approx0.9998790073y_0$

43. The third order model gives the same result as the second, because there is no term of degree three in (8). By the Wallis sine formula, $\displaystyle\int_0^{\pi/2}\sin^4\phi\,d\phi=\frac{1\cdot3}{2\cdot4}\frac{\pi}{2}$, and

$$T\approx4\sqrt{\frac{L}{g}}\int_0^{\pi/2}\left(1+\frac{1}{2}k^2\sin^2\phi+\frac{1\cdot3}{2^2 2!}k^4\sin^4\phi\right)d\phi=4\sqrt{\frac{L}{g}}\left(\frac{\pi}{2}+\frac{k^2}{2}\frac{\pi}{4}+\frac{3k^4}{8}\frac{3\pi}{16}\right)$$

$$=2\pi\sqrt{\frac{L}{g}}\left(1+\frac{k^2}{4}+\frac{9k^4}{64}\right)$$

45. (a) We can differentiate term-by-term:

$$y' = \sum_{k=1}^{\infty} \frac{(-1)^k x^{2k-1}}{2^{2k-1} k!(k-1)!} = \sum_{k=0}^{\infty} \frac{(-1)^{k+1} x^{2k+1}}{2^{2k+1}(k+1)!k!}, \quad y'' = \sum_{k=0}^{\infty} \frac{(-1)^{k+1}(2k+1)x^{2k}}{2^{2k+1}(k+1)!k!}, \text{ and}$$

$$xy'' + y' + xy = \sum_{k=0}^{\infty} \frac{(-1)^{k+1}(2k+1)x^{2k+1}}{2^{2k+1}(k+1)!k!} + \sum_{k=0}^{\infty} \frac{(-1)^{k+1} x^{2k+1}}{2^{2k+1}(k+1)!k!} + \sum_{k=0}^{\infty} \frac{(-1)^k x^{2k+1}}{2^{2k}(k!)^2},$$

$$xy'' + y' + xy = \sum_{k=0}^{\infty} \frac{(-1)^{k+1} x^{2k+1}}{2^{2k}(k!)^2} \left[\frac{2k+1}{2(k+1)} + \frac{1}{2(k+1)} - 1 \right] = 0.$$

(b)　$y' = \displaystyle\sum_{k=0}^{\infty} \frac{(-1)^k (2k+1)x^{2k}}{2^{2k+1} k!(k+1)!}, \quad y'' = \sum_{k=1}^{\infty} \frac{(-1)^k (2k+1)x^{2k-1}}{2^{2k}(k-1)!(k+1)!}.$

Since $J_1(x) = \displaystyle\sum_{k=0}^{\infty} \frac{(-1)^k x^{2k+1}}{2^{2k+1} k!(k+1)!}$ and $x^2 J_1(x) = \displaystyle\sum_{k=1}^{\infty} \frac{(-1)^{k-1} x^{2k+1}}{2^{2k-1}(k-1)!k!}$, it follows that

$$x^2 y'' + xy' + (x^2 - 1)y$$

$$= \sum_{k=1}^{\infty} \frac{(-1)^k (2k+1)x^{2k+1}}{2^{2k}(k-1)!(k+1)!} + \sum_{k=0}^{\infty} \frac{(-1)^k (2k+1)x^{2k+1}}{2^{2k+1}(k!)(k+1)!} + \sum_{k=1}^{\infty} \frac{(-1)^{k-1} x^{2k+1}}{2^{2k-1}(k-1)!k!}$$

$$- \sum_{k=0}^{\infty} \frac{(-1)^k x^{2k+1}}{2^{2k+1} k!(k+1)!}$$

$$= \frac{x}{2} - \frac{x}{2} + \sum_{k=1}^{\infty} \frac{(-1)^k x^{2k+1}}{2^{2k-1}(k-1)!k!} \left(\frac{2k+1}{2(k+1)} + \frac{2k+1}{4k(k+1)} - 1 - \frac{1}{4k(k+1)} \right) = 0.$$

(c)　From Part (a), $J_0'(x) = \displaystyle\sum_{k=0}^{\infty} \frac{(-1)^{k+1} x^{2k+1}}{2^{2k+1}(k+1)!k!} = -J_1(x).$

REVIEW EXERCISES, CHAPTER 9

7.　The series converges for $|x - x_0| < R$ and may or may not converge at $x = x_0 \pm R$.

9.　(a)　always true by Theorem 9.4.2

(b)　sometimes false, for example the harmonic series diverges but $\sum(1/k^2)$ converges

(c)　sometimes false, for example $f(x) = \sin \pi x, a_k = 0, L = 0$

(d)　always true by the comments which follow Example 3(d) of Section 9.1

(e)　sometimes false, for example $a_n = \dfrac{1}{2} + (-1)^n \dfrac{1}{4}$

(f)　sometimes false, for example $u_k = 1/2$

(g)　always false by Theorem 9.4.3

(h)　sometimes false, for example $u_k = 1/k, v_k = 2/k$

(i)　always true by the Comparison Test

(j)　always true by the Comparison Test

(k)　sometimes false, for example $\sum(-1)^k/k$

(l)　sometimes false, for example $\sum(-1)^k/k$

11.　(a)　$a_n = \dfrac{n+2}{(n+1)^2 - n^2} = \dfrac{n+2}{((n+1)+n)((n+1)-n)} = \dfrac{n+2}{2n+1}$, limit $= 1/2$.

(b) $a_n = (-1)^{n-1}\dfrac{n}{2n+1}$, limit does not exist because of alternating signs

13. (a) $a_{n+1}/a_n = (n+1-10)^4/(n-10)^4 = (n-9)^4/(n-10)^4$. Since $n-9 > n-10$ for all n it follows that $(n-9)^4 > (n-10)^4$ and thus that $a_{n+1}/a_n > 1$ for all n, hence the sequence is strictly monotone increasing.

(b) $\dfrac{100^{n+1}}{(2(n+1))!(n+1)!} \times \dfrac{(2n)!n!}{100^n} = \dfrac{100}{(2n+2)(2n+1)(n+1)} < 1$ for $n \geq 3$, so the sequence is ultimately strictly monotone decreasing.

15. (a) geometric, $r = 1/5$, converges $\qquad$ **(b)** $1/(5^k + 1) < 1/5^k$, converges

17. (a) $\dfrac{1}{k^3 + 2k + 1} < \dfrac{1}{k^3}$, $\displaystyle\sum_{k=1}^{\infty} 1/k^3$ converges, so $\displaystyle\sum_{k=1}^{\infty} \dfrac{1}{k^3 + 2k + 1}$ converges by the Comparison Test

(b) Limit Comparison Test, compare with the divergent series $\displaystyle\sum_{k=1}^{\infty} \dfrac{1}{k^{2/5}}$, diverges

19. (a) $\dfrac{9}{\sqrt{k}+1} \geq \dfrac{9}{\sqrt{k}+\sqrt{k}} = \dfrac{9}{2\sqrt{k}}$, $\displaystyle\sum_{k=1}^{\infty} \dfrac{9}{2\sqrt{k}}$ diverges

(b) converges absolutely, because $\left|\dfrac{\cos(1/k)}{k^2}\right| \leq \dfrac{1}{k^2}$ and $\displaystyle\sum_{k=1}^{+\infty} \dfrac{1}{k^2}$ converges

21. $\displaystyle\sum_{k=0}^{\infty} \dfrac{1}{5^k} - \sum_{k=0}^{99} \dfrac{1}{5^k} = \sum_{k=100}^{\infty} \dfrac{1}{5^k} = \dfrac{1}{5^{100}} \sum_{k=0}^{\infty} \dfrac{1}{5^k} = \dfrac{1}{4 \cdot 5^{99}}$

23. (a) $\displaystyle\sum_{k=1}^{\infty} \left(\dfrac{3}{2^k} - \dfrac{2}{3^k}\right) = \sum_{k=1}^{\infty} \dfrac{3}{2^k} - \sum_{k=1}^{\infty} \dfrac{2}{3^k} = \left(\dfrac{3}{2}\right) \dfrac{1}{1-(1/2)} - \left(\dfrac{2}{3}\right) \dfrac{1}{1-(1/3)} = 2$ (geometric series)

(b) $\displaystyle\sum_{k=1}^{n} [\ln(k+1) - \ln k] = \ln(n+1)$, so $\displaystyle\sum_{k=1}^{\infty} [\ln(k+1) - \ln k] = \lim_{n \to +\infty} \ln(n+1) = +\infty$, diverges

(c) $\displaystyle\lim_{n \to +\infty} \sum_{k=1}^{n} \dfrac{1}{2}\left(\dfrac{1}{k} - \dfrac{1}{k+2}\right) = \lim_{n \to +\infty} \dfrac{1}{2}\left(1 + \dfrac{1}{2} - \dfrac{1}{n+1} - \dfrac{1}{n+2}\right) = \dfrac{3}{4}$

(d) $\displaystyle\lim_{n \to +\infty} \sum_{k=1}^{n} \left[\tan^{-1}(k+1) - \tan^{-1} k\right] = \lim_{n \to +\infty} \left[\tan^{-1}(n+1) - \tan^{-1}(1)\right] = \dfrac{\pi}{2} - \dfrac{\pi}{4} = \dfrac{\pi}{4}$

25. Compare with $1/k^p$: converges if $p > 1$, diverges otherwise.

27. (a) $1 \leq k, 2 \leq k, 3 \leq k, \ldots, k \leq k$, therefore $1 \cdot 2 \cdot 3 \cdots k \leq k \cdot k \cdot k \cdots k$, or $k! \leq k^k$.

(b) $\displaystyle\sum \dfrac{1}{k^k} \leq \sum \dfrac{1}{k!}$, converges

(c) $\displaystyle\lim_{k \to +\infty} \left(\dfrac{1}{k^k}\right)^{1/k} = \lim_{k \to +\infty} \dfrac{1}{k} = 0$, converges

29. (a) $p_0(x) = 1, p_1(x) = 1 - 7x, p_2(x) = 1 - 7x + 5x^2, p_3(x) = 1 - 7x + 5x^2 + 4x^3,$ $p_4(x) = 1 - 7x + 5x^2 + 4x^3$

(b) If $f(x)$ is a polynomial of degree n and $k \geq n$ then the Maclaurin polynomial of degree k is the polynomial itself; if $k < n$ then it is the truncated polynomial.

31. $\ln(1+x) = x - x^2/2 + \cdots$; so $|\ln(1+x) - x| \leq x^2/2$ by Theorem 9.6.2.

33. **(a)** $e^2 - 1$ **(b)** $\sin \pi = 0$ **(c)** $\cos e$ **(d)** $e^{-\ln 3} = 1/3$

35. $(27+x)^{1/3} = 3(1+x/3^3)^{1/3} = 3\left(1 + \dfrac{1}{3^4}x - \dfrac{1 \cdot 2}{3^8 2}x^2 + \dfrac{1 \cdot 2 \cdot 5}{3^{12} 3!}x^3 + \cdots\right)$, alternates after first term,

$\dfrac{3 \cdot 2}{3^8 2} < 0.0005$, $\sqrt[3]{28} \approx 3\left(1 + \dfrac{1}{3^4}\right) \approx 3.0370$

37. Both (a) and (b): $x - \dfrac{2}{3}x^3 + \dfrac{2}{15}x^5 - \dfrac{4}{315}x^7$

MAKING CONNECTIONS, CHAPTER 9

1. $P_0P_1 = a\sin\theta,$
$P_1P_2 = a\sin\theta\cos\theta,$
$P_2P_3 = a\sin\theta\cos^2\theta,$
$P_3P_4 = a\sin\theta\cos^3\theta,\ldots$
(see figure)
Each sum is a geometric series.

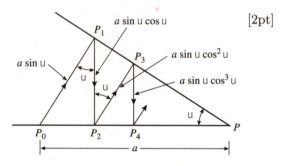

                                                             [2pt]

(a) $P_0P_1 + P_1P_2 + P_2P_3 + \cdots = a\sin\theta + a\sin\theta\cos\theta + a\sin\theta\cos^2\theta + \cdots = \dfrac{a\sin\theta}{1-\cos\theta}$

(b) $P_0P_1 + P_2P_3 + P_4P_5 + \cdots = a\sin\theta + a\sin\theta\cos^2\theta + a\sin\theta\cos^4\theta + \cdots$

$$= \dfrac{a\sin\theta}{1-\cos^2\theta} = \dfrac{a\sin\theta}{\sin^2\theta} = a\csc\theta$$

(c) $P_1P_2 + P_3P_4 + P_5P_6 + \cdots = a\sin\theta\cos\theta + a\sin\theta\cos^3\theta + \cdots$

$$= \dfrac{a\sin\theta\cos\theta}{1-\cos^2\theta} = \dfrac{a\sin\theta\cos\theta}{\sin^2\theta} = a\cot\theta$$

3. $\sum(1/k^p)$ converges if $p > 1$ and diverges if $p \leq 1$, so $\displaystyle\sum_{k=1}^{\infty}(-1)^k\dfrac{1}{k^p}$ converges absolutely if $p > 1$,

and converges conditionally if $0 < p \leq 1$ since it satisfies the Alternating Series Test; it diverges for $p \leq 0$ since $\displaystyle\lim_{k\to+\infty} a_k \neq 0$.

5. $\left(1 - \dfrac{v^2}{c^2}\right)^{-1/2} \approx 1 + \dfrac{v^2}{2c^2}$, so $K = m_0c^2\left[\dfrac{1}{\sqrt{1-v^2/c^2}} - 1\right] \approx m_0c^2(v^2/2c^2) = m_0v^2/2$

CHAPTER 10
Parametric and Polar Curves; Conic Sections

EXERCISE SET 10.1

1. **(a)** $x + 1 = t = y - 1$, $y = x + 2$

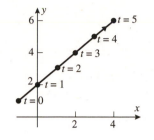

(c)

t	0	1	2	3	4	5
x	-1	0	1	2	3	4
y	1	2	3	4	5	6

3. $t = (x + 4)/3$; $y = 2x + 10$

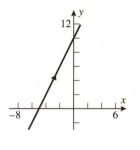

5. $\cos t = x/2$, $\sin t = y/5$; $x^2/4 + y^2/25 = 1$

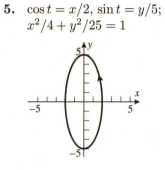

7. $\cos t = (x - 3)/2$, $\sin t = (y - 2)/4$; $(x - 3)^2/4 + (y - 2)^2/16 = 1$

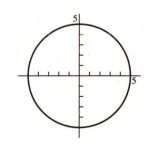

9. $\cos 2t = 1 - 2\sin^2 t$; $x = 1 - 2y^2$, $-1 \le y \le 1$

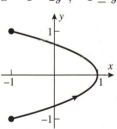

11. $x/2 + y/3 = 1$, $0 \le x \le 2$, $0 \le y \le 3$

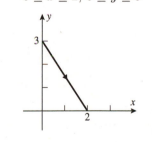

13. $x = 5\cos t$, $y = -5\sin t$, $0 \le t \le 2\pi$

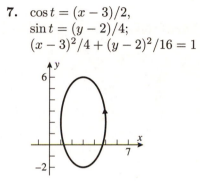

15. $x = 2$, $y = t$

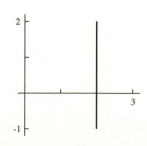

17. $x = t^2$, $y = t$, $-1 \le t \le 1$

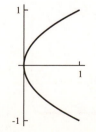

19. (a)

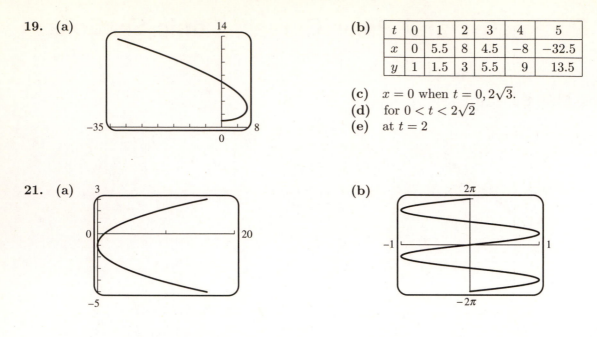

(b)

t	0	1	2	3	4	5
x	0	5.5	8	4.5	-8	-32.5
y	1	1.5	3	5.5	9	13.5

(c) $x = 0$ when $t = 0, 2\sqrt{3}$.
(d) for $0 < t < 2\sqrt{2}$
(e) at $t = 2$

21. (a)

(b)

23. (a) IV, because x always increases whereas y oscillates.

(b) II, because $(x/2)^2 + (y/3)^2 = 1$, an ellipse.

(c) V, because $x^2 + y^2 = t^2$ increases in magnitude while x and y keep changing sign.

(d) VI; examine the cases $t < -1$ and $t > -1$ and you see the curve lies in the first, second and fourth quadrants only.

(e) III, because $y > 0$.

(f) I: Since x and y are bounded, the answer must be I or II; since $x = y = 0$ when $t = \pi/2$, the curve passes through the origin, so it must be I.

25. (a) $|R-P|^2 = (x-x_0)^2+(y-y_0)^2 = t^2[(x_1-x_0)^2+(y_1-y_0)^2]$ and $|Q-P|^2 = (x_1-x_0)^2+(y_1-y_0)^2$, so $r = |R - P| = |Q - P|t = qt$.

(b) $t = 1/2$ **(c)** $t = 3/4$

27. (a) Eliminate $\dfrac{t - t_0}{t_1 - t_0}$ from the parametric equations to obtain $\dfrac{y - y_0}{x - x_0} = \dfrac{y_1 - y_0}{x_1 - x_0}$, which is an equation of the line through the 2 points.

(b) from (x_0, y_0) to (x_1, y_1)

(c) $x = 3 - 2(t - 1)$, $y = -1 + 5(t - 1)$

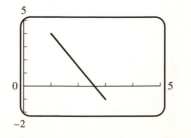

29.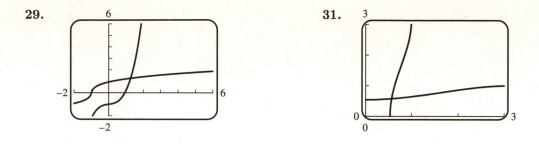

31.

33. False. The parametric curve only gives the part of $y = 1 - x^2$ with $-1 \le x \le 1$.

35. True. By equation (4), $\dfrac{dy}{dx} = \dfrac{dy/dt}{dx/dt} = \dfrac{12t^3 - 6t^2}{x'(t)}$.

37.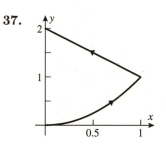

39. **(a)** $x = 4\cos t,\ y = 3\sin t$ **(b)** $x = -1 + 4\cos t,\ y = 2 + 3\sin t$

(c)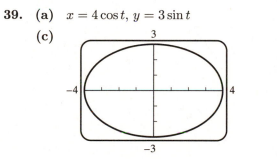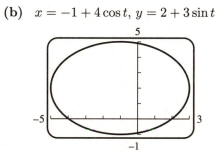

41. **(a)** $dy/dx = \dfrac{2t}{1/2} = 4t;\ dy/dx\big|_{t=-1} = -4;\ dy/dx\big|_{t=1} = 4$

(b) $y = (2x)^2 + 1,\ dy/dx = 8x,\ dy/dx\big|_{x=\pm(1/2)} = \pm 4$

43. From Exercise 41(b), $\dfrac{dy}{dx} = 4t$ so $\dfrac{d^2y}{dx^2} = \dfrac{d}{dt}\left(\dfrac{dy}{dx}\right) \Big/ \dfrac{dx}{dt} = \dfrac{4}{1/2} = 8$. The sign of $\dfrac{d^2y}{dx^2}$ is positive for all t, including $t = \pm 1$.

45. $\dfrac{dy}{dx} = \dfrac{2}{1/(2\sqrt{t})} = 4\sqrt{t},\ \dfrac{d^2y}{dx^2} = \dfrac{d}{dt}\left(\dfrac{dy}{dx}\right)\Big/ \dfrac{dx}{dt} = \dfrac{2/\sqrt{t}}{1/(2\sqrt{t})} = 4,\ \dfrac{dy}{dx}\bigg|_{t=1} = 4,\ \dfrac{d^2y}{dx^2}\bigg|_{t=1} = 4$

47. $\dfrac{dy}{dx} = \dfrac{\sec^2 t}{\sec t \tan t} = \csc t,\ \dfrac{d^2y}{dx^2} = \dfrac{d}{dt}\left(\dfrac{dy}{dx}\right)\Big/ \dfrac{dx}{dt} = \dfrac{-\csc t \cot t}{\sec t \tan t} = -\cot^3 t,$

$\dfrac{dy}{dx}\bigg|_{t=\pi/3} = \dfrac{2}{\sqrt{3}},\ \dfrac{d^2y}{dx^2}\bigg|_{t=\pi/3} = -\dfrac{1}{3\sqrt{3}}$

49. $\dfrac{dy}{dx} = \dfrac{dy/d\theta}{dx/d\theta} = \dfrac{\cos\theta}{1-\sin\theta};$

$\dfrac{d^2y}{dx^2} = \dfrac{d}{d\theta}\left(\dfrac{dy}{dx}\right) \Big/ \dfrac{dx}{d\theta} = \dfrac{(1-\sin\theta)(-\sin\theta)+\cos^2\theta}{(1-\sin\theta)^2}\,\dfrac{1}{1-\sin\theta} = \dfrac{1}{(1-\sin\theta)^2};$

$\dfrac{dy}{dx}\Big|_{\theta=\pi/6} = \dfrac{\sqrt{3}/2}{1-1/2} = \sqrt{3};\ \dfrac{d^2y}{dx^2}\Big|_{\theta=\pi/6} = \dfrac{1}{(1-1/2)^2} = 4$

51. (a) $dy/dx = \dfrac{-e^{-t}}{e^t} = -e^{-2t};$ for $t=1$, $dy/dx = -e^{-2}$, $(x,y) = (e, e^{-1})$; $y - e^{-1} = -e^{-2}(x-e),$

$y = -e^{-2}x + 2e^{-1}$

(b) $y = 1/x, dy/dx = -1/x^2, m = -1/e^2, y - e^{-1} = -\dfrac{1}{e^2}(x-e), y = -\dfrac{1}{e^2}x + \dfrac{2}{e}$

53. $dy/dx = \dfrac{-4\sin t}{2\cos t} = -2\tan t$

(a) $dy/dx = 0$ if $\tan t = 0$; $t = 0, \pi, 2\pi$

(b) $dx/dy = -\dfrac{1}{2}\cot t = 0$ if $\cot t = 0$; $t = \pi/2, 3\pi/2$

55. (a) $a=1,\ b=2$ $a=2,\ b=3$

$a=3,\ b=4$ $a=4,\ b=5$

(b) $x = y = 0$ when $t = 0, \pi;\ \dfrac{dy}{dx} = \dfrac{2\cos 2t}{\cos t};\ \dfrac{dy}{dx}\Big|_{t=0} = 2,\ \dfrac{dy}{dx}\Big|_{t=\pi} = -2,$ the equations of the tangent lines are $y = -2x, y = 2x$.

57. If $x = 4$ then $t^2 = 4$, $t = \pm 2$, $y = 0$ for $t = \pm 2$ so $(4,0)$ is reached when $t = \pm 2$. $dy/dx = (3t^2 - 4)/2t$. For $t = 2$, $dy/dx = 2$ and for $t = -2$, $dy/dx = -2$. The tangent lines are $y = \pm 2(x - 4)$.

59. (a)

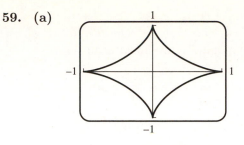

(b) $\dfrac{dx}{dt} = -3\cos^2 t \sin t$ and $\dfrac{dy}{dt} = 3\sin^2 t \cos t$ are both zero when $t = 0, \pi/2, \pi, 3\pi/2, 2\pi$,

so singular points occur at these values of t.

61. (a) From (6), $\dfrac{dy}{dx} = \dfrac{3\sin t}{1 - 3\cos t}$

(b) At $t = 10$, $\dfrac{dy}{dx} = \dfrac{3\sin 10}{1 - 3\cos 10} \approx -0.46402$, $\theta \approx \tan^{-1}(-0.46402) = -0.4345$

63. Eliminate the parameter to get $(x - h)^2/a^2 + (y - k)^2/b^2 = 1$, which is the equation of an ellipse centered at (h, k). Depending on the relative sizes of h and k, the ellipse may be a circle, or may have a horizontal or vertical major axis.

(a) ellipses with a fixed center and varying shapes and sizes

(b) ellipses with varying centers and fixed shape and size

(c) circles of radius 1 with centers on the line $y = x - 1$

65. $L = \displaystyle\int_0^1 \sqrt{(dx/dt)^2 + (dy/dt)^2}\, dt = \int_0^1 \sqrt{(2t)^2 + (t^2)^2}\, dt = \int_0^1 t\sqrt{4 + t^2}\, dt$

Let $u = 4 + t^2$, $du = 2t\, dt$. Then

$$L = \int_4^5 \frac{1}{2}\sqrt{u}\, du = \frac{1}{3}u^{3/2}\bigg]_4^5 = \frac{1}{3}(5\sqrt{5} - 8)$$

67. The curve is a circle of radius 1, traced one and a half times, so the arc length is $\dfrac{3}{2}\cdot 2\pi \cdot 1 = 3\pi$.

69. $L = \displaystyle\int_{-1}^1 \sqrt{(dx/dt)^2 + (dy/dt)^2}\, dt = \int_{-1}^1 \sqrt{[e^{2t}(3\cos t + \sin t)]^2 + [e^{2t}(3\sin t - \cos t)]^2}\, dt$

$$= \int_{-1}^1 \sqrt{10}\, e^{2t}\, dt = \frac{1}{2}\sqrt{10}\, e^{2t}\bigg]_{-1}^1 = \frac{1}{2}\sqrt{10}\, (e^2 - e^{-2})$$

71. (a) $(dx/d\theta)^2 + (dy/d\theta)^2 = (a(1 - \cos\theta))^2 + (a\sin\theta)^2 = a^2(2 - 2\cos\theta)$, so

$$L = \int_0^{2\pi} \sqrt{(dx/d\theta)^2 + (dy/d\theta)^2}\, d\theta = a\int_0^{2\pi} \sqrt{2(1 - \cos\theta)}\, d\theta$$

(b) If you type the definite integral from (a) into your CAS, the output should be something equivalent to "8a". Here's a proof that doesn't use a CAS:

$\cos\theta = 1 - 2\sin^2(\theta/2)$, so $2(1 - \cos\theta) = 4\sin^2(\theta/2)$, and

$$L = a\int_0^{2\pi} \sqrt{2(1 - \cos\theta)}\, d\theta = a\int_0^{2\pi} 2\sin(\theta/2)\, d\theta = -4a\cos(\theta/2)\bigg]_0^{2\pi} = 8a$$

73. (a) The end of the inner arm traces out the circle $x_1 = \cos t, y_1 = \sin t$. Relative to the end of the inner arm, the outer arm traces out the circle $x_2 = \cos 2t, y_2 = -\sin 2t$. Add to get the motion of the center of the rider cage relative to the center of the inner arm:
$x = \cos t + \cos 2t, y = \sin t - \sin 2t$.

(b) Same as part (a), except $x_2 = \cos 2t, y_2 = \sin 2t$, so $x = \cos t + \cos 2t, y = \sin t + \sin 2t$

(c) $L_1 = \int_0^{2\pi} \left[\left(\frac{dx}{dt} \right)^2 + \left(\frac{dy}{dt} \right)^2 \right]^{1/2} dt = \int_0^{2\pi} \sqrt{5 - 4\cos 3t}\, dt \approx 13.36489321$,

$L_2 = \int_0^{2\pi} \sqrt{5 + 4\cos t}\, dt \approx 13.36489322$; L_1 and L_2 appear to be equal, and indeed, with the substitution $u = 3t - \pi$ and the periodicity of $\cos u$,

$L_1 = \frac{1}{3} \int_{-\pi}^{5\pi} \sqrt{5 - 4\cos(u + \pi)}\, du = \int_0^{2\pi} \sqrt{5 + 4\cos u}\, du = L_2$.

75. $x' = 2t, y' = 3, (x')^2 + (y')^2 = 4t^2 + 9$

$S = 2\pi \int_0^2 (3t) \sqrt{4t^2 + 9}\, dt = 6\pi \int_0^4 t\sqrt{4t^2 + 9}\, dt = \frac{\pi}{2}(4t^2 + 9)^{3/2} \Big|_0^2 = \frac{\pi}{2}(125 - 27) = 49\pi$

77. $x' = -2\sin t \cos t, y' = 2\sin t \cos t, (x')^2 + (y')^2 = 8\sin^2 t \cos^2 t$

$S = 2\pi \int_0^{\pi/2} \cos^2 t \sqrt{8\sin^2 t \cos^2 t}\, dt = 4\sqrt{2}\pi \int_0^{\pi/2} \cos^3 t \sin t\, dt = -\sqrt{2}\pi \cos^4 t \Big|_0^{\pi/2} = \sqrt{2}\pi$

79. $x' = -r\sin t, y' = r\cos t, (x')^2 + (y')^2 = r^2, S = 2\pi \int_0^{\pi} r\sin t \sqrt{r^2}\, dt = 2\pi r^2 \int_0^{\pi} \sin t\, dt = 4\pi r^2$

EXERCISE SET 10.2

1.

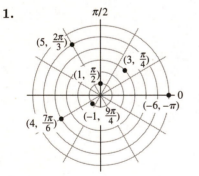

3. (a) $(3\sqrt{3}, 3)$ **(b)** $(-7/2, 7\sqrt{3}/2)$ **(c)** $(3\sqrt{3}, 3)$
(d) $(0, 0)$ **(e)** $(-7\sqrt{3}/2, 7/2)$ **(f)** $(-5, 0)$

5. (a) $(5, \pi), (5, -\pi)$ **(b)** $(4, 11\pi/6), (4, -\pi/6)$ **(c)** $(2, 3\pi/2), (2, -\pi/2)$
(d) $(8\sqrt{2}, 5\pi/4), (8\sqrt{2}, -3\pi/4)$ **(e)** $(6, 2\pi/3), (6, -4\pi/3)$ **(f)** $(\sqrt{2}, \pi/4), (\sqrt{2}, -7\pi/4)$

7. (a) $(5, 0.92730)$ **(b)** $(10, -0.92730)$ **(c)** $(1.27155, -0.66577)$

9. (a) $r^2 = x^2 + y^2 = 4$; circle   **(b)** $y = 4$; horizontal line

(c) $r^2 = 3r\cos\theta$, $x^2 + y^2 = 3x$, $(x - 3/2)^2 + y^2 = 9/4$; circle

(d) $3r\cos\theta + 2r\sin\theta = 6$, $3x + 2y = 6$; line

11. **(a)** $r\cos\theta = 3$ **(b)** $r = \sqrt{7}$

  (c) $r^2 + 6r\sin\theta = 0$, $r = -6\sin\theta$

  (d) $9(r\cos\theta)(r\sin\theta) = 4$, $9r^2\sin\theta\cos\theta = 4$, $r^2\sin 2\theta = 8/9$

13.

$r = 3\sin 2\theta$

15.

$r = 3 - 4\sin\left(\dfrac{\pi}{4}\theta\right)$

17. **(a)** $r = 5$

  (b) $(x - 3)^2 + y^2 = 9$, $r = 6\cos\theta$

  (c) Example 8, $r = 1 - \cos\theta$

19. **(a)** Figure 10.2.19, $a = 3, n = 2, r = 3\sin 2\theta$

  (b) From (8-9), symmetry about the y-axis and Theorem 10.2.1(b), the equation is of the form $r = a \pm b\sin\theta$. The cartesian points $(3, 0)$ and $(0, 5)$ give $a = 3$ and $5 = a + b$, so $b = 2$ and $r = 3 + 2\sin\theta$.

  (c) Example 9, $r^2 = 9\cos 2\theta$

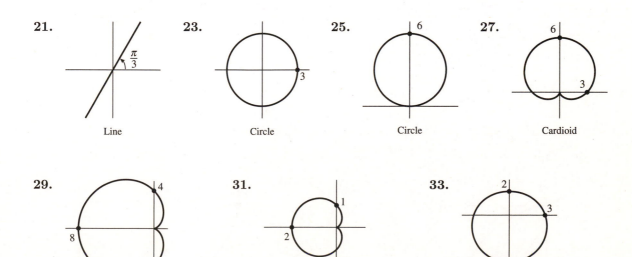

21. Line   **23.** Circle   **25.** Circle   **27.** Cardioid

29. Cardioid   **31.** Cardioid   **33.** Limaçon

35.

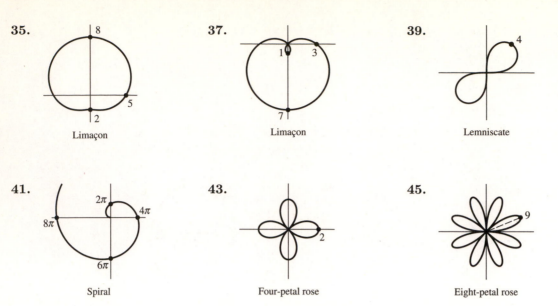

Limaçon

37.

Limaçon

39.

Lemniscate

41.

Spiral

43.

Four-petal rose

45.

Eight-petal rose

47. True. Both have rectangular coordinates $(-1/2, -\sqrt{3}/2)$

49. False. For $\pi/2 < \theta < \pi$, $\sin 2\theta < 0$. Hence the point with polar coordinates $(\sin 2\theta, \theta)$ is in the fourth quadrant.

51. $0 \le \theta < 4\pi$ **53.** $0 \le \theta < 8\pi$ **55.** $0 \le \theta < 5\pi$

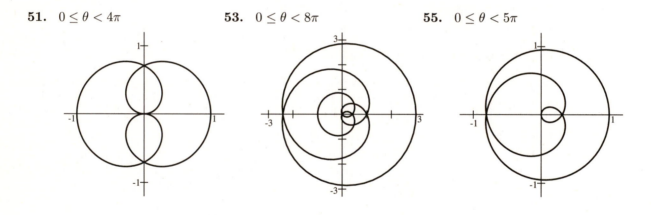

57. (a) $-4\pi \le \theta \le 4\pi$

59. (a) $r = \dfrac{a}{\cos\theta}, r\cos\theta = a, x = a$ (b) $r\sin\theta = b, y = b$

61. (a)

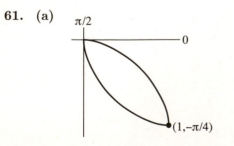

(b)

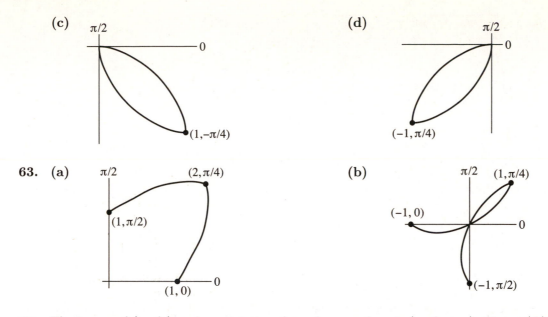

(c) π/2   (d) π/2

(1,−π/4)   (−1,π/4)

63. **(a)** π/2   (2,π/4)   **(b)** π/2 (1,π/4)

(1,π/2)   (−1,0)

(1,0) 0   (−1,π/2)

65. The image of (r_0, θ_0) under a rotation through an angle α is $(r_0, \theta_0 + \alpha)$. Hence $(f(\theta), \theta)$ lies on the original curve if and only if $(f(\theta), \theta + \alpha)$ lies on the rotated curve, i.e. (r, θ) lies on the rotated curve if and only if $r = f(\theta - \alpha)$.

67. **(a)** $r = 1 + \cos(\theta - \pi/4) = 1 + \dfrac{\sqrt{2}}{2}(\cos\theta + \sin\theta)$

  (b) $r = 1 + \cos(\theta - \pi/2) = 1 + \sin\theta$

  (c) $r = 1 + \cos(\theta - \pi) = 1 - \cos\theta$

  (d) $r = 1 + \cos(\theta - 5\pi/4) = 1 - \dfrac{\sqrt{2}}{2}(\cos\theta + \sin\theta)$

69. $y = r\sin\theta = (1 + \cos\theta)\sin\theta = \sin\theta + \sin\theta\cos\theta$,
$dy/d\theta = \cos\theta - \sin^2\theta + \cos^2\theta = 2\cos^2\theta + \cos\theta - 1 = (2\cos\theta - 1)(\cos\theta + 1)$;
$dy/d\theta = 0$ if $\cos\theta = 1/2$ or if $\cos\theta = -1$;
$\theta = \pi/3$ or π (or $\theta = -\pi/3$, which leads to the minimum point).
If $\theta = \pi/3, \pi$, then $y = 3\sqrt{3}/4, 0$ so the maximum value of y is $3\sqrt{3}/4$ and the polar coordinates of the highest point are $(3/2, \pi/3)$.

71. Let (x_1, y_1) and (x_2, y_2) be the rectangular coordinates of the points (r_1, θ_1) and (r_2, θ_2) then
$$d = \sqrt{(x_2 - x_1)^2 + (y_2 - y_1)^2} = \sqrt{(r_2\cos\theta_2 - r_1\cos\theta_1)^2 + (r_2\sin\theta_2 - r_1\sin\theta_1)^2}$$
$$= \sqrt{r_1^2 + r_2^2 - 2r_1r_2(\cos\theta_1\cos\theta_2 + \sin\theta_1\sin\theta_2)} = \sqrt{r_1^2 + r_2^2 - 2r_1r_2\cos(\theta_1 - \theta_2)}.$$
An alternate proof follows directly from the Law of Cosines.

73. The tips occur when $\theta = 0, \pi/2, \pi, 3\pi/2$ for which $r = 1$: $d = \sqrt{1^2 + 1^2 - 2(1)(1)\cos(\pm\pi/2)} = \sqrt{2}$. Geometrically, find the distance between, e.g., the points $(0, 1)$ and $(1, 0)$.

75. **(a)** $0 = (r^2 + a^2)^2 - a^4 - 4a^2r^2\cos^2\theta = r^4 + a^4 + 2r^2a^2 - a^4 - 4a^2r^2\cos^2\theta$
  $= r^4 + 2r^2a^2 - 4a^2r^2\cos^2\theta$, so $r^2 = 2a^2(2\cos^2\theta - 1) = 2a^2\cos 2\theta$.

  (b) The distance from the point (r, θ) to $(a, 0)$ is (from Exercise 73(a))
  $$\sqrt{r^2 + a^2 - 2ra\cos(\theta - 0)} = \sqrt{r^2 - 2ar\cos\theta + a^2},\text{ and to the point }(a, \pi)\text{ is}$$
  $$\sqrt{r^2 + a^2 - 2ra\cos(\theta - \pi)} = \sqrt{r^2 + 2ar\cos\theta + a^2},\text{ and their product is}$$

$$\sqrt{(r^2 + a^2)^2 - 4a^2 r^2 \cos^2 \theta} = \sqrt{r^4 + a^4 + 2a^2 r^2 (1 - 2\cos^2 \theta)}$$
$$= \sqrt{4a^4 \cos^2 2\theta + a^4 + 2a^2 (2a^2 \cos 2\theta)(-\cos 2\theta)} = a^2$$

77. $\displaystyle\lim_{\theta \to 0^\pm} y = \lim_{\theta \to 0^\pm} r \sin \theta = \lim_{\theta \to 0^\pm} \frac{\sin \theta}{\theta^2} = \lim_{\theta \to 0^\pm} \frac{\sin \theta}{\theta} \lim_{\theta \to 0^\pm} \frac{1}{\theta} = 1 \cdot \lim_{\theta \to 0^\pm} \frac{1}{\theta}$, so $\displaystyle\lim_{\theta \to 0^\pm} y$ does not exist.

79. (a)

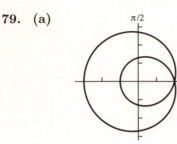

(b) Replacing θ with $-\theta$ changes $r = 2 - \sin(\theta/2)$ into $r = 2 + \sin(\theta/2)$ which is not an equivalent equation. But the locus of points satisfying the first equation, when θ runs from 0 to 4π, is the same as the locus of points satisfying the second equation when θ runs from 0 to 4π, as can be seen under the change of variables (equivalent to reversing direction of θ) $\theta \to 4\pi - \theta$, for which $2 + \sin(4\pi - \theta) = 2 - \sin \theta$.

EXERCISE SET 10.3

1. Substituting $\theta = \pi/6$, $r = 1$, and $dr/d\theta = \sqrt{3}$ in equation (2) gives slope $m = \sqrt{3}$.

3. As in Exercise 1, $\theta = 2$, $dr/d\theta = -1/4$, $r = 1/2$, $m = \dfrac{\tan 2 - 2}{2\tan 2 + 1}$

5. As in Exercise 1, $\theta = \pi/4$, $dr/d\theta = -3\sqrt{2}/2$, $r = \sqrt{2}/2$, $m = 1/2$

7. $m = \dfrac{dy}{dx} = \dfrac{r\cos\theta + (\sin\theta)(dr/d\theta)}{-r\sin\theta + (\cos\theta)(dr/d\theta)} = \dfrac{\cos\theta + 2\sin\theta\cos\theta}{-\sin\theta + \cos^2\theta - \sin^2\theta}$; if $\theta = 0, \pi/2, \pi$,
then $m = 1, 0, -1$.

9. $dx/d\theta = -a\sin\theta(1 + 2\cos\theta)$, $dy/d\theta = a(2\cos\theta - 1)(\cos\theta + 1)$

The tangent line is horizontal if $dy/d\theta = 0$ and $dx/d\theta \neq 0$. $dy/d\theta = 0$ when $\cos\theta = 1/2$ or $\cos\theta = -1$ so $\theta = \pi/3, 5\pi/3$, or π; $dx/d\theta \neq 0$ for $\theta = \pi/3$ and $5\pi/3$. For the singular point $\theta = \pi$ we find that $\lim_{\theta \to \pi} dy/dx = 0$. There are horizontal tangent lines at $(3a/2, \pi/3), (0, \pi)$, and $(3a/2, 5\pi/3)$.

The tangent line is vertical if $dy/d\theta \neq 0$ and $dx/d\theta = 0$. $dx/d\theta = 0$ when $\sin\theta = 0$ or $\cos\theta = -1/2$ so $\theta = 0, \pi, 2\pi/3$, or $4\pi/3$; $dy/d\theta \neq 0$ for $\theta = 0, 2\pi/3$, and $4\pi/3$. The singular point $\theta = \pi$ was discussed earlier. There are vertical tangent lines at $(2a, 0), (a/2, 2\pi/3)$, and $(a/2, 4\pi/3)$.

11. Since $r(\theta + \pi) = -r(\theta)$, the curve is traced out once as θ goes from 0 to π.

$dy/d\theta = (d/d\theta)(\sin^2\theta\cos^2\theta) = (\sin 4\theta)/2 = 0$ at $\theta = 0, \pi/4, \pi/2, 3\pi/4, \pi$. When $\theta = 0, \pi/2$, or π, $r = 0$, so these 3 values give the same point, and we only have 3 points to consider.

$dx/d\theta = (d/d\theta)(\sin\theta\cos^3\theta) = \cos^2\theta(4\cos^2\theta - 3)$ is nonzero when $\theta = 0, \pi/4$, or $3\pi/4$. Hence there are horizontal tangents at all 3 of these points. (There is also a singular point at the origin corresponding to $\theta = \pi/2$.)

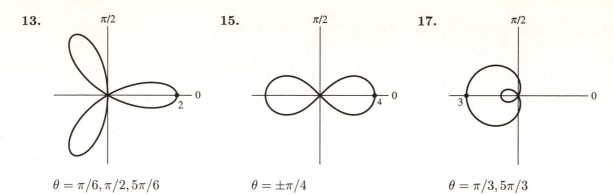

13. $\theta = \pi/6, \pi/2, 5\pi/6$

15. $\theta = \pm\pi/4$

17. $\theta = \pi/3, 5\pi/3$

19. $r^2 + (dr/d\theta)^2 = a^2 + 0^2 = a^2,\ L = \int_0^{2\pi} a\, d\theta = 2\pi a$

21. $r^2 + (dr/d\theta)^2 = [a(1 - \cos\theta)]^2 + [a\sin\theta]^2 = 4a^2 \sin^2(\theta/2),\ L = 2\int_0^{\pi} 2a\sin(\theta/2)\, d\theta = 8a$

23. **(a)** $r^2 + (dr/d\theta)^2 = (\cos n\theta)^2 + (-n\sin n\theta)^2 = \cos^2 n\theta + n^2 \sin^2 n\theta$

$$= (1 - \sin^2 n\theta) + n^2 \sin^2 n\theta = 1 + (n^2 - 1)\sin^2 n\theta,$$

The top half of the petal along the polar axis is traced out as θ goes from 0 to $\pi/(2n)$, so

$$L = 2\int_0^{\pi/(2n)} \sqrt{1 + (n^2 - 1)\sin^2 n\theta}\, d\theta$$

(b) $L = 2\int_0^{\pi/4} \sqrt{1 + 3\sin^2 2\theta}\, d\theta \approx 2.42$

(c)

n	2	3	4	5	6	7	8	9	10	11
L	2.42211	2.22748	2.14461	2.10100	2.07501	2.05816	2.04656	2.03821	2.03199	2.02721

n	12	13	14	15	16	17	18	19	20
L	2.02346	2.02046	2.01802	2.01600	2.01431	2.01288	2.01167	2.01062	2.00971

The limit seems to be 2. This is to be expected, since as $n \to +\infty$ each petal more closely resembles a pair of straight lines of length 1.

25. **(a)** $\displaystyle\int_{\pi/2}^{\pi} \frac{1}{2}(1 - \cos\theta)^2\, d\theta$ **(b)** $\displaystyle\int_0^{\pi/2} 2\cos^2\theta\, d\theta$

(c) $\displaystyle\int_0^{\pi/2} \frac{1}{2}\sin^2 2\theta\, d\theta$ **(d)** $\displaystyle\int_0^{2\pi} \frac{1}{2}\theta^2\, d\theta$

(e) $\displaystyle\int_{-\pi/2}^{\pi/2} \frac{1}{2}(1 - \sin\theta)^2\, d\theta$ **(f)** $\displaystyle\int_{-\pi/4}^{\pi/4} \frac{1}{2}\cos^2 2\theta\, d\theta = \int_0^{\pi/4} \cos^2 2\theta\, d\theta$

27. **(a)** $\displaystyle A = \int_0^{\pi} \frac{1}{2} 4a^2 \sin^2\theta\, d\theta = \pi a^2$ **(b)** $\displaystyle A = \int_{-\pi/2}^{\pi/2} \frac{1}{2} 4a^2 \cos^2\theta\, d\theta = \pi a^2$

29. $\displaystyle A = \int_0^{2\pi} \frac{1}{2}(2 + 2\sin\theta)^2\, d\theta = 6\pi$ **31.** $\displaystyle A = 6\int_0^{\pi/6} \frac{1}{2}(16\cos^2 3\theta)\, d\theta = 4\pi$

33. $\displaystyle A = 2\int_{2\pi/3}^{\pi} \frac{1}{2}(1 + 2\cos\theta)^2\, d\theta = \pi - \frac{3\sqrt{3}}{2}$

35. area $= A_1 - A_2 = \int_0^{\pi/2} \frac{1}{2} 4 \cos^2 \theta \, d\theta - \int_0^{\pi/4} \frac{1}{2} \cos 2\theta \, d\theta = \frac{\pi}{2} - \frac{1}{4}$

37. The circles intersect when $\cos \theta = \sqrt{3} \sin \theta$, $\tan \theta = 1/\sqrt{3}$, $\theta = \pi/6$, so

$$A = A_1 + A_2 = \int_0^{\pi/6} \frac{1}{2} (4\sqrt{3} \sin \theta)^2 \, d\theta + \int_{\pi/6}^{\pi/2} \frac{1}{2} (4 \cos \theta)^2 \, d\theta = 2\pi - 3\sqrt{3} + \frac{4\pi}{3} - \sqrt{3} = \frac{10\pi}{3} - 4\sqrt{3}.$$

39. $A = 2 \int_{\pi/6}^{\pi/2} \frac{1}{2} [9 \sin^2 \theta - (1 + \sin \theta)^2] \, d\theta = \pi$

41. $A = 2 \int_0^{\pi/3} \frac{1}{2} [(2 + 2 \cos \theta)^2 - 9] \, d\theta = \frac{9\sqrt{3}}{2} - \pi$

43. $A = 2 \left[\int_0^{2\pi/3} \frac{1}{2} (1/2 + \cos \theta)^2 \, d\theta - \int_{2\pi/3}^{\pi} \frac{1}{2} (1/2 + \cos \theta)^2 \, d\theta \right] = \frac{\pi + 3\sqrt{3}}{4}$

45. $A = 2 \int_0^{\pi/4} \frac{1}{2} (4 - 2 \sec^2 \theta) \, d\theta = \pi - 2$

47. True. When $\theta = 3\pi$, $r = \cos(3\pi/2) = 0$ so the curve passes through the origin. Also, $\frac{dr}{d\theta} = -\frac{1}{2} \sin(\theta/2) = \frac{1}{2} \neq 0$. Hence, by Theorem 10.3.1, the line $\theta = 3\pi$ is tangent to the curve at the origin. But $\theta = 3\pi$ is the x-axis.

49. False. The area is $\frac{\theta}{2\pi}$ times the area of the circle $= \frac{\theta}{2\pi} \cdot \pi r^2 = \frac{\theta}{2} r^2$, not θr^2.

51. (a) r is not real for $\pi/4 < \theta < 3\pi/4$ and $5\pi/4 < \theta < 7\pi/4$

(b) $A = 4 \int_0^{\pi/4} \frac{1}{2} a^2 \cos 2\theta \, d\theta = a^2$

(c) $A = 4 \int_0^{\pi/6} \frac{1}{2} [4 \cos 2\theta - 2] \, d\theta = 2\sqrt{3} - \frac{2\pi}{3}$

53. $A = \int_{2\pi}^{4\pi} \frac{1}{2} a^2 \theta^2 \, d\theta - \int_0^{2\pi} \frac{1}{2} a^2 \theta^2 \, d\theta = 8\pi^3 a^2$

55. (a) $r^3 \cos^3 \theta - 3r^2 \cos \theta \sin \theta + r^3 \sin^3 \theta = 0$, $r = \dfrac{3 \cos \theta \sin \theta}{\cos^3 \theta + \sin^3 \theta}$

(b) $A = \int_0^{\pi/2} \frac{1}{2} \left(\frac{3 \cos \theta \sin \theta}{\cos^3 \theta + \sin^3 \theta} \right)^2 d\theta = \frac{2 \sin^3 \theta - \cos^3 \theta}{2(\cos^3 \theta + \sin^3 \theta)} \Bigg]_0^{\pi/2} = \frac{3}{2}$

57. If the upper right corner of the square is the point (a, a) then the large circle has equation $r = \sqrt{2}a$ and the small circle has equation $(x - a)^2 + y^2 = a^2$, $r = 2a \cos \theta$, so

area of crescent $= 2 \int_0^{\pi/4} \frac{1}{2} \left[(2a \cos \theta)^2 - (\sqrt{2}a)^2 \right] d\theta = a^2 =$ area of square.

59. $A = \displaystyle\int_0^{\pi/2} \frac{1}{2} 4 \cos^2\theta \sin^4\theta \, d\theta = \pi/16$

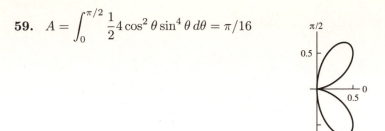

61. $\tan\psi = \tan(\phi - \theta) = \dfrac{\tan\phi - \tan\theta}{1 + \tan\phi\tan\theta} = \dfrac{\dfrac{dy}{dx} - \dfrac{y}{x}}{1 + \dfrac{y}{x}\dfrac{dy}{dx}}$

$= \dfrac{\dfrac{r\cos\theta + (dr/d\theta)\sin\theta}{-r\sin\theta + (dr/d\theta)\cos\theta} - \dfrac{\sin\theta}{\cos\theta}}{1 + \left(\dfrac{r\cos\theta + (dr/d\theta)\sin\theta)}{-r\sin\theta + (dr/d\theta)\cos\theta)}\right)\left(\dfrac{\sin\theta}{\cos\theta}\right)} = \dfrac{r}{dr/d\theta}$

63. $\tan\psi = \dfrac{r}{dr/d\theta} = \dfrac{ae^{b\theta}}{abe^{b\theta}} = \dfrac{1}{b}$ is constant, so ψ is constant.

65. $r^2 + \left(\dfrac{dr}{d\theta}\right)^2 = \cos^2\theta + \sin^2\theta = 1,$

so $S = \displaystyle\int_{-\pi/2}^{\pi/2} 2\pi\cos^2\theta \, d\theta = \pi^2.$

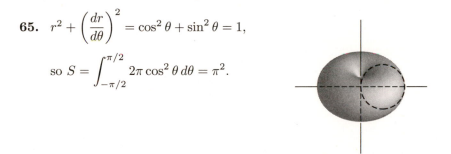

67. $S = \displaystyle\int_0^\pi 2\pi(1 - \cos\theta)\sin\theta\sqrt{1 - 2\cos\theta + \cos^2\theta + \sin^2\theta} \, d\theta$

$= 2\sqrt{2}\pi\displaystyle\int_0^\pi \sin\theta(1 - \cos\theta)^{3/2} \, d\theta = \dfrac{2}{5}2\sqrt{2}\pi(1 - \cos\theta)^{5/2}\Big|_0^\pi = 32\pi/5$

EXERCISE SET 10.4

1. **(a)** $4px = y^2$, point $(1, 1)$, $4p = 1$, $x = y^2$ **(b)** $-4py = x^2$, point $(3, -3)$, $12p = 9$, $-3y = x^2$

(c) $a = 3, b = 2, \dfrac{x^2}{9} + \dfrac{y^2}{4} = 1$ **(d)** $a = 3, b = 2, \dfrac{x^2}{4} + \dfrac{y^2}{9} = 1$

(e) asymptotes: $y = \pm x$, so $a = b$; point $(0, 1)$, so $y^2 - x^2 = 1$

(f) asymptotes: $y = \pm x$, so $b = a$; point $(2, 0)$, so $\dfrac{x^2}{4} - \dfrac{y^2}{4} = 1$

3. (a)

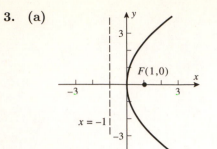

(b)

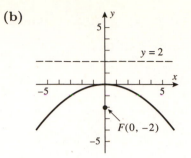

5. (a)

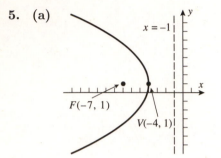

(b)

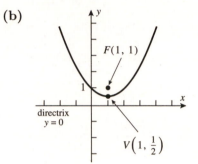

7. (a) $c^2 = 16 - 9 = 7$, $c = \sqrt{7}$

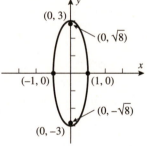

(b) $\dfrac{x^2}{1} + \dfrac{y^2}{9} = 1$

$c^2 = 9 - 1 = 8, c = 2\sqrt{2}$

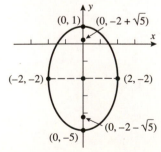

9. (a) $\dfrac{(x+3)^2}{16} + \dfrac{(y-5)^2}{4} = 1$

$c^2 = 16 - 4 = 12, c = 2\sqrt{3}$

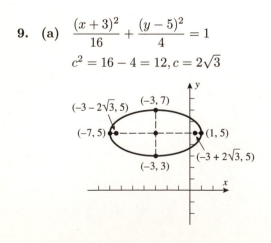

(b) $\dfrac{x^2}{4} + \dfrac{(y+2)^2}{9} = 1$

$c^2 = 9 - 4 = 5, c = \sqrt{5}$

11. **(a)** $c^2 = a^2 + b^2 = 16 + 9 = 25, c = 5$

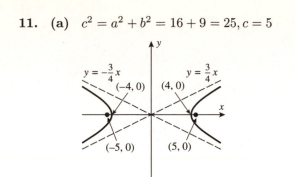

(b) $y^2/4 - x^2/36 = 1$

$c^2 = 4 + 36 = 40, c = 2\sqrt{10}$

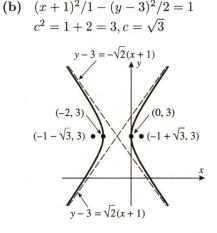

13. **(a)** $c^2 = 3 + 5 = 8, c = 2\sqrt{2}$

(b) $(x+1)^2/1 - (y-3)^2/2 = 1$

$c^2 = 1 + 2 = 3, c = \sqrt{3}$

15. **(a)** $y^2 = 4px, p = 3, y^2 = 12x$

(b) $x^2 = -4py, p = 1/4, x^2 = -y$

17. $y^2 = a(x - h)$, $4 = a(3 - h)$ and $2 = a(2 - h)$, solve simultaneously to get $h = 1$, $a = 2$ so $y^2 = 2(x - 1)$

19. **(a)** $x^2/9 + y^2/4 = 1$

(b) $b = 4, c = 3, a^2 = b^2 + c^2 = 16 + 9 = 25; x^2/16 + y^2/25 = 1$

21. **(a)** $a = 6$, $(-3, 2)$ satisfies $\dfrac{x^2}{a^2} + \dfrac{y^2}{36} = 1$ so $\dfrac{9}{a^2} + \dfrac{4}{36} = 1$, $a^2 = \dfrac{81}{8}$; $\dfrac{x^2}{81/8} + \dfrac{y^2}{36} = 1$

(b) The center is midway between the foci so it is at $(-1, 2)$, thus
$c = 1, b = 2, a^2 = 1 + 4 = 5, a = \sqrt{5}$; $(x+1)^2/4 + (y-2)^2/5 = 1$

23. **(a)** $a = 2, c = 3, b^2 = 9 - 4 = 5; x^2/4 - y^2/5 = 1$

(b) $a = 2, a/b = 2/3, b = 3; y^2/4 - x^2/9 = 1$

25. **(a)** foci along the x-axis: $b/a = 3/4$ and $a^2 + b^2 = 25$, solve to get $a^2 = 16, b^2 = 9$;
$x^2/16 - y^2/9 = 1$
foci along the y-axis: $a/b = 3/4$ and $a^2 + b^2 = 25$ which results in $y^2/9 - x^2/16 = 1$

(b) $c = 3, b/a = 2$ and $a^2 + b^2 = 9$ so $a^2 = 9/5, b^2 = 36/5$; $x^2/(9/5) - y^2/(36/5) = 1$

27. False. The set described is a <u>parabola</u>.

29. False. The distance is $2p$, as shown in Figure 10.4.6.

31. **(a)** $y = ax^2 + b$, $(20, 0)$ and $(10, 12)$ are on the curve so
$400a + b = 0$ and $100a + b = 12$. Solve for b to get
$b = 16$ ft = height of arch.

(b) $\dfrac{x^2}{a^2} + \dfrac{y^2}{b^2} = 1$, $400 = a^2, a = 20$; $\dfrac{100}{400} + \dfrac{144}{b^2} = 1$,

$b = 8\sqrt{3}$ ft = height of arch.

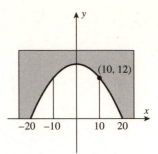

33. We may assume that the vertex is $(0, 0)$ and the parabola opens to the right. Let $P(x_0, y_0)$ be a
point on the parabola $y^2 = 4px$, then by the definition of a parabola, PF = distance from P to
directrix $x = -p$, so $PF = x_0 + p$ where $x_0 \geq 0$ and PF is a minimum when $x_0 = 0$ (the vertex).

35. Use an xy-coordinate system so that $y^2 = 4px$ is an equation of the parabola. Then $(1, 1/2)$ is a
point on the curve so $(1/2)^2 = 4p(1)$, $p = 1/16$. The light source should be placed at the focus
which is $1/16$ ft. from the vertex.

37. **(a)** For any point (x, y), the equation
$y = b \tan t$ has a unique solution t,
$-\pi/2 < t < \pi/2$. On the hyperbola,

$\dfrac{x^2}{a^2} = 1 + \dfrac{y^2}{b^2} = 1 + \tan^2 t = \sec^2 t$,

so $x = \pm a \sec t$.

(b)

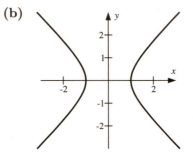

39. $(4, 1)$ and $(4, 5)$ are the foci so the center is at $(4, 3)$ thus $c = 2$, $a = 12/2 = 6$, $b^2 = 36 - 4 = 32$;
$(x - 4)^2/32 + (y - 3)^2/36 = 1$

41. Assume $\dfrac{x^2}{a^2} + \dfrac{y^2}{b^2} = 1$, $A = 4 \displaystyle\int_0^a b\sqrt{1 - x^2/a^2}\, dx = \pi ab$

43. $L = 2a = \sqrt{D^2 + p^2 D^2} = D\sqrt{1 + p^2}$ (see figure), so $a = \dfrac{1}{2}D\sqrt{1 + p^2}$,

but $b = \dfrac{1}{2}D$, $T = c = \sqrt{a^2 - b^2} = \sqrt{\dfrac{1}{4}D^2(1 + p^2) - \dfrac{1}{4}D^2} = \dfrac{1}{2}pD$.

45. As in Exercise 44, $d_2 - d_1 = 2a = vt = (299{,}792{,}458 \text{ m/s})(100 \cdot 10^{-6} \text{ s}) \approx 29979 \text{ m} = 29.979 \text{ km}$.
$a^2 = (vt/2)^2 \approx 224.689 \text{ km}^2$; $c^2 = (50)^2 = 2500 \text{ km}^2$

$b^2 = c^2 - a^2 \approx 2275.311 \text{ km}$, $\dfrac{x^2}{224.688} - \dfrac{y^2}{2275.311} = 1$

But $y = 200$ km, so $x \approx 64.612$ km. The ship is located at $(64.612, 200)$.

47. **(a)** $V = \int_a^{\sqrt{a^2+b^2}} \pi \left(b^2 x^2 / a^2 - b^2\right) dx$

$$= \frac{\pi b^2}{3a^2}(b^2 - 2a^2)\sqrt{a^2 + b^2} + \frac{2}{3}ab^2\pi$$

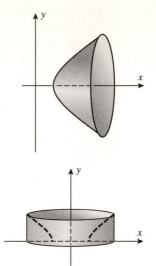

(b) $V = 2\pi \int_a^{\sqrt{a^2+b^2}} x\sqrt{b^2 x^2 / a^2 - b^2}\, dx = (2b^4/3a)\pi$

49. By implicit differentiation, $\dfrac{dy}{dx}\bigg|_{(x_0,y_0)} = -\dfrac{b^2}{a^2}\dfrac{x_0}{y_0}$ if $y_0 \neq 0$, the tangent line is

$y - y_0 = -\dfrac{b^2}{a^2}\dfrac{x_0}{y_0}(x - x_0)$, $a^2 y_0 y - a^2 y_0^2 = -b^2 x_0 x + b^2 x_0^2$, $b^2 x_0 x + a^2 y_0 y = b^2 x_0^2 + a^2 y_0^2$,

but (x_0, y_0) is on the ellipse so $b^2 x_0^2 + a^2 y_0^2 = a^2 b^2$; thus the tangent line is $b^2 x_0 x + a^2 y_0 y = a^2 b^2$,

$x_0 x/a^2 + y_0 y/b^2 = 1$. If $y_0 = 0$ then $x_0 = \pm a$ and the tangent lines are $x = \pm a$ which also follows

from $x_0 x/a^2 + y_0 y/b^2 = 1$.

51. Use $\dfrac{x^2}{a^2} + \dfrac{y^2}{b^2} = 1$ and $\dfrac{x^2}{A^2} - \dfrac{y^2}{B^2} = 1$ as the equations of the ellipse and hyperbola. If (x_0, y_0) is

a point of intersection then $\dfrac{x_0^2}{a^2} + \dfrac{y_0^2}{b^2} = 1 = \dfrac{x_0^2}{A^2} - \dfrac{y_0^2}{B^2}$, so $x_0^2\left(\dfrac{1}{A^2} - \dfrac{1}{a^2}\right) = y_0^2\left(\dfrac{1}{B^2} + \dfrac{1}{b^2}\right)$ and

$a^2 A^2 y_0^2 (b^2 + B^2) = b^2 B^2 x_0^2 (a^2 - A^2)$. Since the conics have the same foci, $a^2 - b^2 = c^2 = A^2 + B^2$,

so $a^2 - A^2 = b^2 + B^2$. Hence $a^2 A^2 y_0^2 = b^2 B^2 x_0^2$. From Exercises 63 and 64, the slopes of the

tangent lines are $-\dfrac{b^2 x_0}{a^2 y_0}$ and $\dfrac{B^2 x_0}{A^2 y_0}$, whose product is $-\dfrac{b^2 B^2 x_0^2}{a^2 A^2 y_0^2} = -1$. Hence the tangent lines are

perpendicular.

53. **(a)** $(x-1)^2 - 5(y+1)^2 = 5$, hyperbola

(b) $x^2 - 3(y+1)^2 = 0, x = \pm\sqrt{3}(y+1)$, two lines

(c) $4(x+2)^2 + 8(y+1)^2 = 4$, ellipse

(d) $3(x+2)^2 + (y+1)^2 = 0$, the point $(-2, -1)$ (degenerate case)

(e) $(x+4)^2 + 2y = 2$, parabola

(f) $5(x+4)^2 + 2y = -14$, parabola

55. distance from the point (x, y) to the focus $(0, -c)$ plus distance to the focus $(0, c) = $ const $= 2a$,

$\sqrt{x^2 + (y+c)^2} + \sqrt{x^2 + (y-c)^2} = 2a, x^2 + (y+c)^2 = 4a^2 + x^2 + (y-c)^2 - 4a\sqrt{x^2 + (y-c)^2}$,

$\sqrt{x^2 + (y-c)^2} = a - \dfrac{c}{a}y$, and since $a^2 - c^2 = b^2$, $\dfrac{x^2}{b^2} + \dfrac{y^2}{a^2} = 1$

57. Assume the equation of the parabola is $x^2 = 4py$. The tangent line at $P = (x_0, y_0)$ (see figure) is given by $(y - y_0)/(x - x_0) = m = x_0/2p$. To find the y-intercept set $x = 0$ and obtain $y = -y_0$. Thus the tangent line meets the y-axis at $Q = (0, -y_0)$. The focus is $F = (0, p) = (0, x_0^2/4y_0)$, so the distance from P to the focus is

$$\sqrt{x_0^2 + (y_0 - p)^2} = \sqrt{4py_0 + (y_0 - p)^2} = \sqrt{(y_0 + p)^2} = y_0 + p \text{ and}$$

the distance from the focus to Q is $p + y_0$. Hence triangle FPQ is isosceles, and angles FPQ and FQP are equal. The angle between the tangent line and the vertical line through P equals angle FQP, so it also equals angle FPQ, as stated in the theorem.

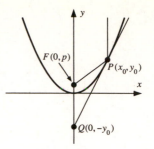

EXERCISE SET 10.5

1. (a) $\sin\theta = \sqrt{3}/2$, $\cos\theta = 1/2$

$x' = (-2)(1/2) + (6)(\sqrt{3}/2) = -1 + 3\sqrt{3}$, $y' = -(-2)(\sqrt{3}/2) + 6(1/2) = 3 + \sqrt{3}$

(b) $x = \dfrac{1}{2}x' - \dfrac{\sqrt{3}}{2}y' = \dfrac{1}{2}(x' - \sqrt{3}y')$, $y = \dfrac{\sqrt{3}}{2}x' + \dfrac{1}{2}y' = \dfrac{1}{2}(\sqrt{3}x' + y')$

$$\sqrt{3}\left[\frac{1}{2}(x' - \sqrt{3}y')\right]\left[\frac{1}{2}(\sqrt{3}x' + y')\right] + \left[\frac{1}{2}(\sqrt{3}x' + y')\right]^2 = 6$$

$$\frac{\sqrt{3}}{4}(\sqrt{3}(x')^2 - 2x'y' - \sqrt{3}(y')^2) + \frac{1}{4}(3(x')^2 + 2\sqrt{3}x'y' + (y')^2) = 6$$

$$\frac{3}{2}(x')^2 - \frac{1}{2}(y')^2 = 6,\ 3(x')^2 - (y')^2 = 12$$

(c)

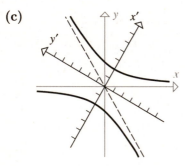

3. $\cot 2\theta = (0 - 0)/1 = 0$, $2\theta = 90°$, $\theta = 45°$
$x = (\sqrt{2}/2)(x' - y')$, $y = (\sqrt{2}/2)(x' + y')$
$(y')^2/18 - (x')^2/18 = 1$, hyperbola

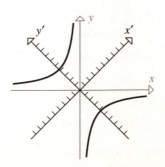

5. $\cot 2\theta = [1 - (-2)]/4 = 3/4$

$\cos 2\theta = 3/5$

$\sin \theta = \sqrt{(1 - 3/5)/2} = 1/\sqrt{5}$

$\cos \theta = \sqrt{(1 + 3/5)/2} = 2/\sqrt{5}$

$x = (1/\sqrt{5})(2x' - y')$

$y = (1/\sqrt{5})(x' + 2y')$

$(x')^2/3 - (y')^2/2 = 1$, hyperbola

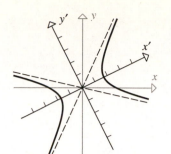

7. $\cot 2\theta = (1 - 3)/(2\sqrt{3}) = -1/\sqrt{3}$,

$2\theta = 120°, \theta = 60°$

$x = (1/2)(x' - \sqrt{3}y')$

$y = (1/2)(\sqrt{3}x' + y')$

$y' = (x')^2$, parabola

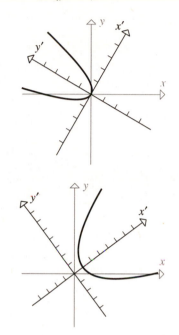

9. $\cot 2\theta = (9 - 16)/(-24) = 7/24$

$\cos 2\theta = 7/25$,

$\sin \theta = 3/5, \qquad \cos \theta = 4/5$

$x = (1/5)(4x' - 3y')$,

$y = (1/5)(3x' + 4y')$

$(y')^2 = 4(x' - 1)$, parabola

11. $\cot 2\theta = (52 - 73)/(-72) = 7/24$

$\cos 2\theta = 7/25, \qquad \sin \theta = 3/5$,

$\cos \theta = 4/5$

$x = (1/5)(4x' - 3y')$,

$y = (1/5)(3x' + 4y')$

$(x' + 1)^2/4 + (y')^2 = 1$, ellipse

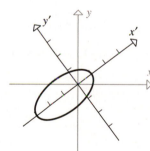

13. $x' = (\sqrt{2}/2)(x + y), y' = (\sqrt{2}/2)(-x + y)$ which when substituted into $3(x')^2 + (y')^2 = 6$ yields $x^2 + xy + y^2 = 3$.

15. Let $x = x' \cos \theta - y' \sin \theta, y = x' \sin \theta + y' \cos \theta$ then $x^2 + y^2 = r^2$ becomes $(\sin^2 \theta + \cos^2 \theta)(x')^2 + (\sin^2 \theta + \cos^2 \theta)(y')^2 = r^2, (x')^2 + (y')^2 = r^2$. Under a rotation transformation the center of the circle stays at the origin of both coordinate systems.

17. Use the Rotation Equations (5).

19. Set $\cot 2\theta = (A - C)/B = 0, 2\theta = \pi/2, \theta = \pi/4, \cos \theta = \sin \theta = 1/\sqrt{2}$. Set $x = (x' - y')/\sqrt{2}$, $y = (x' + y')/\sqrt{2}$ and insert these into the equation to obtain $4y' = (x')^2$; parabola, $p = 1$. In $x'y'$-coordinates: vertex $(0,0)$, focus $(0,1)$, directrix $y' = -1$ In xy-coordinates: vertex $(0,0)$, focus $(-1/\sqrt{2}, 1/\sqrt{2})$, directrix $y = x - \sqrt{2}$

21. $\cot 2\theta = (9 - 16)/(-24) = 7/24$. Use the method of Example 4 to obtain $\cos 2\theta = \dfrac{7}{25}$, so

$\cos\theta = \sqrt{\dfrac{1 + \cos 2\theta}{2}} = \sqrt{\dfrac{1 + \frac{7}{25}}{2}} = \dfrac{4}{5}$, $\sin\theta = \sqrt{\dfrac{1 - \cos 2\theta}{2}} = \dfrac{3}{5}$. Set $x = \dfrac{4}{5}x' - \dfrac{3}{5}y'$, $y = \dfrac{3}{5}x' + \dfrac{4}{5}y'$,

and insert these into the original equation to obtain $(y')^2 = 4(x' - 1)$; parabola, $p = 1$.
In $x'y'$-coordinates: vertex $(1, 0)$, focus $(2, 0)$, directrix $x' = 0$
In xy-coordinates: vertex $(4/5, 3/5)$, focus $(8/5, 6/5)$, directrix $y = -4x/3$

23. $\cot 2\theta = (288 - 337)/(-168) = 49/168 = 7/24$; proceed as in Exercise 21 to obtain $\cos\theta = 4/5$, $\sin\theta = 3/5$. Set $x = (4x' - 3y')/5$, $y = (3x' + 4y')/5$ to get $(x')^2/16 + (y')^2/9 = 1$; ellipse, $a = 4$, $b = 3$, $c = \sqrt{7}$.
In $x'y'$-coordinates: foci $(\pm\sqrt{7}, 0)$, vertices $(\pm 4, 0)$, minor axis endpoints $(0, \pm 3)$
In xy-coordinates: foci $\pm(4\sqrt{7}/5, 3\sqrt{7}/5)$, vertices $\pm(16/5, 12/5)$,
           minor axis endpoints $\pm(-9/5, 12/5)$

25. $\cot 2\theta = (31 - 21)/(10\sqrt{3}) = 1/\sqrt{3}$, $2\theta = \pi/3$, $\theta = \pi/6$, $\cos\theta = \sqrt{3}/2$, $\sin\theta = 1/2$. Set $x = \sqrt{3}x'/2 - y'/2$, $y = x'/2 + \sqrt{3}y'/2$ and obtain $(x')^2/4 + (y' + 2)^2/9 = 1$; ellipse, $a = 3$, $b = 2$, $c = \sqrt{9 - 4} = \sqrt{5}$.
In $x'y'$-coordinates: foci $(0, -2 \pm \sqrt{5})$, vertices $(0, 1)$ and $(0, -5)$, ends of minor axis $(\pm 2, -2)$
In xy-coordinates: foci $\left(1 - \dfrac{\sqrt{5}}{2}, -\sqrt{3} + \dfrac{\sqrt{15}}{2}\right)$ and $\left(1 + \dfrac{\sqrt{5}}{2}, -\sqrt{3} - \dfrac{\sqrt{15}}{2}\right)$,

vertices $\left(-\dfrac{1}{2}, \dfrac{\sqrt{3}}{2}\right)$ and $\left(\dfrac{5}{2}, -\dfrac{5\sqrt{3}}{2}\right)$,

ends of minor axis $\left(1 + \sqrt{3}, 1 - \sqrt{3}\right)$ and $\left(1 - \sqrt{3}, -1 - \sqrt{3}\right)$

27. $\cot 2\theta = (1 - 11)/(-10\sqrt{3}) = 1/\sqrt{3}$, $2\theta = \pi/3$, $\theta = \pi/6$, $\cos\theta = \sqrt{3}/2$, $\sin\theta = 1/2$. Set $x = \sqrt{3}x'/2 - y'/2$, $y = x'/2 + \sqrt{3}y'/2$ and obtain $(x')^2/16 - (y')^2/4 = 1$; hyperbola, $a = 4$, $b = 2$, $c = \sqrt{20} = 2\sqrt{5}$.
In $x'y'$-coordinates: foci $(\pm 2\sqrt{5}, 0)$, vertices $(\pm 4, 0)$, asymptotes $y' = \pm x'/2$
In xy-coordinates: foci $\pm(\sqrt{15}, \sqrt{5})$, vertices $\pm(2\sqrt{3}, 2)$, asymptotes $y = \dfrac{5\sqrt{3} \pm 8}{11}x$

29. $\cot 2\theta = ((-7) - 32)/(-52) = 3/4$; proceed as in Example 4 to obtain $\cos 2\theta = 3/5$,
$\cos\theta = \sqrt{\dfrac{1 + \cos 2\theta}{2}} = \dfrac{2}{\sqrt{5}}$, $\sin\theta = \dfrac{1}{\sqrt{5}}$. Set $x = \dfrac{2x' - y'}{\sqrt{5}}$, $y = \dfrac{x' + 2y'}{\sqrt{5}}$ and the equation

becomes $\dfrac{(x')^2}{9} - \dfrac{(y' - 4)^2}{4} = 1$; hyperbola, $a = 3$, $b = 2$, $c = \sqrt{13}$.
In $x'y'$-coordinates: foci $(\pm\sqrt{13}, 4)$, vertices $(\pm 3, 4)$, asymptotes $y' = 4 \pm 2x'/3$
In xy-coordinates: foci $\left(\dfrac{-4 + 2\sqrt{13}}{\sqrt{5}}, \dfrac{8 + \sqrt{13}}{\sqrt{5}}\right)$ and $\left(\dfrac{-4 - 2\sqrt{13}}{\sqrt{5}}, \dfrac{8 - \sqrt{13}}{\sqrt{5}}\right)$,

vertices $\left(\dfrac{2}{\sqrt{5}}, \dfrac{11}{\sqrt{5}}\right)$ and $(-2\sqrt{5}, \sqrt{5})$,

asymptotes $y = \dfrac{7x}{4} + 3\sqrt{5}$ and $y = -\dfrac{x}{8} + \dfrac{3\sqrt{5}}{2}$

31. $(\sqrt{x} + \sqrt{y})^2 = 1 = x + y + 2\sqrt{xy}$, $(1 - x - y)^2 = x^2 + y^2 + 1 - 2x - 2y + 2xy = 4xy$, so $x^2 - 2xy + y^2 - 2x - 2y + 1 = 0$. Set $\cot 2\theta = 0$, then $\theta = \pi/4$. Change variables by the Rotation Equations to obtain $2(y')^2 - 2\sqrt{2}x' + 1 = 0$, which is the equation of a parabola. The original equation implies that x and y are in the interval $[0, 1]$, so we only get part of the parabola.

33. It suffices to show that the expression $B'^2 - 4A'C'$ is independent of θ. Set

$g = B' = B(\cos^2 \theta - \sin^2 \theta) + 2(C - A) \sin \theta \cos \theta$

$f = A' = (A \cos^2 \theta + B \cos \theta \sin \theta + C \sin^2 \theta)$

$h = C' = (A \sin^2 \theta - B \sin \theta \cos \theta + C \cos^2 \theta)$

It is easy to show that

$g'(\theta) = -2B \sin 2\theta + 2(C - A) \cos 2\theta,$

$f'(\theta) = (C - A) \sin 2\theta + B \cos 2\theta$

$h'(\theta) = (A - C) \sin 2\theta - B \cos 2\theta$ and it is a bit more tedious to show that

$\dfrac{d}{d\theta}(g^2 - 4fh) = 0.$

It follows that $B'^2 - 4A'C'$ is independent of θ and by taking $\theta = 0$, we have $B'^2 - 4A'C' = B^2 - 4AC$.

35. If $A = C$ then $\cot 2\theta = (A - C)B = 0$, so $2\theta = \pi/2$, and $\theta = \pi/4$.

EXERCISE SET 10.6

1. **(a)** $r = \dfrac{3/2}{1 - \cos \theta}, e = 1, d = 3/2$
                  (b) $r = \dfrac{3/2}{1 + \frac{1}{2} \sin \theta}, e = 1/2, d = 3$

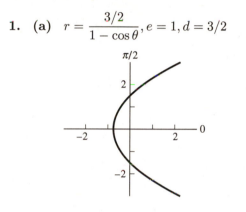
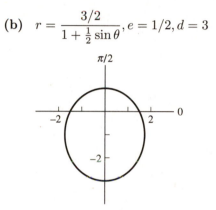

3. **(a)** $e = 1, d = 8,$ parabola, opens up
               (b) $r = \dfrac{4}{1 + \frac{3}{4} \sin \theta}, e = 3/4, d = 16/3,$

                                               ellipse, directrix $16/3$ units
                                               above the pole

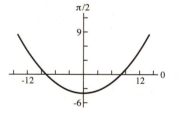
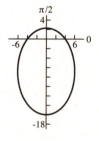

5. **(a)** $d = 2, r = \dfrac{ed}{1 + e \cos \theta} = \dfrac{3/2}{1 + \frac{3}{4} \cos \theta} = \dfrac{6}{4 + 3 \cos \theta}$

    (b) $e = 1, d = 1, r = \dfrac{ed}{1 + e \cos \theta} = \dfrac{1}{1 + \cos \theta}$

(c) $e = 4/3, d = 3, r = \dfrac{ed}{1 + e\sin\theta} = \dfrac{4}{1 + \frac{4}{3}\sin\theta} = \dfrac{12}{3 + 4\sin\theta}$

7. (a) $r = \dfrac{3}{1 + \frac{1}{2}\sin\theta}, e = 1/2, d = 6$, directrix 6 units above pole; if $\theta = \pi/2 : r_0 = 2$;

if $\theta = 3\pi/2 : r_1 = 6, a = (r_0 + r_1)/2 = 4, b = \sqrt{r_0 r_1} = 2\sqrt{3}$, center $(0, -2)$ (rectangular

coordinates), $\dfrac{x^2}{12} + \dfrac{(y+2)^2}{16} = 1$

(b) $r = \dfrac{1/2}{1 - \frac{1}{2}\cos\theta}, e = 1/2, d = 1$, directrix 1 unit left of pole; if $\theta = \pi : r_0 = \dfrac{1/2}{3/2} = 1/3$;

if $\theta = 0 : r_1 = 1, a = 2/3, b = 1/\sqrt{3}$, center $= (1/3, 0)$ (rectangular coordinates),

$\dfrac{9}{4}(x - 1/3)^2 + 3y^2 = 1$

9. (a) $r = \dfrac{3}{1 + 2\sin\theta}, e = 2, d = 3/2$, hyperbola, directrix 3/2 units above pole, if $\theta = \pi/2 :$

$r_0 = 1; \theta = 3\pi/2 : r = -3$, so $r_1 = 3$, center $(0, 2), a = 1, b = \sqrt{3}, -\dfrac{x^2}{3} + (y - 2)^2 = 1$

(b) $r = \dfrac{5/2}{1 - \frac{3}{2}\cos\theta}, e = 3/2, d = 5/3$, hyperbola, directrix 5/3 units left of pole, if $\theta = \pi :$

$r_0 = 1; \theta = 0 : r = -5, r_1 = 5$, center $(-3, 0)$, $a = 2, b = \sqrt{5}, \dfrac{1}{4}(x + 3)^2 - \dfrac{1}{5}y^2 = 1$

11. (a) $r = \dfrac{\frac{1}{2}d}{1 + \frac{1}{2}\cos\theta} = \dfrac{d}{2 + \cos\theta}$, if $\theta = 0 : r_0 = d/3; \theta = \pi, r_1 = d$,

$8 = a = \dfrac{1}{2}(r_1 + r_0) = \dfrac{2}{3}d, d = 12, \quad r = \dfrac{12}{2 + \cos\theta}$

(b) $r = \dfrac{\frac{3}{5}d}{1 - \frac{3}{5}\sin\theta} = \dfrac{3d}{5 - 3\sin\theta}$, if $\theta = 3\pi/2 : r_0 = \dfrac{3}{8}d; \theta = \pi/2, r_1 = \dfrac{3}{2}d$,

$4 = a = \dfrac{1}{2}(r_1 + r_0) = \dfrac{15}{16}d, d = \dfrac{64}{15}, r = \dfrac{3(64/15)}{5 - 3\sin\theta} = \dfrac{64}{25 - 15\sin\theta}$

13. For a hyperbola, both vertices and the directrix lie between the foci. So if one focus is at the origin and one vertex is at (5,0), then the directrix must lie to the right of the origin. By Theorem 10.6.2, the equation of the hyperbola has the form $r = \dfrac{ed}{1 + e\cos\theta}$.

Since the hyperbola is equilateral, $a = b$, so $c = \sqrt{2}a$ and $e = c/a = \sqrt{2}$. Since $(5, 0)$ lies on the hyperbola, either $r(0) = 5$ or $r(\pi) = -5$. In the first case the equation is $r = \dfrac{5\sqrt{2} + 5}{1 + \sqrt{2}\cos\theta}$; in the

second case it is $r = \dfrac{5\sqrt{2} - 5}{1 + \sqrt{2}\cos\theta}$.

15. (a) From Figure 10.4.22, $\dfrac{x^2}{a^2} - \dfrac{y^2}{b^2} = 1, \dfrac{x^2}{a^2} - \dfrac{y^2}{c^2 - a^2} = 1, \left(1 - \dfrac{c^2}{a^2}\right)x^2 + y^2 = a^2 - c^2,$

$c^2 + x^2 + y^2 = \left(\dfrac{c}{a}x\right)^2 + a^2, (x - c)^2 + y^2 = \left(\dfrac{c}{a}x - a\right)^2,$

$\sqrt{(x - c)^2 + y^2} = \dfrac{c}{a}x - a$ for $x > a^2/c$.

(b) From part (a) and Figure 10.6.1, $PF = \dfrac{c}{a}PD, \dfrac{PF}{PD} = \dfrac{c}{a}$.

17. **(a)** $e = c/a = \dfrac{\frac{1}{2}(r_1 + r_0)}{\frac{1}{2}(r_1 - r_0)} = \dfrac{r_1 + r_0}{r_1 - r_0}$

(b) $e = \dfrac{r_1/r_0 + 1}{r_1/r_0 - 1}, e(r_1/r_0 - 1) = r_1/r_0 + 1, \dfrac{r_1}{r_0} = \dfrac{e + 1}{e - 1}$

19. True. A non-circular ellipse can be described by the focus-directrix characterization as shown in Figure 10.6.1, so its eccentricity satisfies $0 < e < 1$ by part (b) of Theorem 10.6.1.

21. False. The eccentricity is determined by the ellipse's shape, not its size.

23. **(a)** $T = a^{3/2} = 39.5^{1.5} \approx 248$ yr

(b) $r_0 = a(1 - e) = 39.5(1 - 0.249) = 29.6645$ AU $\approx 4,449,675,000$ km
$r_1 = a(1 + e) = 39.5(1 + 0.249) = 49.3355$ AU $\approx 7,400,325,000$ km

(c) $r = \dfrac{a(1 - e^2)}{1 + e\cos\theta} \approx \dfrac{39.5(1 - (0.249)^2)}{1 + 0.249\cos\theta} \approx \dfrac{37.05}{1 + 0.249\cos\theta}$ AU

(d)

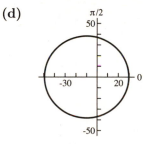

25. **(a)** $a = T^{2/3} = 2380^{2/3} \approx 178.26$ AU

(b) $r_0 = a(1 - e) \approx 0.8735$ AU, $r_1 = a(1 + e) \approx 355.64$ AU

(c) $r = \dfrac{a(1 - e^2)}{1 + e\cos\theta} \approx \dfrac{1.74}{1 + 0.9951\cos\theta}$ AU

(d)

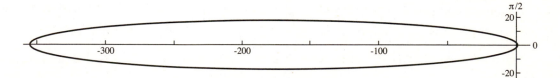

27. $r_0 = a(1 - e) \approx 7003$ km, $h_{\min} \approx 7003 - 6440 = 563$ km,
$r_1 = a(1 + e) \approx 10,726$ km, $h_{\max} \approx 10,726 - 6440 = 4286$ km

REVIEW EXERCISES, CHAPTER 10

1. $x(t) = \sqrt{2}\cos t, \quad y(t) = -\sqrt{2}\sin t, \quad 0 \le t \le 3\pi/2$

3. **(a)** $dy/dx = \dfrac{1/2}{2t} = 1/(4t); \ dy/dx\big|_{t=-1} = -1/4; \ dy/dx\big|_{t=1} = 1/4$

(b) $x = (2y)^2 + 1, dx/dy = 8y, dy/dx\big|_{y=\pm(1/2)} = \pm 1/4$

5. $dy/dx = \dfrac{4\cos t}{-2\sin t} = -2\cot t$

    (a) $dy/dx = 0$ if $\cot t = 0$, $t = \pi/2 + n\pi$ for $n = 0, \pm 1, \cdots$

    (b) $dx/dy = -\dfrac{1}{2}\tan t = 0$ if $\tan t = 0$, $t = n\pi$ for $n = 0, \pm 1, \cdots$

7. **(a)** $(-4\sqrt{2}, -4\sqrt{2})$ **(b)** $(7/\sqrt{2}, -7/\sqrt{2})$ **(c)** $(4\sqrt{2}, 4\sqrt{2})$

    (d) $(5, 0)$ **(e)** $(0, -2)$ **(f)** $(0, 0)$

9. **(a)** $(5, 0.6435)$ **(b)** $(\sqrt{29}, 5.0929)$ **(c)** $(1.2716, 0.6658)$

11. **(a)** $r = 2a/(1 + \cos\theta), r + x = 2a, x^2 + y^2 = (2a - x)^2, y^2 = -4ax + 4a^2$, parabola

    (b) $r^2(\cos^2\theta - \sin^2\theta) = x^2 - y^2 = a^2$, hyperbola

    (c) $r\sin(\theta - \pi/4) = (\sqrt{2}/2)r(\sin\theta - \cos\theta) = 4, y - x = 4\sqrt{2}$, line

    (d) $r^2 = 4r\cos\theta + 8r\sin\theta, x^2 + y^2 = 4x + 8y, (x - 2)^2 + (y - 4)^2 = 20$, circle

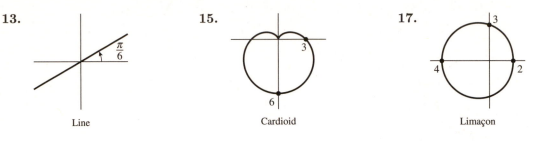

13.   **15.**   **17.**

       Line   Cardioid   Limaçon

19. **(a)** $x = r\cos\theta = \cos\theta - \cos^2\theta, dx/d\theta = -\sin\theta + 2\sin\theta\cos\theta = \sin\theta(2\cos\theta - 1) = 0$ if $\sin\theta = 0$ or $\cos\theta = 1/2$, so $\theta = 0, \pi, \pi/3, 5\pi/3$; maximum $x = 1/4$ at $\theta = \pi/3, 5\pi/3$, minimum $x = -2$ at $\theta = \pi$

    (b) $y = r\sin\theta = \sin\theta - \sin\theta\cos\theta, dy/d\theta = \cos\theta + 1 - 2\cos^2\theta = 0$ at $\cos\theta = 1, -1/2$, so $\theta = 0, 2\pi/3, 4\pi/3$; maximum $y = 3\sqrt{3}/4$ at $\theta = 2\pi/3$, minimum $y = -3\sqrt{3}/4$ at $\theta = 4\pi/3$

21. **(a)** As t runs from 0 to π, the upper portion of the curve is traced out from right to left; as t runs from π to 2π the bottom portion is traced out from right to left, except for the bottom part of the loop. The loop is traced out counterclockwise for $\pi + \sin^{-1}\dfrac{1}{4} < t < 2\pi - \sin^{-1}\dfrac{1}{4}$.

    (b) $\lim\limits_{t\to 0^+} x = +\infty, \lim\limits_{t\to 0^+} y = 1; \lim\limits_{t\to\pi^-} x = -\infty, \lim\limits_{t\to\pi^-} y = 1; \lim\limits_{t\to\pi^+} x = +\infty, \lim\limits_{t\to\pi^+} y = 1;$

        $\lim\limits_{t\to 2\pi^-} x = -\infty, \lim\limits_{t\to 2\pi^-} y = 1;$ the horizontal asymptote is $y = 1$.

    (c) horizontal tangent line when $dy/dx = 0$, or $dy/dt = 0$, so $\cos t = 0, t = \pi/2, 3\pi/2$; vertical tangent line when $dx/dt = 0$, so $-\csc^2 t - 4\sin t = 0, t = \pi + \sin^{-1}\dfrac{1}{\sqrt[3]{4}}, 2\pi - \sin^{-1}\dfrac{1}{\sqrt[3]{4}}, t \approx 3.823, 5.602$

    (d) Since $\tan\theta = \dfrac{y}{x} = \tan t$, we may take $\theta = t$. $r^2 = x^2 + y^2 = x^2(1 + \tan^2 t) = x^2\sec^2 t = (4 + \csc t)^2 = (4 + \csc\theta)^2$, so $r = 4 + \csc\theta$.

        $r = 0$ when $\csc\theta = -4, \sin\theta = -\dfrac{1}{4}$. The tangent lines at the pole are $\theta = \pi + \sin^{-1}\dfrac{1}{4}$ and $\theta = 2\pi - \sin^{-1}\dfrac{1}{4}$.

23. $A = 2 \int_0^\pi \frac{1}{2}(2 + 2\cos\theta)^2 d\theta = 6\pi$

25. $A = \int_0^{\pi/6} \frac{1}{2}(2\sin\theta)^2 d\theta + \int_{\pi/6}^{\pi/3} \frac{1}{2} \cdot 1^2 d\theta + \int_{\pi/3}^{\pi/2} \frac{1}{2}(2\cos\theta)^2 d\theta$

The first and third integrals are equal, by symmetry, so

$A = \int_0^{\pi/6} 4\sin^2\theta\, d\theta + \frac{1}{2}\left(\frac{\pi}{3} - \frac{\pi}{6}\right)$

$= \int_0^{\pi/6} 2(1 - \cos 2\theta)\, d\theta + \frac{\pi}{12}$

$= (2\theta - \sin 2\theta)\Big]_0^{\pi/6} + \frac{\pi}{12}$

$= \frac{\pi}{3} - \frac{\sqrt{3}}{2} + \frac{\pi}{12} = \frac{5\pi}{12} - \frac{\sqrt{3}}{2}$

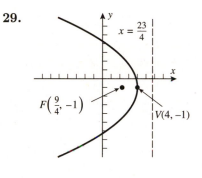

27.

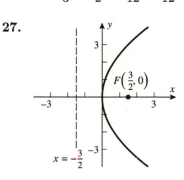

29.

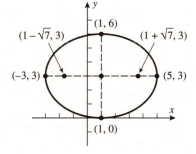

31. $c^2 = 25 - 4 = 21, c = \sqrt{21}$

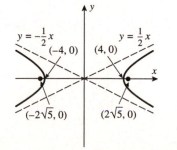

33. $\dfrac{(x-1)^2}{16} + \dfrac{(y-3)^2}{9} = 1$

$c^2 = 16 - 9 = 7, c = \sqrt{7}$

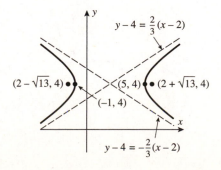

35. $c^2 = a^2 + b^2 = 16 + 4 = 20, c = 2\sqrt{5}$

37. $c^2 = 9 + 4 = 13, c = \sqrt{13}$

39. $x^2 = -4py$, $p = 4$, $x^2 = -16y$　　　　**41.** $a = 3$, $a/b = 1$, $b = 3$; $y^2/9 - x^2/9 = 1$

43. **(a)** $y = y_0 + (v_0 \sin\alpha)\dfrac{x}{v_0\cos\alpha} - \dfrac{g}{2}\left(\dfrac{x}{v_0\cos\alpha}\right)^2 = y_0 + x\tan\alpha - \dfrac{g}{2v_0^2\cos^2\alpha}x^2$

(b) $\dfrac{dy}{dx} = \tan\alpha - \dfrac{g}{v_0^2\cos^2\alpha}x$, $dy/dx = 0$ at $x = \dfrac{v_0^2}{g}\sin\alpha\cos\alpha$,

$y = y_0 + \dfrac{v_0^2}{g}\sin^2\alpha - \dfrac{g}{2v_0^2\cos^2\alpha}\left(\dfrac{v_0^2\sin\alpha\cos\alpha}{g}\right)^2 = y_0 + \dfrac{v_0^2}{2g}\sin^2\alpha$

45. $\cot 2\theta = \dfrac{A-C}{B} = 0$, $2\theta = \pi/2$, $\theta = \pi/4$, $\cos\theta = \sin\theta = \sqrt{2}/2$, so
$x = (\sqrt{2}/2)(x' - y')$, $y = (\sqrt{2}/2)(x' + y')$, $5(y')^2 - (x')^2 = 6$, hyperbola

47. $\cot 2\theta = (4\sqrt{5} - \sqrt{5})/(4\sqrt{5}) = 3/4$, so $\cos 2\theta = 3/5$ and thus $\cos\theta = \sqrt{(1+\cos 2\theta)/2} = 2/\sqrt{5}$
and $\sin\theta = \sqrt{(1-\cos 2\theta)/2} = 1/\sqrt{5}$. Hence the transformed equation is $5\sqrt{5}(x')^2 - 5\sqrt{5}y' = 0$,
$y' = (x')^2$, parabola

49. **(a)** $r = \dfrac{1/3}{1 + \frac{1}{3}\cos\theta}$, ellipse, right of pole, distance $= 1$

(b) hyperbola, left of pole, distance $= 1/3$

(c) $r = \dfrac{1/3}{1 + \sin\theta}$, parabola, above pole, distance $= 1/3$

(d) parabola, below pole, distance $= 3$

51. **(a)** $e = 4/5 = c/a$, $c = 4a/5$, but $a = 5$ so $c = 4$, $b = 3$, $\dfrac{(x+3)^2}{25} + \dfrac{(y-2)^2}{9} = 1$

(b) directrix $y = 2$, $p = 2$, $(x+2)^2 = -8y$

(c) center $(-1,5)$, vertices $(-1,7)$ and $(-1,3)$, $a = 2$, $a/b = 8$, $b = 1/4$, $\dfrac{(y-5)^2}{4} - 16(x+1)^2 = 1$

53. $a = 3$, $b = 2$, $c = \sqrt{5}$, $C = 4(3)\displaystyle\int_0^{\pi/2}\sqrt{1 - (5/9)\cos^2 u}\, du \approx 15.86543959$

MAKING CONNECTIONS, CHAPTER 10

1. **(a)**

(b) As $t \to +\infty$, the curve spirals in toward a point P in the first quadrant. As $t \to -\infty$, it spirals in toward the reflection of P through the origin. (It can be shown that $P = (1/2, 1/2)$.)

(c) $L = \displaystyle\int_{-1}^{1} \left[\cos^2 \left(\frac{\pi t^2}{2} \right) + \sin^2 \left(\frac{\pi t^2}{2} \right) \right] dt = 2$

3. Let P denote the pencil tip, and let $R(x, 0)$ be the point below Q and P which lies on the line L. Then $QP + PF$ is the length of the string and $QR = QP + PR$ is the length of the side of the triangle. These two are equal, so $PF = PR$. But this is the definition of a parabola according to Definition 10.4.1.

5. (a) Position the ellipse so its equation is $\dfrac{x^2}{a^2} + \dfrac{y^2}{b^2} = 1$. Then $y = \dfrac{b}{a} \sqrt{a^2 - x^2}$, so

$$V = 2 \int_0^a \pi y^2 \, dx = 2 \int_0^a \pi \frac{b^2}{a^2} (a^2 - x^2) \, dx = \frac{4}{3} \pi a b^2.$$

Also, $\dfrac{dy}{dx} = -\dfrac{bx}{a\sqrt{a^2 - x^2}}$ so $1 + \left(\dfrac{dy}{dx} \right)^2 = \dfrac{a^4 - (a^2 - b^2)x^2}{a^2(a^2 - x^2)} = \dfrac{a^4 - c^2 x^2}{a^2(a^2 - x^2)}$, where

$c = \sqrt{a^2 - b^2}$. Then $S = 2 \displaystyle\int_0^a 2\pi y \sqrt{1 + (dy/dx)^2} \, dx = \dfrac{4\pi b}{a} \int_0^a \sqrt{a^2 - x^2} \sqrt{\dfrac{a^4 - c^2 x^2}{a^2(a^2 - x^2)}} \, dx$

$$= \frac{4\pi bc}{a^2} \int_0^a \sqrt{\frac{a^4}{c^2} - x^2} \, dx = \frac{4\pi bc}{a^2} \left[\frac{x}{2} \sqrt{\frac{a^4}{c^2} - x^2} + \frac{a^4}{2c^2} \sin^{-1} \frac{cx}{a^2} \right]_0^a = 2\pi ab \left(\frac{b}{a} + \frac{a}{c} \sin^{-1} \frac{c}{a} \right),$$

by Endpaper Integral Table Formula 74.

(b) In part (a) interchange a and b to obtain the result.

APPENDIX A

Graphing Functions Using Calculators and Computer Algebra Systems

EXERCISE SET A

1. (a)   (b)   (c)

(d)   (e)

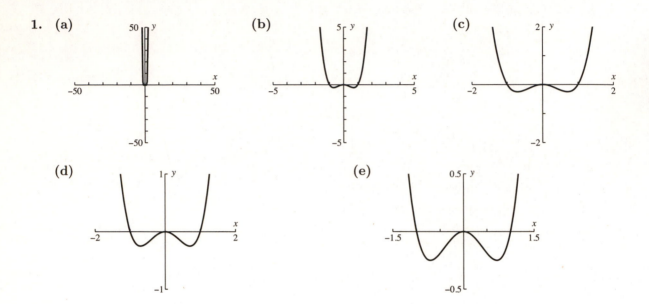

3. (a) Only two points of the graph, $(-1, 13)$ and $(1, 13)$, are in the window.

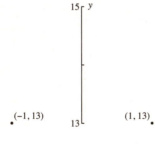

(b)

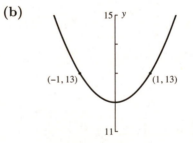

(c)

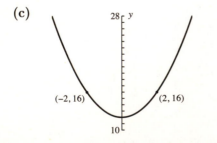

(d) This graph uses the window $[-4, 4] \times [-4, 28]$.

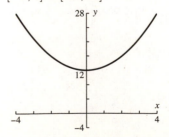

5. The domain is $[-\sqrt{8}, \sqrt{8}]$ and the range is $[0,4]$. The graph below uses the window $[-3,3] \times [-1,5]$.

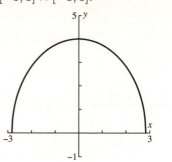

7.

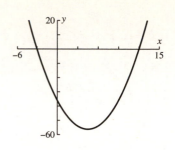

9.

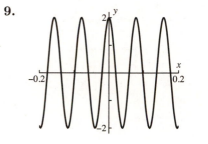

11.

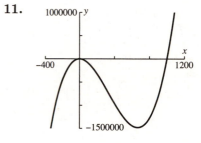

13.

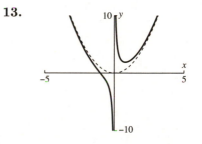

15.

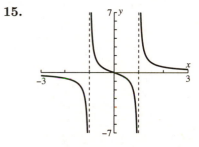

17. (a) $f(x) = \sqrt{16 - x^2}$

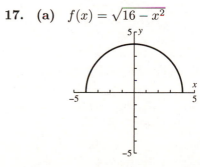

(b) $f(x) = -\sqrt{16 - x^2}$

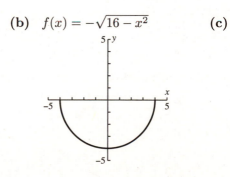

(c)

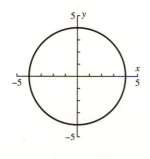

(d)

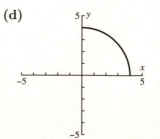

(e) No; it fails the vertical line test.

19. **(a)**

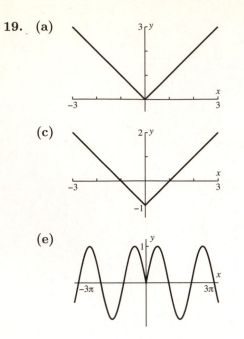

(b)

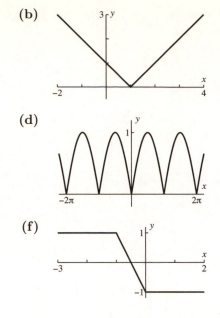

(c)

(d)

(e)

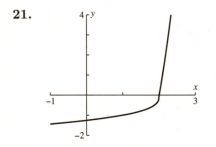

(f)

21.

23. **(a)** The two graphs should be identical:

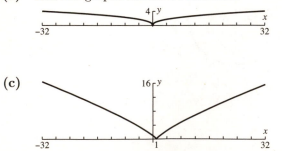

(b) The two graphs should be identical:

(c)

(d)

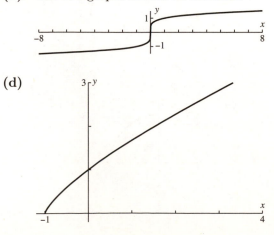

25. **(a)**

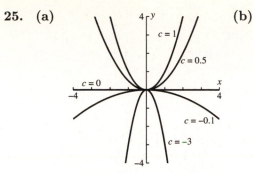

Positive values of c make the parabola open upward; negative values make it open downward. Large values (positive or negative) make it pointier.

(b)

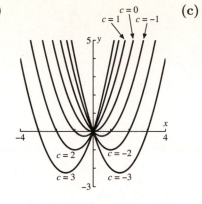

The parabola keeps the same shape, but its vertex is moved to $(-c/2, -c^2/4)$.

(c)

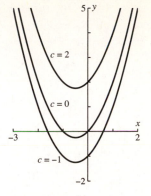

The parabola keeps the same shape, but is translated vertically.

27.

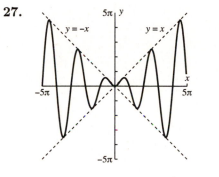

29. **(a)**

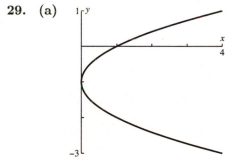

(b)

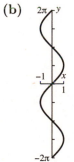

31.

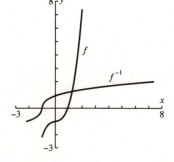

33.

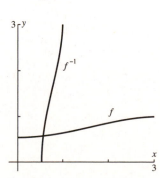

35. **(a)** One possible answer is $x = 4\cos t$, $y = 3\sin t$.

(b) One possible answer is $x = -1 + 4\cos t$, $y = 2 + 3\sin t$.

(c)

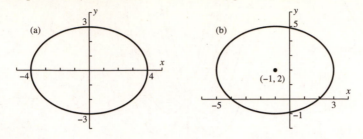

APPENDIX B
Trigonometry Review

EXERCISE SET B

1. (a) $5\pi/12$ (b) $13\pi/6$ (c) $\pi/9$ (d) $23\pi/30$

3. (a) $12°$ (b) $(270/\pi)°$ (c) $288°$ (d) $540°$

5.

	$\sin\theta$	$\cos\theta$	$\tan\theta$	$\csc\theta$	$\sec\theta$	$\cot\theta$
(a)	$\sqrt{21}/5$	$2/5$	$\sqrt{21}/2$	$5/\sqrt{21}$	$5/2$	$2/\sqrt{21}$
(b)	$3/4$	$\sqrt{7}/4$	$3/\sqrt{7}$	$4/3$	$4/\sqrt{7}$	$\sqrt{7}/3$
(c)	$3/\sqrt{10}$	$1/\sqrt{10}$	3	$\sqrt{10}/3$	$\sqrt{10}$	$1/3$

7. $\sin\theta = 3/\sqrt{10}$, $\cos\theta = 1/\sqrt{10}$ **9.** $\tan\theta = \sqrt{21}/2$, $\csc\theta = 5/\sqrt{21}$

11. Let x be the length of the side adjacent to θ, then $\cos\theta = x/6 = 0.3$, $x = 1.8$.

13.

	θ	$\sin\theta$	$\cos\theta$	$\tan\theta$	$\csc\theta$	$\sec\theta$	$\cot\theta$
(a)	$225°$	$-1/\sqrt{2}$	$-1/\sqrt{2}$	1	$-\sqrt{2}$	$-\sqrt{2}$	1
(b)	$-210°$	$1/2$	$-\sqrt{3}/2$	$-1/\sqrt{3}$	2	$-2/\sqrt{3}$	$-\sqrt{3}$
(c)	$5\pi/3$	$-\sqrt{3}/2$	$1/2$	$-\sqrt{3}$	$-2/\sqrt{3}$	2	$-1/\sqrt{3}$
(d)	$-3\pi/2$	1	0	$-$	1	$-$	0

15.

	$\sin\theta$	$\cos\theta$	$\tan\theta$	$\csc\theta$	$\sec\theta$	$\cot\theta$
(a)	$4/5$	$3/5$	$4/3$	$5/4$	$5/3$	$3/4$
(b)	$-4/5$	$3/5$	$-4/3$	$-5/4$	$5/3$	$-3/4$
(c)	$1/2$	$-\sqrt{3}/2$	$-1/\sqrt{3}$	2	$-2\sqrt{3}$	$-\sqrt{3}$
(d)	$-1/2$	$\sqrt{3}/2$	$-1/\sqrt{3}$	-2	$2/\sqrt{3}$	$-\sqrt{3}$
(e)	$1/\sqrt{2}$	$1/\sqrt{2}$	1	$\sqrt{2}$	$\sqrt{2}$	1
(f)	$1/\sqrt{2}$	$-1/\sqrt{2}$	-1	$\sqrt{2}$	$-\sqrt{2}$	-1

17. (a) $x = 3\sin 25° \approx 1.2679$ (b) $x = 3/\tan(2\pi/9) \approx 3.5753$

19.

	$\sin\theta$	$\cos\theta$	$\tan\theta$	$\csc\theta$	$\sec\theta$	$\cot\theta$
(a)	$a/3$	$\sqrt{9-a^2}/3$	$a/\sqrt{9-a^2}$	$3/a$	$3/\sqrt{9-a^2}$	$\sqrt{9-a^2}/a$
(b)	$a/\sqrt{a^2+25}$	$5/\sqrt{a^2+25}$	$a/5$	$\sqrt{a^2+25}/a$	$\sqrt{a^2+25}/5$	$5/a$
(c)	$\sqrt{a^2-1}/a$	$1/a$	$\sqrt{a^2-1}$	$a/\sqrt{a^2-1}$	a	$1/\sqrt{a^2-1}$

21. (a) $\theta = 3\pi/4 \pm n\pi$, $n = 0, 1, 2, \ldots$
     (b) $\theta = \pi/3 \pm 2n\pi$ and $\theta = 5\pi/3 \pm 2n\pi$, $n = 0, 1, 2, \ldots$

23. **(a)** $\theta = \pi/6 \pm n\pi$, $n = 0, 1, 2, \ldots$

(b) $\theta = 4\pi/3 \pm 2n\pi$ and $\theta = 5\pi/3 \pm 2n\pi$, $n = 0, 1, 2, \ldots$

25. **(a)** $\theta = 3\pi/4 \pm n\pi$, $n = 0, 1, 2, \ldots$ **(b)** $\theta = \pi/6 \pm n\pi$, $n = 0, 1, 2, \ldots$

27. **(a)** $\theta = \pi/3 \pm 2n\pi$ and $\theta = 2\pi/3 \pm 2n\pi$, $n = 0, 1, 2, \ldots$

(b) $\theta = \pi/6 \pm 2n\pi$ and $\theta = 11\pi/6 \pm 2n\pi$, $n = 0, 1, 2, \ldots$

29. $\sin\theta = 2/5$, $\cos\theta = -\sqrt{21}/5$, $\tan\theta = -2/\sqrt{21}$, $\csc\theta = 5/2$, $\sec\theta = -5/\sqrt{21}$, $\cot\theta = -\sqrt{21}/2$

31. **(a)** $\theta = \pm n\pi$, $n = 0, 1, 2, \ldots$ **(b)** $\theta = \pi/2 \pm n\pi$, $n = 0, 1, 2, \ldots$

(c) $\theta = \pm n\pi$, $n = 0, 1, 2, \ldots$ **(d)** $\theta = \pm n\pi$, $n = 0, 1, 2, \ldots$

(e) $\theta = \pi/2 \pm n\pi$, $n = 0, 1, 2, \ldots$ **(f)** $\theta = \pm n\pi$, $n = 0, 1, 2, \ldots$

33. **(a)** $s = r\theta = 4(\pi/6) = 2\pi/3$ cm   **(b)** $s = r\theta = 4(5\pi/6) = 10\pi/3$ cm

35. $\theta = s/r = 2/5$

37. **(a)** $2\pi r = R(2\pi - \theta)$, $r = \dfrac{2\pi - \theta}{2\pi}R$

(b) $h = \sqrt{R^2 - r^2} = \sqrt{R^2 - (2\pi - \theta)^2 R^2/(4\pi^2)} = \dfrac{\sqrt{4\pi\theta - \theta^2}}{2\pi}R$

39. Let h be the altitude as shown in the figure, then

$h = 3\sin 60° = 3\sqrt{3}/2$ so $A = \dfrac{1}{2}(3\sqrt{3}/2)(7) = 21\sqrt{3}/4$.

41. Let x be the distance above the ground, then $x = 10\sin 67° \approx 9.2$ ft.

43. From the figure, $h = x - y$ but $x = d\tan\beta$,

$y = d\tan\alpha$ so $h = d(\tan\beta - \tan\alpha)$.

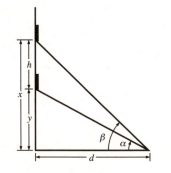

45. **(a)** $\sin 2\theta = 2\sin\theta\cos\theta = 2(\sqrt{5}/3)(2/3) = 4\sqrt{5}/9$

(b) $\cos 2\theta = 2\cos^2\theta - 1 = 2(2/3)^2 - 1 = -1/9$

47. $\sin 3\theta = \sin(2\theta + \theta) = \sin 2\theta\cos\theta + \cos 2\theta\sin\theta = (2\sin\theta\cos\theta)\cos\theta + (\cos^2\theta - \sin^2\theta)\sin\theta$

$= 2\sin\theta\cos^2\theta + \sin\theta\cos^2\theta - \sin^3\theta = 3\sin\theta\cos^2\theta - \sin^3\theta$; similarly, $\cos 3\theta = \cos^3\theta - 3\sin^2\theta\cos\theta$

49. $\dfrac{\cos\theta\tan\theta + \sin\theta}{\tan\theta} = \dfrac{\cos\theta(\sin\theta/\cos\theta) + \sin\theta}{\sin\theta/\cos\theta} = 2\cos\theta$

51. $\tan\theta + \cot\theta = \dfrac{\sin\theta}{\cos\theta} + \dfrac{\cos\theta}{\sin\theta} = \dfrac{\sin^2\theta + \cos^2\theta}{\sin\theta\cos\theta} = \dfrac{1}{\sin\theta\cos\theta} = \dfrac{2}{2\sin\theta\cos\theta} = \dfrac{2}{\sin 2\theta} = 2\csc 2\theta$

53. $\dfrac{\sin\theta+\cos 2\theta-1}{\cos\theta-\sin 2\theta}=\dfrac{\sin\theta+(1-2\sin^2\theta)-1}{\cos\theta-2\sin\theta\cos\theta}=\dfrac{\sin\theta(1-2\sin\theta)}{\cos\theta(1-2\sin\theta)}=\tan\theta$

55. Using (47), $2\cos 2\theta\sin\theta=2(1/2)[\sin(-\theta)+\sin 3\theta]=\sin 3\theta-\sin\theta$

57. $\tan(\theta/2)=\dfrac{\sin(\theta/2)}{\cos(\theta/2)}=\dfrac{2\sin(\theta/2)\cos(\theta/2)}{2\cos^2(\theta/2)}=\dfrac{\sin\theta}{1+\cos\theta}$

59. From the figures, area $=\dfrac{1}{2}hc$ but $h=b\sin A$

so area $=\dfrac{1}{2}bc\sin A$. The formulas

area $=\dfrac{1}{2}ac\sin B$ and area $=\dfrac{1}{2}ab\sin C$

follow by drawing altitudes from vertices B and C, respectively.

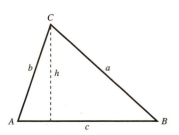

61. **(a)** $\sin(\pi/2+\theta)=\sin(\pi/2)\cos\theta+\cos(\pi/2)\sin\theta=(1)\cos\theta+(0)\sin\theta=\cos\theta$
 (b) $\cos(\pi/2+\theta)=\cos(\pi/2)\cos\theta-\sin(\pi/2)\sin\theta=(0)\cos\theta-(1)\sin\theta=-\sin\theta$
 (c) $\sin(3\pi/2-\theta)=\sin(3\pi/2)\cos\theta-\cos(3\pi/2)\sin\theta=(-1)\cos\theta-(0)\sin\theta=-\cos\theta$
 (d) $\cos(3\pi/2+\theta)=\cos(3\pi/2)\cos\theta-\sin(3\pi/2)\sin\theta=(0)\cos\theta-(-1)\sin\theta=\sin\theta$

63. **(a)** Add (34) and (36) to get $\sin(\alpha-\beta)+\sin(\alpha+\beta)=2\sin\alpha\cos\beta$ so
 $\sin\alpha\cos\beta=(1/2)[\sin(\alpha-\beta)+\sin(\alpha+\beta)]$.
 (b) Subtract (35) from (37). **(c)** Add (35) and (37).

65. $\sin\alpha+\sin(-\beta)=2\sin\dfrac{\alpha-\beta}{2}\cos\dfrac{\alpha+\beta}{2}$, but $\sin(-\beta)=-\sin\beta$ so

$\sin\alpha-\sin\beta=2\cos\dfrac{\alpha+\beta}{2}\sin\dfrac{\alpha-\beta}{2}$.

67. Consider the triangle having a, b, and d as sides. The angle formed by sides a and b is $\pi-\theta$ so
from the law of cosines, $d^2=a^2+b^2-2ab\cos(\pi-\theta)=a^2+b^2+2ab\cos\theta$, $d=\sqrt{a^2+b^2+2ab\cos\theta}$.

69. **(a)** $\tan^{-1}(-1/2)\approx-27°$ so angle of inclination $\approx 180°-27°=153°$
 (b) angle of inclination $=\tan^{-1}1=45°$
 (c) $\tan^{-1}(-2)\approx-63°$ so angle of inclination $\approx 180°-63°=117°$
 (d) angle of inclination $=\tan^{-1}57\approx 89°$

71. **(a)** angle of inclination $=\tan^{-1}\sqrt{3}=60°$
 (b) $y=-2x-5$. $\tan^{-1}(-2)\approx-63°$ so angle of inclination $\approx 180°-63°=117°$

APPENDIX C
Solving Polynomial Equations

EXERCISE SET C

1. **(a)** $q(x) = x^2 + 4x + 2, r(x) = -11x + 6$
   (b) $q(x) = 2x^2 + 4, r(x) = 9$
   (c) $q(x) = x^3 - x^2 + 2x - 2, r(x) = 2x + 1$

3. **(a)** $q(x) = 3x^2 + 6x + 8, r(x) = 15$
   (b) $q(x) = x^3 - 5x^2 + 20x - 100, r(x) = 504$
   (c) $q(x) = x^4 + x^3 + x^2 + x + 1, r(x) = 0$

5.

x	0	1	-3	7
$p(x)$	-4	-3	101	5001

7. **(a)** $q(x) = x^2 + 6x + 13, r = 20$ **(b)** $q(x) = x^2 + 3x - 2, r = -4$

9. Assume $r = a/b$ where a and b are integers with $a > 0$:
   (a) b divides 1, $b = \pm 1$; a divides 24, $a = 1, 2, 3, 4, 6, 8, 12, 24$;
   the possible candidates are $\{\pm 1, \pm 2, \pm 3, \pm 4, \pm 6, \pm 8, \pm 12, \pm 24\}$
   (b) b divides 3 so $b = \pm 1, \pm 3$; a divides -10 so $a = 1, 2, 5, 10$;
   the possible candidates are $\{\pm 1, \pm 2, \pm 5, \pm 10, \pm 1/3, \pm 2/3, \pm 5/3, \pm 10/3\}$
   (c) b divides 1 so $b = \pm 1$; a divides 17 so $a = 1, 17$;
   the possible candidates are $\{\pm 1, \pm 17\}$

11. $(x + 1)(x - 1)(x - 2)$ 13. $(x + 3)^3(x + 1)$

15. $(x + 3)(x + 2)(x + 1)^2(x - 3)$ 17. -3 is the only real root.

19. $x = -2, -2/3, -1 \pm \sqrt{3}$ are the real roots.   21. $-2, 2, 3$ are the only real roots.

23. If $x - 1$ is a factor then $p(1) = 0$, so $k^2 - 7k + 10 = 0$, $k^2 - 7k + 10 = (k - 2)(k - 5)$, so $k = 2, 5$.

25. If the side of the cube is x then $x^2(x - 3) = 196$; the only real root of this equation is $x = 7$ cm.

27. Use the Factor Theorem with x as the variable and y as the constant c.
   (a) For any positive integer n the polynomial $x^n - y^n$ has $x = y$ as a root.
   (b) For any positive even integer n the polynomial $x^n - y^n$ has $x = -y$ as a root.
   (c) For any positive odd integer n the polynomial $x^n + y^n$ has $x = -y$ as a root.